Mobile and Wireless: Communications and Technology

Mobile and Wireless: Communications and Technology

Edited by
Timothy Kolaya

Willford Press
www.willfordpress.com

Published by Willford Press,
118-35 Queens Blvd., Suite 400,
Forest Hills, NY 11375, USA

ISBN: 978-1-68285-346-7

Cataloging-in-Publication Data

Mobile and wireless : communications and technology / edited by Timothy Kolaya.
p. cm.
Includes bibliographical references and index.
ISBN 978-1-68285-346-7
1. Mobile computing. 2. Wireless communication systems. 3. Mobile communication systems.
4. Electronic data processing--Distributed processing. 5. Communication and technology. I. Kolaya, Timothy.

QA76.59 .M63 2017
004--dc23

For information on all Willford Press publications
visit our website at www.willfordpress.com

Willford Press

Printed in the United States of America.

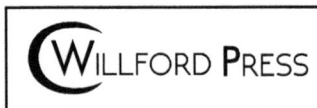

Contents

Preface

I am honored to present to you this unique book which encompasses the most up-to-date data in the field. I was extremely pleased to get this opportunity of editing the work of experts from across the globe. I have also written papers in this field and researched the various aspects revolving around the progress of the discipline. I have tried to unify my knowledge along with that of stalwarts from every corner of the world, to produce a text which not only benefits the readers but also facilitates the growth of the field.

This book on mobile computing delves into the myriad areas that mobile and wireless communications technology has been applied to. Mobile computing is a compound of inter-related technical innovations that are reflected in the fields of mobile software, mobile hardware and mobile communications. Technological breakthroughs in this field allow for greater transmission of data and information. This book is a vital tool for all researching or studying mobile and wireless communications and technology as it gives incredible insights into emerging trends and concepts. This book is a collective contribution of a renowned group of international experts. Coherent flow of topics, student-friendly language and extensive use of examples make this book an invaluable source of knowledge.

Finally, I would like to thank all the contributing authors for their valuable time and contributions. This book would not have been possible without their efforts. I would also like to thank my friends and family for their constant support.

Editor

Dependable Community-Cloud Framework for Smartphones

Arnold Adimabua Ojugo[1], Fidelis Obukowho Aghware[2], Rume Elizabeth Yoro[3], Mary Oluwatoyin Yerokun[4], Andrew Okonji Eboka[4], Christiana Nneamaka Anujeonye[4], Fidelia Ngozi Efozia[5]

[1]Dept. of Math/Computer, Federal University of Petroleum Resources Effurun, Delta State, Nigeria
[2]Dept. of Computer Science Education, College of Education, Agbor, Delta State, Nigeria
[3]Dept. of Computer Sci., Delta State Polytechnic, Ogwashi-Uku, Delta State, Nigeria
[4]Dept. of Computer Sci. Education, Federal College of Education (Technical), Asaba, Delta State, Nigeria
[5]Prototype Engineering Development Institute, Fed. Ministry of Science Technology, Osun State, Nigeria

Email address:
arnoldojugo@yahoo.com (A. A. Ojugo), ojugo_arnold@yahoo.com (A. A. Ojugo), aghwarefo@yahoo.com (F. O. Aghware), rumerisky@yahoo.com (R. E. Yoro), an_drey2k@yahoo.com (A. O. Eboka), agapenexus@hotmail.co.uk (M. O. Yerokun), anujeonyechristy@gmail.com (C. N. Anujeonye), fenngo31@yahoo.com (F. N. Efozia)

Abstract: Cloud computing enable users to access ubiquitous, on-demand, convenient and shared resource (apps and storage) – as rapidly released by a provider with minimal managed effort. The increased growth of user access to mobile smartphones from 42.5% in 2013 to 78.9% by 2013 and the advent of Androids has made smartphones a preferred choice over PCs due to its design, portability, speed, functionality and Internet access ease – all of which continues to pose significant risk to user data security with high vulnerability to attacks. With its implication to work related functions and biz issues, it exposes sensitive data to adversaries. The study thus, describes a support tool named PushCloud that lets users account the ability to sign-in and perform backup functions on contacts, messages, picture files, documents, videos and recorded voice amongst others. Its other benefit is in the fact that it pools together cloud service providers and allows users a cross platform with minimal price difference. The system helps address security related issue from a user's end via AES-256 encryption on an integrated cloud model, explores its storage capability to guarantee data recovery with a remote server (BDC) for back- and front-end data storage ease.

Keywords: Stochastic, Immunize, Network, Vertices, SIS, SIR, Function, Search Space, Solution, Models

1. Introduction

Today, Android is become a leading platform for mobile devices with its open source feat that distinguishes it from most other mobile platforms such Blackberry, Windows Phone and iOS (Morril, 2010). It is not a specification or distribution of traditional Linux, neither is it a collection of replaceable components or chunk of software ported on a device. Its open source platform is built by Google with OS, middleware, and apps for mobile systems based on Linux kernel that enables developers to write apps majorly in Java with support for C/C++ (Bray, 2010). Its major success is its license that allows third-party porting developments to it. Since its release, it has been constantly improved either in feats, supported hardware, and also extended to new device types besides the originally intended ones (Maia et al, 2010). Recent efforts are to enhance real-time capabilities as employed in a variety of embedded systems (Tapas Kumar and Kolin, 2010).

1.1. Android Platform (AP)

Pernel et al (2013) and Agam (2011) in "Google Android and Linaro Android SDK" note AP is an eco-system layer of app component implemented on mobile (smartphone) hardware as thus (see fig 1):

a. Linux OS provides basic functionality such as security,

process/memory management and networking to support vast device drivers. It handles human machine interfaces, file systems, network access etc. Its kernel is modified by Google to use low memory killer, specific inter-process communication system, kernel log feats, shared memory system and many other changes as developed. It runs on standard Vanilla Linux, merging specific changes into its kernel. Recent release aimed at real-time Linux kernel is v4.0.3 (Ice Cream Sandwich).

b. Library with Google's *libc* called *Bionic*, media/graphics (OpenGL|ES), browser-webkit and light-database SQLite. DVM (Dalvik Virtual Machine) completely differs from Sun's JVM and uses register based byte code to conserves memory, max performance and can instantiate many of its apps multiple times, with each app having its own private copy running. DVM uses Linux for memory management and multi-threading to support the Java language.

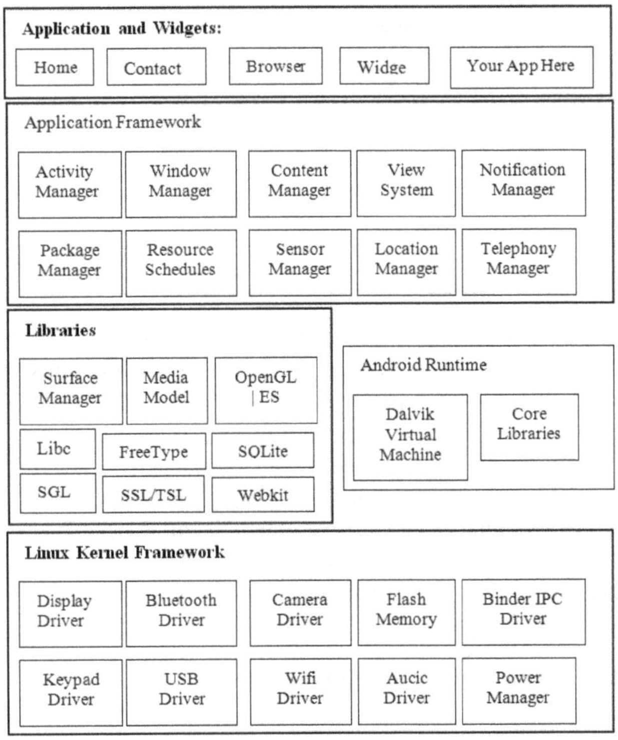

Fig. 1. *Android OS Platform.*

DVM uses *bionic* (not compatible with *glibc*) so that its native libraries are faster to implement with small custom *pthread* to support services such as system and logging capabilities. Writable data segments are small so as to be loaded into memory with each process. This keeps code size small so that Linux loads only once, all read-only pages. *Bionic* is used: (a) to avoid inclusion of GPL code at user space level in its platform where BSD is used, and (b) for small memory footprint devices with high speed CPUs at relatively low frequencies. *Bionic libc* does not handle C++ exceptions (though omitting such lower level exceptions pose no problem as Java is Android's primary language. It handles exceptions internally). *Bionic* has no priority inheritance for mutexex as implemented in *glibc*. Available in its kernel and accessed via own library in system calls, its lack of priority inversion disqualifies it for real-time capability as applied in robotics/automotive. Google's reason for a complete new VM from scratch as accomplished with DVM's register-based byte code is to reduce patent infringement risk. Thus, existing real-time apps modified for JVM cannot easily be ported to DVM.

c. Application Framework provides higher-level services to apps such as Java classes amongst others. Its use can vary between/with varying implementation.

d. Application/Widget are Android routine distributed apps such as email, SMS, calendar, contacts and Web browser.

1.2. Literature Review

In embedded systems (automotive or robotic), its ability to meet deadlines, time constraints is a critical specification part in its design as such systems must response to stimuli within a certain pre-specified real-time constraints. Thus, the reliability of software has not to focus only on the functional failures but require and detailed evaluation of the ability of the system to meet these timing specifications (Bhupinder and Vijay, 2010).

From a device mainly used for phone calls and messages, the mobile phone (smartphone) is become a multi-purpose device. Though favored by its size, there exists thermal constraints, battery consumption and computational powers that limits it usage and capabilities. Cloud computing has the potential to transform the IT-industry. Thus, Harmen (2012) investigated the possible increase in speed of smartphones by offloading computational heavy app functions via cloud computing. He developed an app that was used to conduct computational heavy tests, and the results showed that it is not beneficial to use cloud computing to carry out these types of tasks; it is faster to use the smartphone.

Pernel et al (2013) In their test for real time behavior and performance on the Android platform – so as to make clear if Android usage be advised in open real-time environments – used for evaluation, a test suite of four performance tests namely: thread switch latency, interrupt latency, sustained interrupt frequency, and semaphore acquire-release timing in contention case, and one behavior test to checks the mutex locking behavior. Their test results showed that the Android in its current state cannot be qualified to be used in real-time environments. Finally we provide some potential solutions for using Android in such environments.

2. Cloud Computing Technology

Cloud computing is the underlying infrastructure that helps scale services exponentially and flex resources rapidly in response to variable supply and demand. Hurwitz et al (2010) It hinges on terminologies and technologies such as:

a. Cloud Services are the actual apps employed by a user to perform one or two tasks. Such as using Snapfish to share photo online, Force.com to create niche market services, NetSuite for ERP services, amongst others.

b. Multi-Tenancy Cloud services are either at software or at infrastructure layer. Thus, many instances of software and platform it runs are made available to serves many clients. With shared resources, providers have access controls and security for a protected environment for each user.

c. Enterprise-Services (software/infrastructure) is designed to serve an enterprise' specific internal needs not limited to and includes data security, integration, configurability, access, reliability and availability.
d. Global-Services (software/infrastructure) designed for external, arbitrary and non-secure user. Software is native, multi-tenant and designed with Web 2.0 to be scalable and relies on software-based resiliency.
e. Private/Internal cloud connotes enterprise-class service with virtualized and automated infrastructure. While different from cloud-based infrastructures, they both share similar feats, and benefits from same technologies that help cloud services providers rapidly scale.
f. Elasticity allows flexibility to meet user preferences and needs on a near real-time basis, in response to supply and demand triggers. It is also ability of a service or infrastructure to adjust to users' fluctuating demands in service by automatically provisioning of resources as well as by moving the service to be executed on another part of the system.

2.1. Cloud Computing Services

Hamrén (2012) Cloud services are grouped as thus:

a. On-demand self-service: A user can unilaterally provision computing services such as network storage and server time as needed automatically without interference.
b. Network access is available of network accessed through standard mechanisms that promote use by heterogeneous thin or thick client platforms (e.g. mobile phones, tablets, laptops).
c. Pooling allows a provider's resources to serve many users on a multi-tenant model, with various physical and virtual resources, dynamically assigned/reassigned to meet users' needs irrespective of location dependence. Users have no control over the exact location of provided resources but are able to specify location at a higher level of abstraction (like as country or datacenter). Resources include storage, processing, memory, and network bandwidth.
d. Elasticity is the release of services as automatically scaled rapidly outward and inward on users' demand. To users, such capability should appear to be unlimited and can be appropriated in any quantity at any time.
e. Service resource measure can be automatic, controlled and optimized to leverage metering capability at some level of abstraction appropriate to service type such as storage, processing, bandwidth and active user accounts. Internet resource usage can be monitored, controlled and reported to provide the needed transparency for both provider and users of the utilized service.

2.2. Service Model

Hamrén (2012) Service models are grouped into three as:

a. Platform as Service (PaaS) is the capability provided to the user to deploy onto a cloud infrastructure, user-created or acquired apps using programming languages, libraries, services and tools as supported by a cloud provider. The user only has control over his deployed apps and possibly configuration settings for app-hosting environment.
b. Software as Service (SaaS) are services and apps provided to users on cloud infrastructure, accessible from client's devices via a thin client interface (email, web browser), or via a program interface. User has no control of underlying infrastructure such as network, operating system, servers, storage, or individual apps; but is limited to user-specific application configuration settings.
c. Infrastructure as Service (IaaS) is capability provided to user such as processing, storage and other resources, so he can deploy and run arbitrary software to include operating systems and apps. User only has control over operating systems, storage, deployed apps and limited control of selecting network devices (like firewalls and SSH embedded within the organization wishing to engage in cloud services. All of which is aimed at improved data security and integrity from the client-end).

2.3. Deployment Models

Deployment models deal with various forms of intrusion from adversaries with malicious intent towards data. Thus, data security must be ensured. Ureigho (2012) and Ojugo et al (2012a) various methods to improve intrusion detection on cloud infrastructure exist but clouds are deployed as:

a. Private cloud is exclusive to an organization with multiple users. It is owned, operated and managed by organization, third party, or both; and may exist on or off premises.
b. Community cloud is exclusive to a specific community of users from an organization with shared concerns such as mission, security requirements, policy, and compliance considerations. It may be owned, managed, and operated by one or more of the organizations in the community, a third party, or both; and may exist on or off premises.
c. Public cloud is made for open use by the public. It may be owned, managed, and operated by a business, academic, or government organization, or some combination of them. It exists on the premises of the cloud provider.
d. Hybrid cloud combines two/more cloud infrastructures (private, community or public) with unique entities, but bound together by standardized or proprietary technology to enables data and application portability. For example, cloud bursts for load balancing between clouds.

3. Advanced Encryption Standard (AES)

AES is a specification for the encryption of electronic data established by the U.S. National Institute of Standards and Technology (NIST) in 2001. It is based on the Rijndael cipher developed by two Belgian cryptographers, Joan Daemen and Vincent Rijmen via proposal to NIST at the AES selection process. Rijndael is a family of ciphers with different key and block sizes. For AES, NIST selected three members of the Rijndael family, each with a block size of 128-bits, but three different key lengths: 128, 192 and 256 bits. AES 192/256 is approved for top-secret data by most Governments as closely aligned with public crypto.

Ureigho (2012) Crypto knowledge in the public and foreign intelligence domains has skyrocketed, and a vulnerability that the NSA can exploit is possibly a vulnerability that someone else can exploit. Thus, drafting of AES focuses on choosing a candidate standard that though may be broken given any amount of time and data, but will prove intractable for a time. Adversaries can only break crypto when they have the keys no matter how mathematically secure the crypto is. Most adversaries focus more on key retrieval via methods like brute force by attacking the endpoints that generate the keys. Though not as hard as it seems if we consider how many user and corporate machines get infected with malware alongside the sort/range of key-related backdoors are planted in popular software), and a simple subpoena may get keys in some situations. As more user data moves toward cloud, backdoors in public services (voluntarily provided or not) are going to make the job of key recovery even easier.

Hurtwiz et al (2010) AES was initiated in 1997 by National Institute of Standards and Technology (NIST), a unit of U.S. Commerce Department search to find a robust replacement for the Data Encryption Standard (DES) and to a lesser degree Triple DES. The specification called for symmetric algorithm (same key for encryption/decryption) using block encryption of 128-bits size, supporting key sizes of 128, 192 and 256 bits, as a minimum. The algorithm was required to be royalty-free for use worldwide and offer security of a sufficient level to protect data for the next 20 to 30 years. It was to be easy to implement in hardware and software, as well as in restricted environments (for example, in a smart card) and offer good defenses against various attack techniques.

Its selection process fully subjected to public scrutiny and preliminary analysis by the world cryptographic community to decide. This will ensure the best possible analysis design via its full visibility. On this submission that saw Rijndael were other cipher families also subjected to more extensive analysis namely: (a) MARS by IBM Research team, (b) RC6 by RSA Security, (c) Serpent by Ross Andersen, Eli Biham and Lars Knudsen, (d) Twofish by a large team of researchers including Counterpane's respected cryptographer, Bruce Schneier.

Implementations were tested extensively in C and Java for speed, reliability in cryptosystem, key usage, algorithm set-up time and its resistance to various attacks (both in hardware- and software-centric systems). Detailed analysis was provided by global crypto-community, and in 2000, NIST announced Rijndael (standardized in 2001 by the Secretary of Commerce and approved Federal Information Processing Standard). Thus, all sensitive, unclassified documents will use Rijndael as AES. Table 1 shows a number of key combinations and key size.

Table 1. *Key combination and size.*

Key Size	Possible Combinations
16-bits	65536
32-bits	$4.2 * 10^{9}$
56-bits (DES)	$7.2 * 10^{16}$
64-bits	$1.8 * 10^{19}$
128-bits (AES)	$3.4 * 10^{38}$
192-bits (AES)	$6.2 * 10^{57}$
256-bits (AES)	$* 10^{77}$

There is an exponential increase in possible combinations as key size increases. DES is a symmetric crypto algorithm with a key size of 56 bits that has been cracked in the past using brute force attack. The argument that a 128-bit symmetric key is computationally secure against brute-force attack is proved thus: if (a) fastest supercomputer of 10.51 Pentaflops = 10.51 x 10^{15} flops (floating point operations per second), (b) flops per combination check is 1000, (c) combination checks per second = (10.51 x 10^{15}) / 1000 = 10.51 x 10^{12}, and (d) number of seconds in one Year = 365 x 24 x 60 x 60 = 31536000, then the number of year required to crack 128-AES is given by:

$$Years\ to\ crack\ 128bit\ AES = \frac{3.4 * 10^{38}}{31536000(10.51 * 10^{12})}$$
$$= 1.02 * 10^{18} = 1\ Billion - Billion\ years$$

Table 2. *Key combination and time to crack.*

Key Size	Time to Crack
56-bits (DES)	399secs
128-bits (AES)	$1.02 * 10^{18}$ years
192-bits (AES)	$1.872 * 10^{37}$ years
256-bits (AES)	$3.31 * 10^{18}$ years

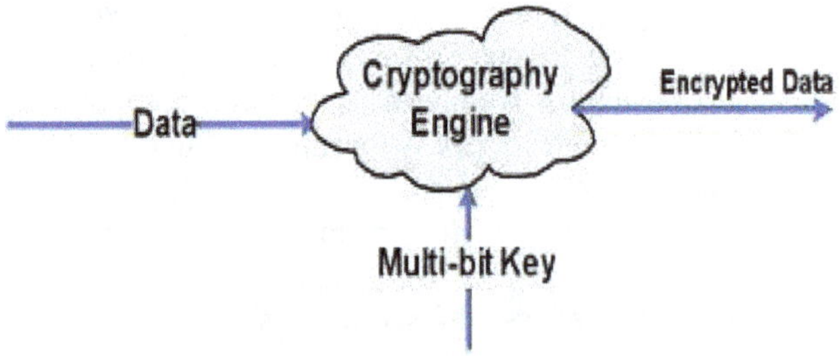

Fig. 2. *Multi-bit key cryptographic algorithm.*

AES's strength is generally expressed in length of the numeric 'key' used to scramble/unscramble messages. The study aims to display the strength of the AES against brute force attacks with different key sizes and the time it takes to successfully mount a brute force attack factoring future advancements in processing speeds. A cryptographic algorithm requires multi-bit key to encrypt the data as shown in Fig. 2.

3.1. *Brute Force Attack on AES*

The key length used in the encryption determines the practical feasibility of performing a brute-force attack, with longer keys exponentially more difficult to crack than shorter ones. Brute-force attack systematically checks all possible keys till correct key is found. Brute force attack on a 5-bit key is as in fig 3. This shows it takes max of 32-rounds to check every possible combination starting with 00000. Given sufficient time, a brute force attack is capable of cracking any known algorithm.

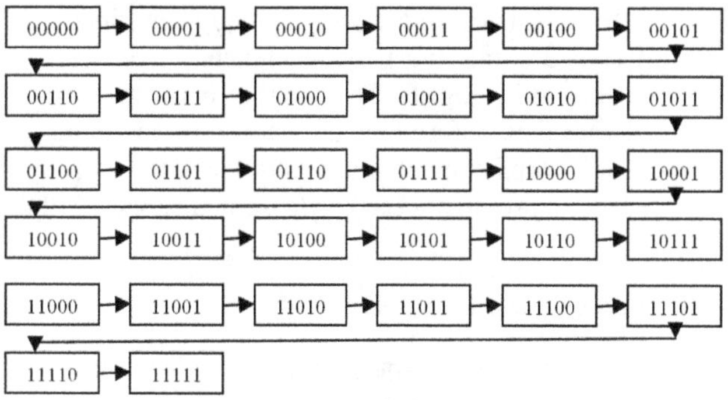

Fig. 3. *Brute force attack on 5-bit key.*

3.2. *Advanced Encryption Standard in Browsers*

The cipher AES-256 is used among other places in TSL/SSL across the Internet. It's considered amongst the top ciphers and in theory, it is less penetrable with its extensive combinations of keys, which is and remains massive.

Kangas (2012) SSL/TLS provide the majority of security in the data transmitted over the Internet today. Most users are unaware of the degree of security and privacy inherent in a secure connection ranging from almost none to a really good enough for US government TOPSECRET data. The cipher and encryption technique is what varies and thus provides the variable level of security needed. There are a large number of different ciphers. Some are very fast and very insecure. Some are slower and very secure. Others are weak (export-grade).

AES is a successor cipher and encryption technique to DES and was standardized in 2001 after a 5 year review. Currently, one of the most popular algorithms used in symmetric key cryptography (used for actual data transfer in SSL and TSL). It is also the gold-standard encryption method and many security-conscious organizations require its employees to use AES-256 (256-bit AES) for all communications.

This study highlights AES role in SSL, which web browsers and email programs support it, how to implement the 256-bit AES encryption on all secure communications, and more.

AES is FIPS (Federal Information Processing Standard) certified with no known brute-force attack success (except some side channel timing attacks on processing of AES that are not feasible over a network environment, not applicable to SSL in general). AES security is strong enough to be certified for use by most governments for top secret information.

Its design and strength of all key lengths (128, 192 and 256 bits) are sufficient to protect classified information up to the SECRET level. TOPSECRET data requires the use of either 192 or 256 key lengths. AES in products intended to protect national security systems and data must be reviewed/certified by NSA prior to their acquisition and use. (Hathaway, 2003)

It is often debated if 128-bit AES is computationally secure against brute-force attack. Governments and businesses place a great deal of faith in the belief that AES is so secure that its security key can never be broken. From table 2, it takes the fastest supercomputer, 1 billion-billion years to crack AES 128-bit via brute force attack. If we assume that a computing system existed that can recover DES key in 1sec, it take same machine approximately 149 trillion years to crack 128-bit AES. Though, difference in cracking AES-128bit and AES-256 is minimal as any breakthrough in 128-bit will probably render 256-bit tractable also. AES remains safe against brute force attacks contrary to belief and arguments. Its key size for encryption always be large enough despite the considerable advancements in processor speeds based on Moore's law as in fig. 3.

4. Framework and Implementation

Data stored in cloud receives malicious attempts. Clients may not understand security feats provided by Cloud Service Providers. Thus, study proposes a reliable AES encryption, employed at a client's end on community cloud to help protect data, separate from firewalls and other infrastructure in place by cloud providers. We models a security framework to make cloud "dependable" and achieve these:

a. Implement an integrated community-cloud that allows a user choice at sign-in unto the cloud infrastructure with AES-256 encryption at the client's end for improved data integrity against adversary to yield a dependable cloud.
b. Storage support for the integrated cloud with a remote server (completely transparent to user).
c. Data, at client's end is secured via AES-256 (to protect user data and message contents at end-to-end connection within the community cloud). SSL protects the username and password – alongside NAT, firewall and gateway that are implemented within the (Intranet) framework.
d. Sync, selected content in mobile devices to any available cloud technology on the model and ensure they are available anywhere and anytime you want to access or manage them which also takes out anxiety of losing important files, if device is damaged, lost or stolen.
e. Recommend suitable security models from existing benchmarks to improve user confidence in cloud computing services.

4.1. *The Nigerian Front for Integrated Mobile Cloud*

The framework will bring see many cloud service providers (infrastructure and services) brought together into one single user platform via mobile computing. It is an extension of the miniaturization process and faster computing on Moore's law, bringing about dependable and secure data storage capability to users via portable mobile devices.

4.2. Experimental Model Design Overview

The study provides a community-cloud model and support for users at Federal University of Petroleum Resources Effurun Nigeria. The model achieves this via a native app for mobile device (operable from Android v2.2) with support for web-service to allow Internet connectivity ease and connection to remote server and cloud-provider services via an API call (adapter). The framework will masks all technical nuances between application-model and data-models such as session management, connectivity, authentication and authorization. Its security is handled via AES-256 implemented on SSL/TSL applied to all data as backed by the cloud firewall to ensure security. Its client end-to-end encryption solution uses AES-256 to protect its data; while SSL protects username and password (see fig 4).

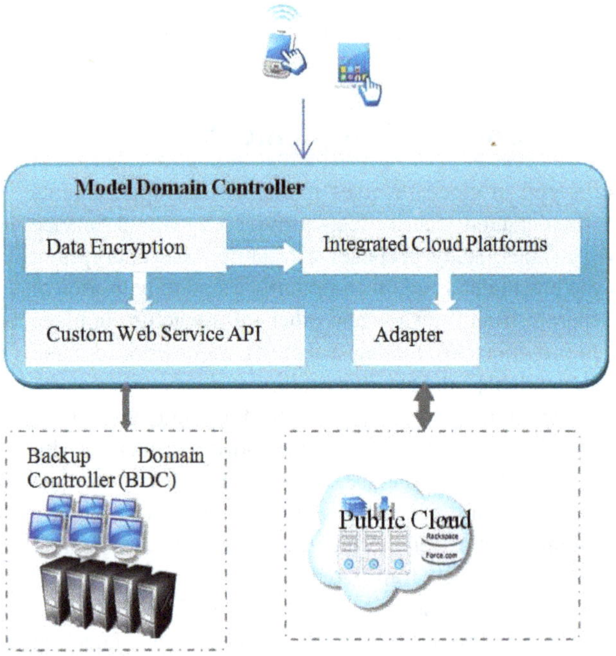

Fig. 4. Application and Data Model for FUPRE Community-Cloud.

Tools used for the development of this native app include Android SDK, Apache XAMPP and Google's Android Studio. This native app is enabled and ported on any Android platform from v2.2 with forward compatibility.

The system implements AES-256 encryption as supported by SSL/TSL. It is adopted because: (a) it is computationally, mathematically secure against brute force attacks, (b) quite flexible, (c) its small-size Java codes and support for C(++/#), (d) memory size required is small as ported on AP. Thus, has no effect on smartphone speed and performance, and (e) ease of integration as implemented with Java and support for C-language into its web browsers with ease of connectivity.

Web-browser used includes Safari, Firefox and Netscape – all of which enable AES-256 encryption on SSL/TLS protect data transfer between user and server. However, data transfer over the Internet between the sender and recipient remains unprotected, no matter how good SSL in use is.

4.3. System Implementation / Snapshots

Implementing the app on Android emulator with many of its snapshots as in Appendix, lets a user to download the native app and create an account so as to be able to sign-in and perform backup functions on contacts, messages, picture files, documents, videos and recorded voice amongst other backups and/or synchronization. After which, the user on first backup is notified to choose the cloud platform (whichever choice is made comes with its many benefits but the price difference is minimal). The technical nuances are not discussed such as billing etc. The app then performs AES encryption of user data before they are backed up too either to the remote server (backup domain controller – BDC) or to the integrated cloud provider's server via a cloud service API call (adapter).

4.4. Rationale for Cloud Implementation

AES is secure, its data encryption is more mathematically efficient, is elegant cryptographic algorithm with a model whose strength resides in the key-length options. Time required for an intruder to crack algorithm is directly proportional to length of key used to secure data transfer or communication. AES allows a choice between 192 and 256-bit implementation, making it exponentially stronger. There are no significant tradeoffs in functionality, speed and memory – as implementation on Android Studio makes it quite portable for target device. Required memory is relatively small and does not affect device speed even with extra functionalities. Target Android OS is v2.2 with forward compatibility to v4.0. This will bring closer to users, cloud technology with its many benefits at a cheap price.

5. Conclusion

The incessant need of users to protect stored, online data continues to foster the field of Data Forensic, which aims at measures to help detect network intrusion alongside keeping such adversaries off-bay via biometrics and cryptography so as to achieve the needed data non-repudiation, confidentiality, security and integrity for client end-to-end transaction. Cloud infrastructure is a system that enables 5-essential feats namely: self-provision, pay-per-use, on-demand resources availability, scalability and resource pooling. It consists of a physical layer of hardware resources necessary to support cloud services (such server, storage and network devices) and an abstraction layer of software-deployed on physical layer to manifests as cloud feats such as virtualization, grid computing, outsourcing and utility computing. Thus, study yields an integrated cloud on Android platform for smartphones as motivated by the need to proffer clients' transaction the much required security.

Appendixes

Appendix A. *Snapshot of PushCloud App installed on Android.*

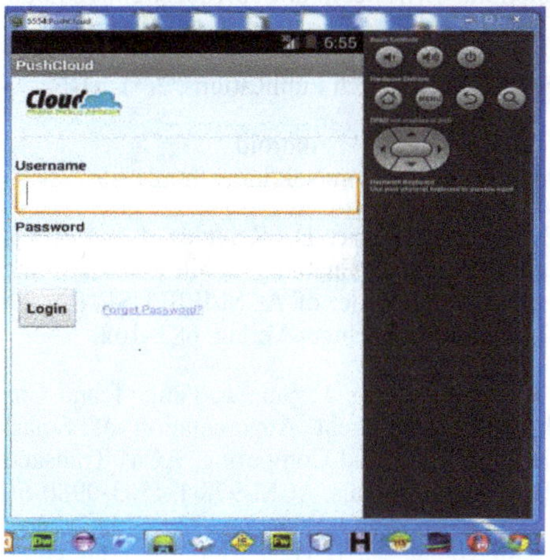

Appendix B. *User Account Login on PushCloud.*

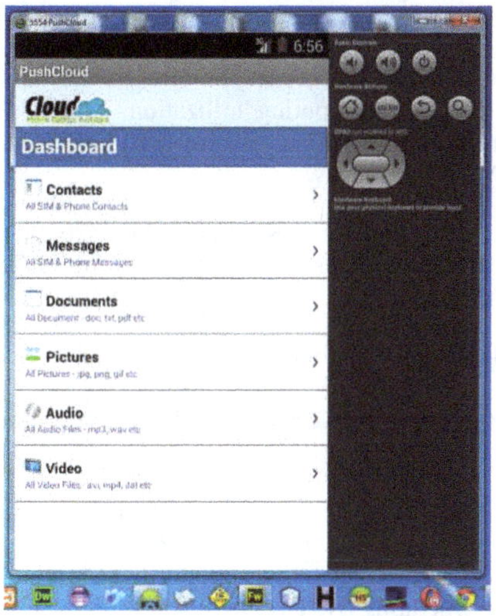

Appendix C. *PushCloud Dashboard with Menu for Contacts.*

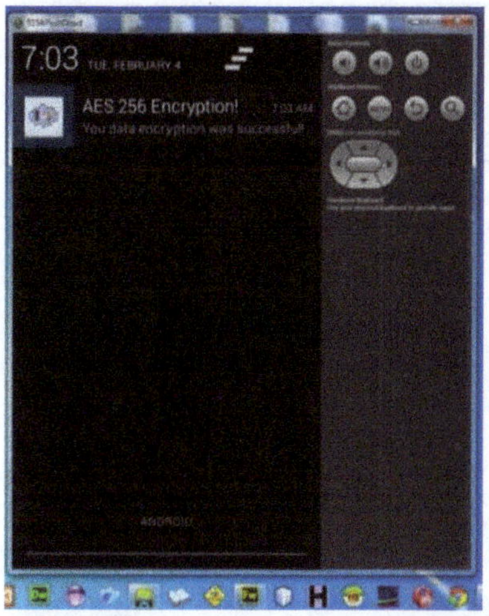

Appendix D. *Contact backup with AES Encryption first.*

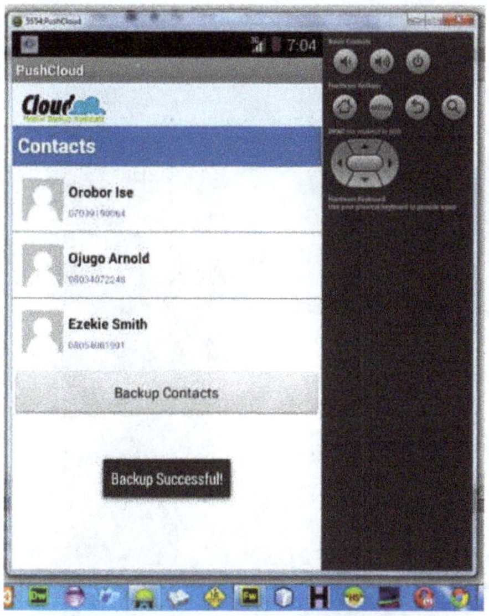

Appendix E. *Contacts Backup successful on PushCloud.*

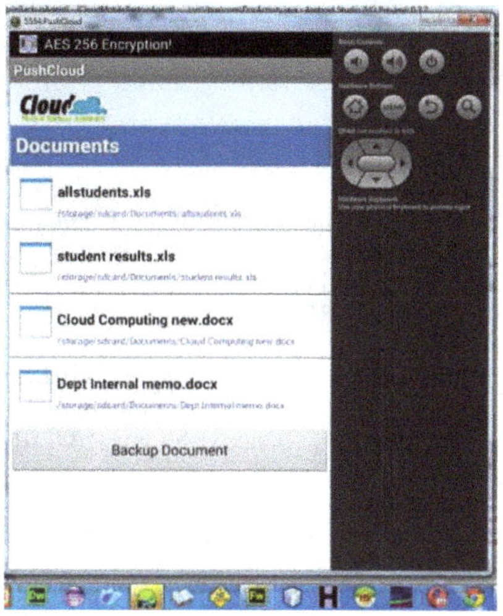

Appendix F. *PushCloud Integrated setting for User Choice of Cloud Infrastructure to backup Data.*

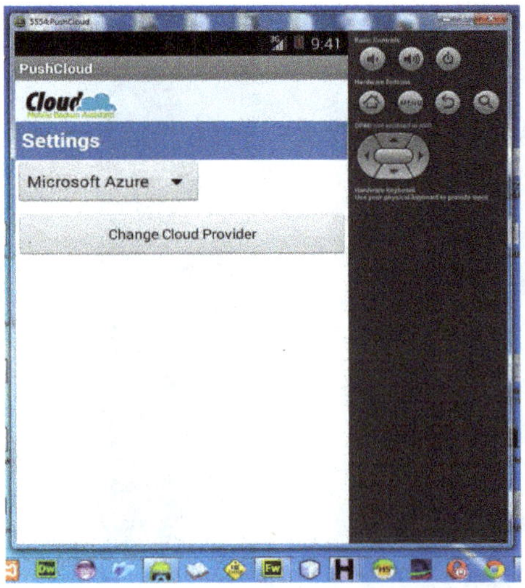

Appendix G. *Dropdown Menu of PushCloud Integrated setting for User Choice of Cloud Infrastructure to backup Data.*

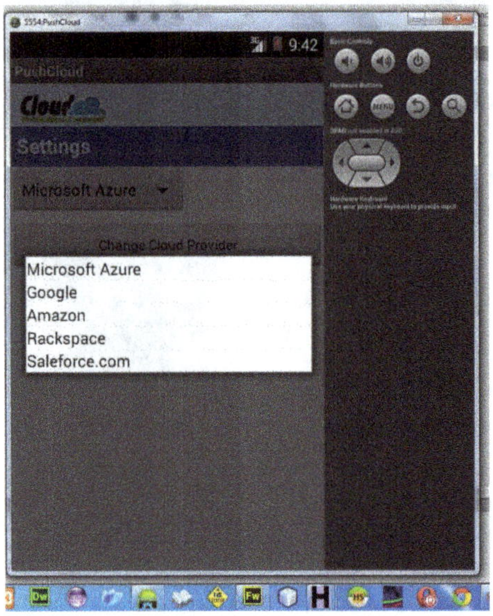

Appendix H. *Documents List with Backup Option.*

References

[1] Agam, S., (2011). Google's Android 4.0 ported to x86 processors, retrieved online via: http://www.computerworld.com/s/article/9222323/Google_s_Android_4.0_ported_to_x86_processors.

[2] Alonso, G., Rellermeyer, J and Roscoe, T., (2011). R-osgi: Distributed applications through software modularization, IFIP Lecture Notes in Computer Science (LNCS), 4834, p1–20.

[3] Android, Linaro Android Build Service, https://android-build.linaro.org/ last accessed January 2014.

[4] Bhupinder, S.M and Vijay, K.M (2010). Reliable Real-Time Applications on Android OS, retrieved online via: www.users.ece.gatech.edu/~vkm/Android_Real_Time.pdf.

[5] Bray, T., (2010). Ongoing by Tim Bray-What Android Is, www.tbray.org/ongoing/When/201x/2010/11/14/What-Android-Is.

[6] Chun, B., Ihm, S., Maniatis, P., Naik, M and Patti, A., (2011). Clonecloud: Elastic execution between mobile device and cloud, Proc. of 6th ACM conference on Computer systems, p301–314.

[7] Cohen, R., (2010). The cloud computing opportunity by the numbers. www.elasticvapor.com/2010/05/cloud-computing-opportunity-by-numbers.html.

[8] Cole, B., (2012). Real-time Android: real possibility, really hard to do - or just plain impossible? www.embedded.com/electronics-blogs/cole-bin/4372870/Real time-Android/real-possibility-really-really-hard-to-do-or-just-plain-impossible.html.

[9] Crysta X, Crysta X. NET, [online] www.crystax.net/nl/android/ndk/7. Last accessed January 2014.

[10] Dinh, H.T., Lee, C., Niyato, D and Wang, P., (2012). A survey of mobile cloud computing: architecture, applications and approaches, Wiley Wireless Communications and Mobile Computing, http://onlinelibrary.wiley.com/doi/10.1002/wcm.1203/abstract.

[11] Divya, V.L., (2012). Mobile application with cloud computing, Int. J. of Scientific Research Publications, 2(4), p1-6.

[12] Google, Android SDK, http://developer.android.com/sdk/index.html.

[13] Giurgiu, I., Riva, O., Juric, D., Krivulev, I and Alonso, G., (2009). Calling the cloud: Enabling mobile phones as interfaces to cloud applications, Proc. of ACM/IFIP/USENIX 10th Int. Conf. on Middleware. Springer-Verlag, p83–102.

[14] Guo, Y., Zhang, L., Kong, J., Sun, J., Feng, T and Chen, X., (2011). Jupiter: Transparent Augmentation of Smartphone Capabilities through Cloud Computing, ACM Transaction on Mobiheld, Portugal: Cascais, ACM-978-1-4503-0980-6/11/10.

[15] Gupta, P and Gupta, S., (2012). Mobile Cloud computing: future of cloud, Int. J. Adv. Res. in Electrical, Electronics and Instrumentation Engineering, 1(3), p134.

[16] Hamrén, O., (2012). Mobile phones and cloud computing: A quantitative research paper on mobile phone application offloading by cloud computing utilization, Master's Thesis, Dept of informatics, Human Computer Interaction SPM 2012.07, UMEA University.

[17] Hurwitz, J., Bloor, R and Kaufman, M., (2010). "Cloud computing for dummies: HP special edition", Wiley publications, New York.

[18] IBM, (2014). Inside the Linux 2.6 Completely Fair Scheduler, [Online]. www.ibm.com/developerworks/library/lcompletely-fair-scheduler/.

[19] Jeong, S., Zhang, X., Kunjithapatham, A and Gibbs, S., (2010) Towards an elastic application model for augmenting computing capabilities of mobile platforms, Mobile Wireless Middleware, Operating Systems, and Applications, p161–174.

[20] Kalkov,I., Franke, B and Schommer, J., (2012). A Real-time Extension to the Android Platform, In Proceedings of the 10th International Workshop on Java Technologies for Real-time and Embedded Systems, Copenhagen, Denmark.

[21] Krishnan, S., (2010). Programming Windows Azure. O'Reilly Media, Inc., 1005 Gravenstein Highway North, Sebastopol, CA 95472.

[22] Kumar, K. & Lu, Y. (2010). Cloud computing for mobile users: can offloading computation save energy? IEEE Computer Society.

[23] Lu, Y., Li, S and Shen, H., (2011). Virtualized screen: A third element for cloud-mobile convergence, IEEE Multimedia, 18(2), p4–11.

[24] Maia, C., Nogueira, L and Pinho, L.M., (2010). Evaluating Android OS for Embedded Real-Time Systems, Proceedings of 6th International Workshop on Operating Systems Platforms for Embedded Real-Time Applications, Brussels, Belgium.

[25] Marinelli, E.E., (2009). Hydrax: Cloud computing on mobile devices using MapReduce, Masters' Thesis, School of Computer Science, Carnegie Mellon University, Pittsburg, CMU-CS-09-164.

[26] Marrapese, B., (2010). Google ceo: a few years later, the mobile phone becomes a super computer. http://www.itnews-blog.com/it/21320.html.

[27] Mei, L., Chan, W and Tse, T., (2008). A tale of clouds: paradigm comparisons and some thoughts on research issues, IEEE Asia-Pacific Services Computing Conference APSCC'08, p464.

[28] Ojugo. A.A., Orobor, A.I., Yoro, R.E and Aghware, F.O., (2012a). Dependable community cloud model implemented using model view controller: a case of FUPRE, Technical report on cloud Technologies (FUPRE-Tech-03-2012), p11–24.

[29] Ojugo. A.A., Eboka. A.O, Okonta, E.O, Yoro, R.E and Aghware, F.O., (2012b). Implementation issues of VoIP for rural telephony in Nigeria, J. Emerging Trends in Computing and Info System, 4(2), p172.

[30] Ojugo, A.A., Eboka, A.O and Yoro, R.E., (2013a). Technical issues for IP-based network in Nigeria, J. of Wireless Communications and Mobile Computing, 2(2) p43-50.

[31] Ojugo, A.A., Eboka, A.O., Yerokun, M.O., Iyawa, I.J.B and Yoro, R.E., (2013b). Cryptography: salvaging exploitation against data integrity, Int. J. Networks and Communications, 2(2), p47-55, doi: 10.11648/j.ajnc.20130202.14.

[32] Pernel, L, Fayyad-Kazan, H and Timmerman, M., (2013). Android and real time application: take care, J. Emerging Trends in Computing and Info. Systems, 4, Special Issue ICCSII, ISSN 2079-8407.

[33] Rittinghouse, W.J and Ransome, F.J., (2010). Cloud Computing implementation, management and security. Boca Raton, FL: CRC Press

[34] Reese, G. (2009). Cloud Application Architectures. O'Reilly Media, Inc., 1005 Gravenstein Highway North, Sebastopol, CA 95472.

[35] Sarna, D.E.Y., (2011). Implementing and developing cloud computing applications. Taylor and Francis Group, Boca Raton, FL: CRC Press.

[36] Satyanarayanan, M., Bahl, P., Caceres, R and Davies, N., (2009). The case for vm-based cloudlets in mobile computing, IEEE Pervasive Computing, 8(4), p14–23.

[37] Shetty, K and Singh, S., (2011). Cloud Based Application Development for Accessing Restaurant Information on Mobile Device using LBS, Int. J. UbiComp, 2(4), DOI:10.5121/iju.2011.2404 37.

[38] Sung, A., Xu, J., Chavez, P., Mukkamala, S., (2004). Static analyzer of vicious executables, Proceedings of 20th Annual Computer Security Applications Conf., IEEE Computer Society, p326-334.

[39] Tapas Kumar, K and Kolin, P., (2010). Android on Mobile Devices: An Energy Perspective, IEEE 10th International Conference on Computer and Information Technology, Kuala-Lumpur, Malaysia.

[40] Ureigho, R.O.J., (2012). Dependable cloud computing: a framework for secure cloud, Unpublished PhD thesis, Department of Computer Science, Ebonyi State University Abakiliki, Ebonyi State.

[41] Vinutha, S., Raju, C.K and Siddappa, M., (2012). Development of hospital management system utilizing cloud computing and Android OS using VPN connections, Int. J. Sci. Tech. Res, 1(6), p59.

[42] Zhang, X., Kunjithapatham, A., Jeong, S and Gibbs, S., (2011). Towards an elastic application model for augmenting the computing capabilities of mobile devices with cloud computing, Mobile Networks and Applications, 16(3), p270–284.

2

Spherical Grid Protocol to Enhance Quality of Service to Resource Constrained Wireless Sensor Networks

Jai Prakash Prasad[1], Suresh Chandra Mohan[2]

[1]Visvesvaraya Technological University, Research Resource Centre, Belgaum, Karnataka, India
[2]Department of ECE, Bapuji Institute of Engineering & Technology, Davangere, Karnataka, India

Email address:
jaiaasu@gmail.com (J. P. Prasad)

Abstract: Wireless Sensor Network (WSN) is an emerging wireless communication networks to provide potential secure optimized data routing between source and destination. Presently the performance analysis of the Wireless Sensor Network routing technique and security protocols is the major research issues. A major issue in wireless sensor network (WSN) is the energy constraint in a node and its limited computing resources, which may pose an operational hazard or limitations on the network lifetime. Therefore, analyses of innovative secure routing techniques are required to utilize the resources of WSN to improve the life time of network in WSN. To design & develop any innovative wireless sensor network routing and security protocol, the network characteristics & its design issues are important consideration. The proposed spherical Grid Routing Protocol (SGRP) is designed and developed to enhance Quality of service to improve WSN performance using NS2 simulator compare to compared to modified LEACH techniques. The packet transmitted, packet received, average energy, throughput and packet delivery ratio are the main common performance measures that are used for performance analysis of proposed SGRP protocols.

Keywords: Wireless Sensor Network, Routing Protocols, NS2 Simulator, Packet Drop, Average Energy, Throughput, Packet Delivery Ratio

1. Introduction

In Wireless Sensor Network (WSN) each sensor node component is mainly consists of sensors, processor, memory and radio trans-receiver. Each sensor node is responsible to sense input attribute such as temperature, humidity, or pressure depending on the application involved and forward the sensed data to the destination using optimized energy efficient routing path. The figure 1 shows the general structure of WSN.

Presently there are plenty of algorithms for routing sensor sensed data in Wireless Sensor Network applications. Sensor nodes can be used for communication purposes with efficient use of their energy in various domains based on the requirement & utilization of resources more effectively to perform a specific task. Due to this WSN are extensively used in environmental monitoring, distributed control system, detection of radioactive sources, agricultural & farm practices, internet, military and surveillance. The characteristics of wireless sensor networks are summarized below:

- *Dense sensor node deployment:* Sensor nodes deployed for a specific application can be several orders of magnitudes.
- *Battery-powered sensor nodes:* Sensor nodes are deployed in an application where it is very difficult to replace or recharge the batteries.
- *Limited energy:* Sensor node batteries having limited energy.
- *Computation & Storage constraint:* Sensor nodes are having low computational & limited storage capabilities.
- *Self-configurable:* Any changes in network autonomously configure themselves into a communication network.
- *Unreliable sensor nodes:* Sensor node may fail due to its deployment in harsh or hostile environment.
- *Data redundancy:* Multiple sensor node deployed in a region have a certain level of redundancy w. r. to data.

The main design objectives of sensor networks are:-

- *Small node size:* Small node size reduces the power consumption and cost of sensor nodes.

- *Low node cost:* By designing low node cost result into the cost reduction of whole network.
- *Low power consumption:* Low power consumption of sensor node increases the lifetime of the sensor network.
- *Scalability:* Design requirement for sensor network must be scalable to grow in network size.
- *Reliability:* Proper error control and correction scheme will ensure reliable data delivery over noisy channel.
- *Self-configurability:* Sensor network must reconfigure themselves in case of topology changes and node failures.
- *Adaptability:* Network protocol designed should be adaptive to node density and topology changes.
- *Channel utilization:* Effective use of bandwidth improves channel utilization.
- *Fault tolerance:* Sensor nodes should have the abilities of self testing, self-calibrating, self-repairing, and self-recovering.
- *Security:* Security protocol which protect interception of sensor signal which causes loss of message confidentiality.
- *QoS support:* QoS is measured in terms of delivery latency and packet loss.

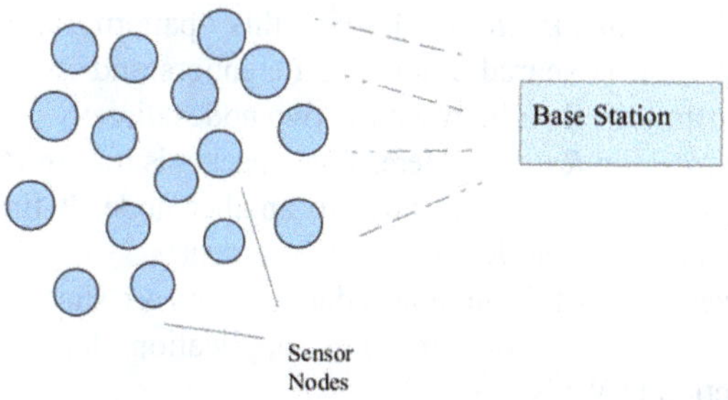

Fig. 1. *A general WSN architecture.*

The implementation of wireless sensor network in remote area and its maintenances is challenging task due to resource constrained nature of wireless sensor networks. To achieve data confidentiality and integrity in wireless sensor network there are presently available varieties of schemes in symmetric and asymmetric encryption scheme. To provide security to routed data while it is on the way from the unauthorized access by hacker there is need to design and develop optimized energy efficient secure routing protocol to improve quality of service as well as network life time compare to existing algorithm. A general public key asymmetric encryption scheme is shown in figure 2.

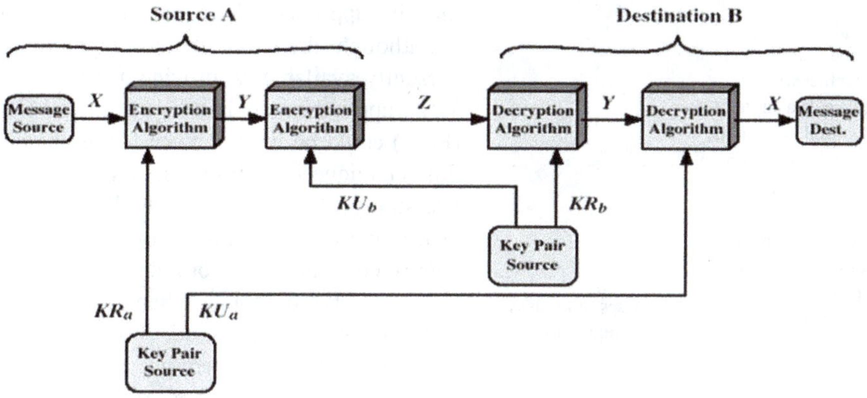

Fig. 2. *A Public Key Encryption Scheme.*

2. Traffic Patterns in WSN

There are varieties of wireless sensor network communication topological patterns as shown in figure 3. These Patterns are used to form a topology for the WSN.

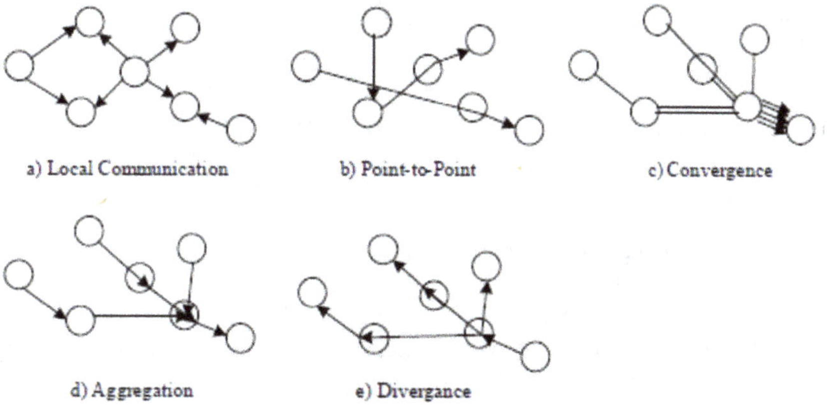

Fig. 3. *Types of Traffic Patterns in WSN.*

- *Local Communication:* Using this pattern anode information is shared among its neighbors and is used to transmit the data between the two nodes directly.
- *Point-to-Point Routing:* Here, a data packet is transmitted from a randomly chosen node to another node. It finds application in a wireless LAN environment.
- *Convergence:* Multiple nodes data packets are shared to a single base node. It finds application for data collection in WSNs.
- *Aggregation:* Processing of data packet happens in the intermediate nodes and the aggregated packet is routed to the base node.
- *Divergence:* It is used to send a request or data from a base node to another sensor node.

***Table 1.** A Summary on Comparison of Grid Routing.*

Grid Routing	Present Work	Scope for future work
GBDD	• First sink appearing in the sensor field triggers grid construction with sufficiently large lifetime. • Sink constructs new grid only when no valid grid is present. • Significant overall energy savings Improvements. • Smaller average packet delay.	Packet delay is more.
GRASS	• It combines the ideas of fixed cluster-based routing together with application-specific data aggregation. • It uses optimal as well as heuristic algorithms that solve the joint problem of optimal routing with data aggregation.	Latency is more.
ARA	• It taking into account, the residual energy of the sensor nodes and creating an adaptive route path. • ARA provides a network lifetime growth of about 20% from other algorithms. • ARA can be implemented also in WSN with randomly deployed nodes.	Uses of lossless compression algorithms.
GBDAS	• It constructs a chain by linking all cell heads so that sensed data can be disseminated along the chain. • The energy consumption of sensor nodes is evenly distributed to maximize their lifetimes. • Lifetime of network is better.	Redundancy is more.
CBDAS	• Each node in a cell takes turn to be cell head and each cell head takes turn to be cycle leader • The energy depletion is evenly distributed.	Extra cell selection Overheads

3. WSN Routing Techniques: A Review

Secure Routing technique uses a strategy to ensure connectivity between the different nodes in the WSN. A number of studies as shown in table 1 that compares Grid based Routing scheme. The GRID routing comparison is made to analyze & understand its strength and weaknesses. The popular grid based routing protocols are GBDD, GRASS, ARA, GBDAS & CBDAS.

4. Elliptic Curve Cryptography [ECC]

Elliptic curves are used in public key cryptography for securing information from unauthorized access. Let assume d is a private key which is randomly selected from [1, n-1], where n is integer no. Assuming Q being public key is computed by dP, where P, Q elliptic curve points. Once the key pair (d, Q) is generated, a variety of cryptosystems such as signature, encryption/decryption, and key management system can be set up. Then dP is calculated which is known as scalar multiplication. The term dP is also for the calculations of signature, encryption, and key agreement in the ECC system.

Intuitive approach:

$$dP = P+P+\ldots+P$$

It requires d-1 times point addition over the elliptic curve. For an example, to compute *17 P*, we could start with $2P$, double that, and that two more times, finally add P, i.e. $17P=2(2(2(2P)))+P$. This needs only 4 point doublings and one point addition instead of 16 point additions in the intuitive approach. This is called Double-and-Add algorithm.

Although there are many cryptography techniques are presently available, to provide a better security especially in WSN applications public key Elliptical Curve Cryptography (ECC) could be better choice for researchers. The benefit of this technique is that they uses smaller size key which need less storage, less bandwidth and less energy, thereby reducing processing and communication overhead, which is ideal for energy-constrained sensor nodes.

In the cryptographic schemes, elliptic curves over two finite fields are mostly used.

- Prime field F_p, where p is a prime.
- Binary field $F_2{}^m$, where m is a positive integer.

The equation of the elliptic curve over F_p is defined as:

$y^2 \bmod p = (x^3 + ax + b) \bmod p$

Where: $(4a^3 + 27b^2) \bmod p \neq 0$

x, y, a, b $\in [0, p-1]$

- Point addition for EC over F_p

$x_R = (\lambda^2 - x_P - x_Q) \bmod p$

$y_R = (\lambda(x_P - x_R) - y_P) \bmod p$

Where: $\lambda = ((y_Q - y_P)/(x_Q - x_P)) \bmod p$

- Point doubling for EC over F_P

$x_R = (\lambda^2 - 2x_P) \bmod p$

$y_R = (\lambda(x_P - x_R) - y_P) \bmod p$

Where: $\lambda = ((3x^2_P + a)/(2y_P)) \bmod p$

A elliptic curve E over the finite field $F_2{}^m$is given through the following equation,

$y^2 + xy = x^3 + ax^2 + b$

Where $x, y, a, b \in F_2{}^m$

- Point Addition and Doubling over $F_2{}^m$

Let $P=(x_P, y_P)$, $Q=(x_Q, y_Q)$ on the curve $y^2 + xy = x^3 + ax^2 + b$

The $R=P+Q$ can be computed:

$x_R = \lambda^2 + \lambda + x_P + x_Q + a$

$y_R = \lambda(x_P + x_R) + x_R + y_P$

Where: $\lambda = ((y_Q + y_P)/(x_Q + x_P))$

Then $R=2P$ can be computed:

$x_R = \lambda^2 + \lambda + a$

$y_R = x^2_P \lambda x_R + x_R$

Where: $\lambda = ((x_P + y_P)/(x_P))$

The implementation of ECC using Deffie-Hellmen algorithm as shown in figure 4 is explained below.

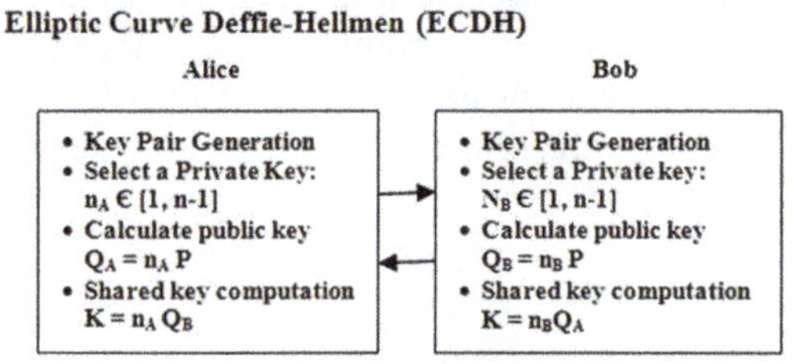
Elliptic Curve Deffie-Hellmen (ECDH)
Alice
Bob
• Key Pair Generation
• Select a Private Key:
$n_A \in [1, n-1]$
• Calculate public key
$Q_A = n_A P$
• Shared key computation
$K = n_A Q_B$
• Key Pair Generation
• Select a Private key:
$N_B \in [1, n-1]$
• Calculate public key
$Q_B = n_B P$
• Shared key computation
$K = n_B Q_A$

Fig. 4. *ECDH Algorithm.*

Consistency: $K=n_AQ_B=n_An_BP=n_BQ_A$

An Example of ECDH:

- Alice and Bob make a key agreement over the following prime, curve, and point.

  p=3851, E: $y^2=X^3+324x+1287$, P = (920, 303) $\in E$ (F3851)
- Alice chooses the private key $n_A \in 1194$,
- Computes $Q_A=1194P$= (2067, 2178) $\in E$ (F3851), and sends it to Bob.
- Bob chooses the private key $n_B=1759$
- Computes $Q_A=1759P$= (3684, 3125) $\in E$ (F3851), and sends it to Alice.
- Alice computes n_AQ_B=1194(3684, 3125) = (3347, 1242) $\in E$ (F3851)
- Bob computes n_BQ_A=1759(2067, 2178) = (3347, 1242) $\in E$ (F3851)

5. Spherical GRID Routing Protocol (SGRP): A Proposed Method

The Spherical GRID routing protocol architecture is shown in figure 5. Here sensor nodes are uniformly distributed over a field to monitor its environment. All sensor nodes transmit its data to its neighbor nodes using chain route in spherical fashion.

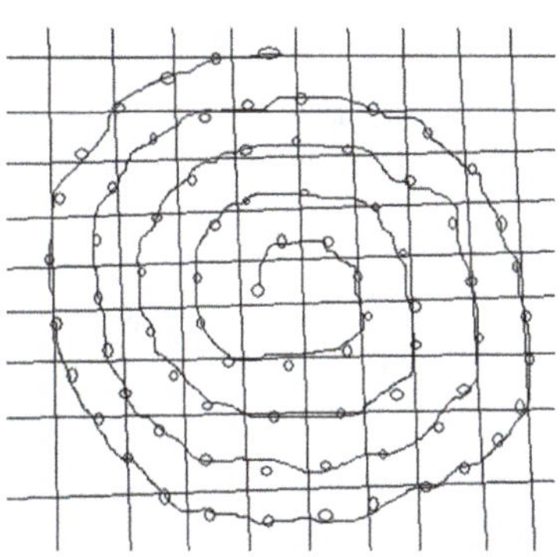

Fig. 5. *A General SGRP Networks.*

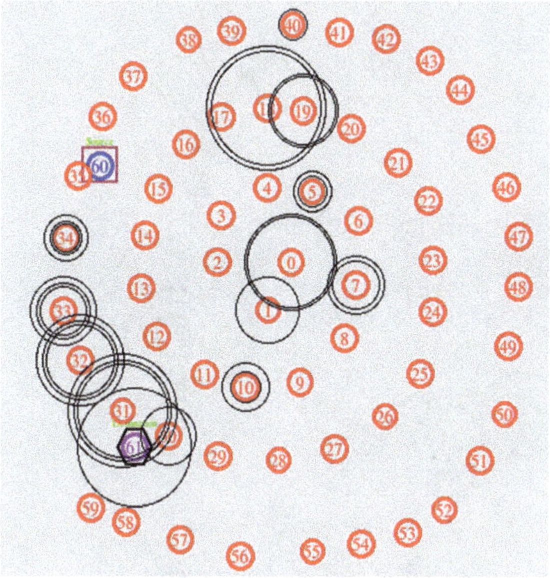

Fig. 6. *60 Nodes SGRP Network Scenario-I.*

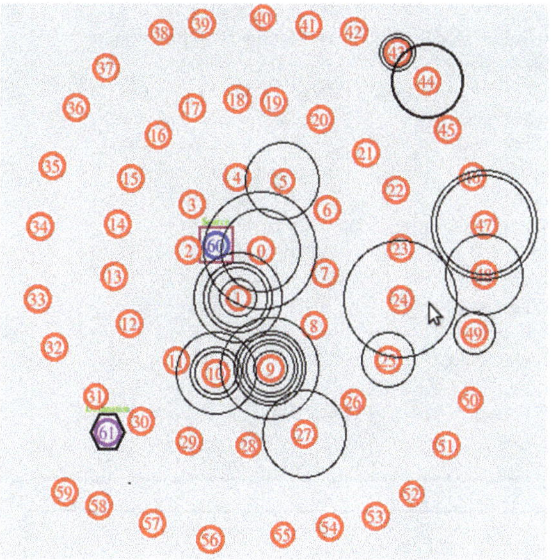

Fig. 7. *60 Nodes SGRP Network Scenario-II.*

A 60 nodes SGRP WSN network is simulated using NS-2 simulator. In network scenario-I & II as shown in figure 6 & 7 respectively, 60 nodes are arranged and distributed in spherical fashion. The source node no. 60 is indicated here as target nodes and its movement are traces by its nearest sensor nodes. The nearest node to source node 60 informs about it to the destination node no. 61 using spherical path. The performance efficiency of network is evaluated using performance metric such as transmitted packet, received packet, packet delivery ratio, average throughput and average residual energy. The simulation parameter setup of NS-2 is shown in table 2. The modified leach routing protocol & its topological setup is shown in figure 8. In modified LEACH Protocol whole network is divided into no. of clustered networks & exchange of packet takes place among clusters. Also the performance evaluation of SGRP algorithm versus modified LEACH protocol is shown in figure 9.

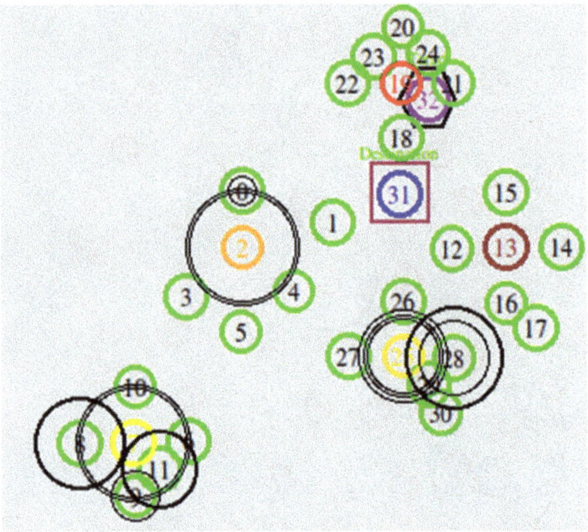

Fig. 8. *Modified Leach Protocol.*

Table 2. *Simulation parameters for WSN.*

Simulation Parameters	Value
Channel type	Wireless Channel
Radio-propagation model	Propagation/Two Ray Ground
Network interface type	Phy /WirelessPhy
MAC type	Mac/802_11
Interface queue type	Queue/DropTail /PriQueue
Link layer type	LL
Antenna model	Antenna/Omni Antenna
Max packet in ifq	50
Number of mobile nodes	16/25/36/49
Routing protocol	AODV
X dimension of topography	2000
Y dimension of topography	1000
Time of simulation end	80
Initial energy in Joules	80
Network Type	Mobile
Connection Pattern	Random
Packet Size	512 bytes
Connection type	CBR/UDP/TCP

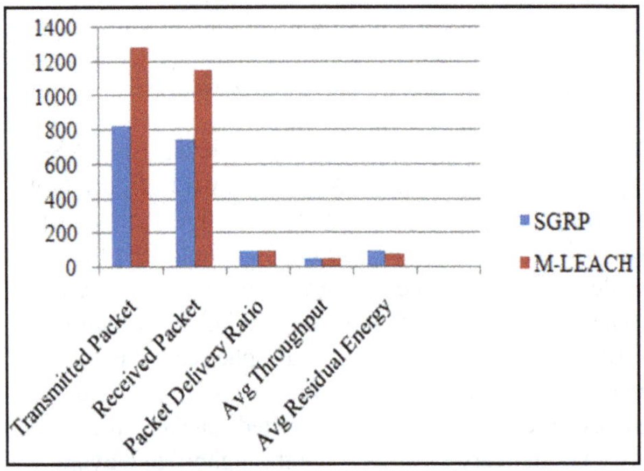

Fig. 9. *SGRP v/s Modified LEACH.*

6. Conclusion

The proposed Spherical Grid Routing protocol (SGRP) performance metrics are evaluated as i.e. transmitted packet, received packet, packet delivery ratio, average throughput and average residual energy and compared with the performance metrics calculations of modified LEACH Protocol. The conclusion from the analysis of results is that SGRP protocol achieves better performance compare to popular WSN modified LEACH protocol. Using these performances analysis the researchers gets better ideas to design and develop improved routing protocol by overcoming the limitations such as complete network failure due to a node energy exhaust of SPRG that can offer better PDR, Throughput, low packet drops & low power consumption in highly random mobility network. The outcomes from the result of performance metrics can adds extra life time in a node for providing better QoS in secure routing applications in real time practical applications using proposed method.

References

[1] W. R. Heinzelman, A. Chandrakasan, and H. Balakrishnan, “Energy efficient communication protocol for wireless micro sensor networks,” in Proceedings of the 33rd Annual Hawaii International Conference on System Sciences (HICSS), pp. 10–20, January 2000.

[2] Kemal Akkaya and Mohamed Younis, A Survey on Routing Protocols for Wireless Sensor Networks, Ad hoc Networks, vol. 3, no. 3, pp. 325-349, May 2005.

[3] Ananthram Swami et al., “Wireless Sensor Networks: Signal Processing and Communication Perspectives”, John Wiley, 2007.

[4] T. P. Sharma, R. C. Joshi, Manoj Misra, “GBDD: Grid Based Data Dissemination in Wireless Sensor Networks,” In Proc. 16th International Conference on Advanced Computing and Communications (ADCOM 2008), Chennai, India, 2008, pp. 234-240.

[5] Jamal N. Al-Karaki Raza Ul-Mustafa Ahmed E. Kamal, “Data Aggregation and Routing in Wireless Sensor Networks: Optimal And Heuristic Algorithms”, Computer Networks, Volume 53, Issue 7, Pages 945–960, 13 May 2009.

[6] Dragoş I. Săcăleanu, Dragoş M. Ofrim, Rodica Stoian, Vasile Lăzărescu, “Increasing lifetime in grid wireless sensor networks through routing algorithm and data aggregation techniques”, International Journal Of Communications, Issue 4, Volume 5, 2011.

[7] Neng-Chung Wang, Yung-Kuei Chiang, Chih-Hung Hsieh, and Young-Long Chen, “Grid-Based Data Aggregation for Wireless Sensor Networks”, Journal of Advances in Computer Networks, Vol. 1, No. 4, December 2013.

[8] Yung-Kuei Chiang, Neng-Chung Wang and Chih-Hung Hsieh, “A Cycle-Based Data Aggregation Scheme for Grid-Based Wireless Sensor Networks”, Sensors 2014, 14, 8447-8464; doi: 10.3390/s140508447.

[9] I. F. Akyildiz et al., A Survey on Sensor Setworks, IEEE Communication Mag., vol. 40, no. 8, Aug. 2002, pp. 102–114.

[10] W. R. Heinzelman, A. Chandrakasan, and H. Balakrishnan, Energy efficient communication protocol for wireless micro sensor networks, in Proceedings of the 33rd Annual Hawaii International Conference on System Sciences (HICSS), pp. 10–20, January 2000.

[11] T. P. Sharma, R. C. Joshi, Manoj Misra, "GBDD: Grid Based Data Dissemination in Wireless Sensor Networks," In Proc. 16th International Conference on Advanced Computing and Communications (ADCOM 2008), Chennai, India, 2008, pp. 234-240.

[12] Jamal N. Al-Karaki Raza Ul-Mustafa Ahmed E. Kamal, "Data Aggregation and Routing in Wireless Sensor Networks: Optimal And Heuristic Algorithms", Computer Networks, Volume 53, Issue 7, Pages 945–960, 13 May 2009.

[13] Dragoş I. Săcăleanu, Dragoş M. Ofrim, Rodica Stoian, Vasile Lăzărescu, "Increasing lifetime in grid wireless sensor networks through routing algorithm and data aggregation techniques", International Journal Of Communications, Issue 4, Volume 5, 2011.

[14] Neng-Chung Wang, Yung-Kuei Chiang, Chih-Hung Hsieh, and Young-Long Chen, "Grid-Based Data Aggregation for Wireless Sensor Networks", Journal of Advances in Computer Networks, Vol. 1, No. 4, December 2013.

[15] The network simulator - ns-2, http://www.isi.edu/nsnam/ns/

[16] Pritam Gajkumar Shah, Xu Huang, Dharmendra Sharma, "Analytical study of implementation issues of Elliptical Curve Cryptography for Wireless Sensor Networks", IEEE 24th International Conference on Advanced Information Networking and Application Workshops, 2010, pp. 589-592, IEEE, 2010, DOI 10.1109/WAINA.2010.47.

[17] N. Koblitz, "Elliptical curve cryptosystems", Mathematics of Computation, Vol. 48. pp. 203-209, 1987.

[18] Y. Shou, H. Guyennet, and M. Lehsaini, "Parallel Scalar Multiplication on Elliptic Curves in Wireless Sensor Networks", 14th Int. Conf. on Distributed Computing and Networking (ICDCN), LNCS 7730, pp. 300-314, Bombay, India, Jan 2013.

[19] Wenbo Shi and Peng Gong, "A New User Authentication Protocol for Wireless Sensor Networks using Elliptical Curves Cryptography" in proceedings of Hindawi Publishing Corporation International Journal of Distributed Sensor Networks, Vol. 730831, 1-7, 2013.

3

Evaluation of SNR for AWGN, Rayleigh and Rician Fading Channels Under DPSK Modulation Scheme with Constant BER

Deepak K. Chy[1], Md. Khaliluzzaman[2]

[1]Department of Electrical & Electronic Engineering, University of Information Technology & Sciences (UITS), Dhaka, Bangladesh
[2]Department of Computer Science & Engineering, University of Information Technology & Sciences (UITS), Dhaka, Bangladesh

Email address:
dk_chy53@yahoo.de (D. K. Chy), khalil_021@yahoo.co.in (M. Khaliluzzaman)

Abstract: With the growing demand in modern communication, it has become necessary to give better and efficient service to users by using better technique. Technique such as Amplitude Shift Keying (ASK), Frequency Shift Keying (FSK), Phase Shift Keying (PSK), Differential Phase shift Keying (DPSK) and Quadrature Amplitude Modulation (QAM) are very important parts of the implementation of modern communications systems in which DPSK is the simplest and most robust of all techniques. In this paper, evaluation of SNR in terms of constant bit error rate is performed on AWGN, Rayleigh and Rician fading channels. Among these channels, Rician is showing better performance as compared to AWGN and Rayleigh.

Keywords: DPSK, SNR, BER, AWGN, Rayleigh, Rician

1. Introduction

Mobile communications and wireless network have experienced massive growth and commercial success in the recent years. However, the radio channels in mobile radio systems are usually not amiable as the wired one. Unlike wired channels that are stationary and predictable whereas wireless channels are extremely random and time-variant.

In general Communication can be defined simply as 'sending and receiving messages', or 'the transmission of messages from one person to another'. Effective communication occurs only when the receiver understands the exact message sent by the transmitter [1].

The next generation of wireless communication systems faces the demand for increased data rates, higher mobility, larger carrier frequencies, and more link reliability. Wireless channels are characterized by fading, multipath, limited bandwidth, and frequency and time selectivity which make system design a challenge. It is therefore crucial to have an understanding of the behavior of wireless channels in order to know their performance limits and to be able to design efficient communication systems for them. This paper considers the analysis of the performance of digital communication systems with different coding and modulation schemes. Although digital communication is much better than the analog communication, still it has certain issues that need to be addressed. Especially when it comes to wireless communication, one of the major research considerations becomes the effect of multipath propagation. A thorough analysis is necessary for strategic planning of any system design by doing comparative study of different modulation techniques via different multipath communication channels. To study and draw the graph in terms Bit Error Rate (BER) versus E_b / No in multipath communication channels for modulation schemes. Therefore, understand the system could go for more suitable modulation technique to suit the channel quality and can suggest better modulation schemes [2].

The performance of data transmission over wireless channels is well captured by observing their BER, which is a function of Signal to Noise Ratio (SNR) [5] at the receiver. In wireless channels, several models have been proposed and investigated to calculate SNR. All the models are a function of the distance between the sender and the receiver, the path loss exponent and the channel gain. Several probability distributed functions are available to model a time-variant parameter i.e. channel gain. We describe the three important and frequently used distributions. Those are Additive White Gaussian Noise (AWGN), Rayleigh and Rician models [9].

The remaining paper is organized as follows. In section II, channel models are described. In section III, fading is given. In section IV Modulation Techniques are explained and simulation results are given in section V. The paper is concluded in section VI.

2. Channel Model

Wireless communication is now become an important part in our daily life and it is widely used in the technology development areas. Assembling of the various channels can be done accurately because the performance and the design of the channels depend upon the accuracy of the simulation. In the wireless communication field, fading is the important consideration because it tells about the fading patterns in the various conditions. There is no such model which tells about the environment. A signal that has chosen should be error free, or close to being error free [3]. If the signal is error free then the high quality of voice and data transmission can be done. The main issue arises while the development of the application is that the selection of the fading model. The analysis based on the DPSK will give the idea which helps for the application development in the market [4].

There are three main basic fading channel models i.e. Additive White Gaussian Noise (AWGN), Line of Sight (Rician) and Non Line of Sight (Rayleigh) Fading Channel models.

2.1. Rayleigh Fading Channel

Rayleigh fading occurs due to the multilink reception. In Rayleigh fading model the effect of the environment spreading to a larger area on a radio signal. It is one of the cheapest models of the signal propagation (i.e. for ionosphere and troposphere). Rayleigh fading is most applicable when there is no dominant propagation along a line of sight between transmitter and receiver. If the channel impulse response will be modeled as a Gaussian process with respect to the distribution of the individual components and if the process has zero mean and phase lie between 0 to 2π radians [6]. Then, the probability density function can be given by:

$$P_R(R) = \frac{R}{e^2}\, e - \frac{R^2}{2\sigma^2}\ ,\ 0 \le R < \alpha \qquad (1)$$

2.2. Rician Fading Channel

Rician Fading is a part of Rayleigh fading with the introduction of a strong line of sight path in the Rayleigh fading environment. Rician fading is worthy for satellite communications and is acceptable for some urban scenarios. Rician fading is a type of small-scale fading because the probability of deep fades is less than that in the Rayleigh-fading case [6]. The probability density function of the amplitude is a Rician distribution and is mathematically expressed as follows:

$$P_R(R) = \frac{R}{e^2}\, e - \frac{R^2+A^2}{\sigma^2}\, I_0\left(\frac{RA}{\sigma^2}\right),\ 0 \le R < \alpha \qquad (2)$$

2.3. Additive White Gaussian Noise Channel

For the case of Doppler Effect between a moving source and stationary receiver, narrowband data model is used to model the received signal at the antenna arrays. It presumes that the enclosure of the signal wave front inseminating across the antenna array necessarily remains constant [6]. This model is valid for the signals having bandwidth much smaller than the carrier frequency. According to above hypothesis, the received signal can be written as

$$H(t) = A(\Theta)b(t) + N(t) \qquad (3)$$

Where, $A(\Theta)$ is the array manifold vector and $N(t)$ is AWGN with zero mean and two-sided power spectral density given by N_0 /2.

3. Fading

Fading refers to the distortion that a carrier-modulated telecommunication signal experiences over certain propagation media. In wireless systems, fading is due to multipath propagation and is sometimes referred to as multipath induced fading. To understand fading, it is essential to understand multipath. In wireless telecommunications, multipath is the propagation phenomenon that results in radio signals' reaching the receiving antenna by two or more paths. Causes of multipath include atmospheric ducting, ionosphere reflection and refraction, and reflection from terrestrial objects, such as mountains and buildings. The effects of multipath include constructive and destructive interference, and phase shifting of the signal. This distortion of signals caused by multipath is known as fading. In other words it can be said that in the real world, multipath occurs when there is more than one path available for radio signal propagation. The phenomenon of reflection, diffraction and scattering all give rise to additional radio propagation paths beyond the direct optical LOS[7] (Line of Sight) path between the radio transmitter and receiver.

A Fading Channel is known as communications channel which has to face different fading phenomenon's, during signal transmission. In real world environment, the radio propagation effects combine together and multipath is generated by these fading channels. Due to multiple signal propagation paths, multiple signals will be received by receiver and the actual received signal level is the vector sum of the all signals. These signal incident from any direction or angle of arrival. In multipath, some signals aid the direct path and some others subtract it.

3.1. Fading on the Basis of Effect of Multipath

Large scale fading is defined as the fading which depends upon the location with respect to objects or it shows clearly in case of the short distance of the transmitter or the receiver. A continuous variation in the phase and amplitude occurs when a signal moves from a distance in the order of wavelength or it can also say that the small scale fading refer to the changes occur in the position of the transmitter and receiver in order of wavelength [8]. These changes are very small.

3.2. Fading on the Basis of the Doppler Spread Effect

Slow fading occurs when the minimum time required for the channel is large to change its magnitude from its previous value relative to the delay behavior of the channel. Slow fading can also be formed by shadowing. In shadowing, when large buildings or hills create problem for the path of the main signal of the transmitter and receiver, the received power is obtained by shadowing can be modeled by using log -distance path loss or log-normal distribution. The minimum time required for the channel is to change its magnitude from its previous value relative to the delay behavior of the channel is known as Fast fading.

4. Modulation

One way to communicate a message signal whose frequency spectrum does not fall within that fixed frequency range, or one that is unsuitable for the channel, is to change a transmittable signal according to the information in the message signal. This alteration is called *modulation*, and it is the modulated signal that is transmitted. The receiver then recovers the original signal through a process called *demodulation.*

Good bit error rate performance, less power consumption and good spectral efficiency are the Properties of modulation techniques. In digital system, the message signal is to be transmitted is digital in nature or we can say that the transmission of the information in digital form. There are various types of modulating schemes involves in the communication system like Phase shift keying(PSK), Frequency shift keying(FSK), Minimum shift keying(MSK), Quadrature phase shift keying(QPSK), Quadrature amplitude modulation(QAM) and Differential Phase Shift Keying (DPSK).

4.1. Differential Phase Shift Keying

Differential phase shift keying (DPSK) [9], a common form of phase modulation conveys data by changing the phase of carrier wave. In Phase shift keying, High state contains only one cycle but DPSK contains one and half cycle. Differential Shift Keying is a modulation technique that codes information by using the phase difference between two neighboring symbols. In the transmitter, each symbol is modulated relative to the previous symbol and modulating signal, for instance in BPSK 0 represents no change and 1 represents +180 degrees. In the receiver, the current symbol is demodulated using the previous symbol as a reference. The previous symbol serves as an estimate of the channel. A no change condition causes the modulated signal to remain to remain at the same 0 or 1 state of the previous symbol. Differential modulation is theoretically 3 dB poorer than coherent. This is because the differential system has 2 sources of error: a corrupted symbol, and a corrupted reference.

In DPSK, the transmitter, each symbol is modulated relative to the phase of the immediately preceding signal element and the data being transmitted. In this paper, we choose M-DPSK where M=2, 4, 8, 16, 32, 64 scheme to analyze SNR with constant BER in different fading channels.

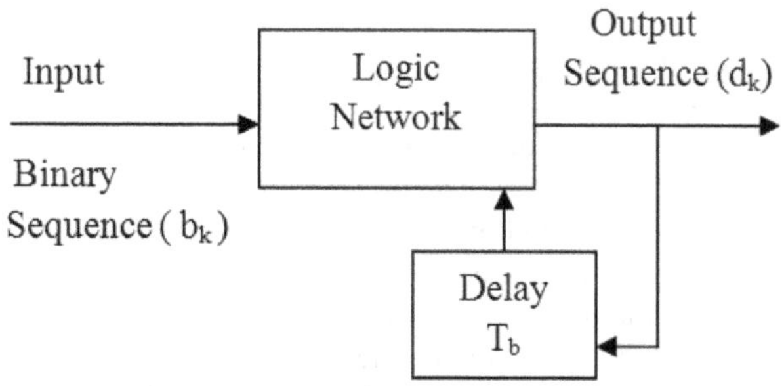

Figure 1. *Block Diagram of DPSK Modulation.*

4.2. Bit Error Rate (BER)

The BER or quality of the digital link is calculated from the number of bits received in error divided by the number of bits transmitted.

BER= (Bits in Error) / (Total bits received).

In digital transmission, the number of bit errors is the number of received bits of a data stream over a communication channel that has been altered due to noise, interference, distortion or bit synchronization errors. The BER is the number of bit errors divided by the total number of transferred bits during a particular time interval. BER is a unit less performance measure, often expressed as a percentage.

For example, N erroneous bits out of 1000 bits transmitted would be expressed as $N*10^{-3}$. In this paper we assume that one erroneous bit out of 1000 bits would be transmitted. That is the bit error rate is $1*10^{-3}$.

4.3. Signal to Noise Ratio (SNR)

SNR is the ratio of the received signal strength over the noise strength in the frequency range of the operation. It is an important parameter of the physical layer of Local Area Wireless Network (LAWN). Noise strength, in general, can include the noise in the environment and other unwanted signals (interference). BER is inversely related to SNR, that is high BER causes low SNR. High BER causes increases packet loss, increase in delay and decreases throughput. The exact relation between the SNR and the BER is not easy to determine in the multi channel environment. Signal to noise ratio (SNR) is an indicator commonly used to evaluate the quality of a communication link and measured in decibels and represented by Eq. (4).

$$SNR = 10 \log_{10} (\text{Signal Power} / \text{Noise Power}) \text{ dB} \quad (4)$$

SNR measures the quality of a transmission over a network channel. The grater the signal to noise ratio, the easier it is to identify and subsequently isolate and eliminate the source of noise. A SNR of zero indicates that the desired signal is virtually indistinguishable from the unwanted noise.

In this paper, we tested various fading channel in accordance with SNR with constant bit error rate, in our case which is 10^{-3}.

5. Simulation and Results

In this paper, one of the important topics in wireless communication, which is the concept of fading, is demonstrated by the approach available in MATLAB [10].

A wireless Communication system was designed in MATLAB. It was assumed the data was first encoded with linear block coding and then was transmitted thought the channel. The transmitted signal is distorted by noise which was assumed as Additive white Gaussian noise. The fading effect was also taken in consideration. So, different fading channel mode was considered.

The important topic in wireless communications, that is the concept of fading, is demonstrated by the approach available in MATLAB. One of the important aspects of the path between the transmitter and receiver is the occurrence of fading. MATLAB provides a simple and easy way to demonstrate fading taking place in wireless systems. Statistical testing can subsequently be used to establish the validity of the fading models frequently used in wireless systems. The different fading models and MATLAB based simulation approaches will now be described.

In order to be statistically significant, each simulation must generate some number of errors. If any channel does not contain any noise the channel capacity will be infinite (according to Shannon's channels capacity theorem) in practice no channel can be noise free. In this simulation the test signal contains 1000 bits and receiver will identify only one bit error out of 1000 bits, that is bit error rate is 0.1% since a bit error rate of one percent is considered quiet high, that's why we consider BER much lower than 1%.

The simulation is followed by using m file. In this approach, the simulation is successfully done using DPSK modulation technique. The desired BER versus SNR are obtained for simulation in AWGN, Rayleigh and Rician channels.

5.1. Transmission through AWGN Channel

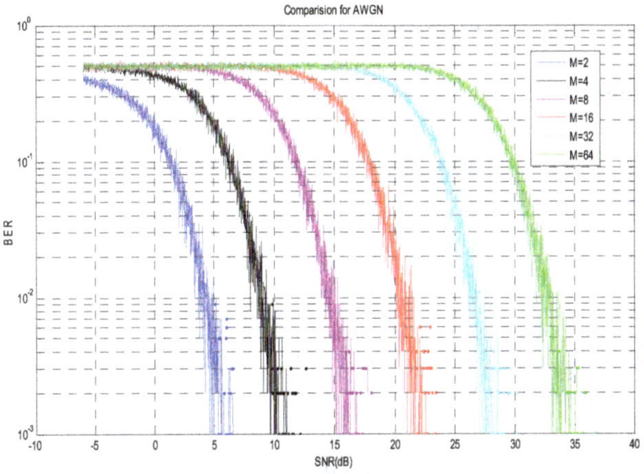

Figure 2. *Bit Error Rate vs. SNR over AWGN channel for 2, 4, 8, 16, 32, 64 DPSK.*

Figure 2 shows Bit Error Rate vs. SNR over AWGN channel for 2, 4, 8, 16, 32, 64 DPSK.

5.2. Transmission through Rayleigh Channel

Figure 3 shows Bit Error Rate vs. SNR over Rayleigh channel for 2, 4, 8, 16, 32, 64 DPSK.

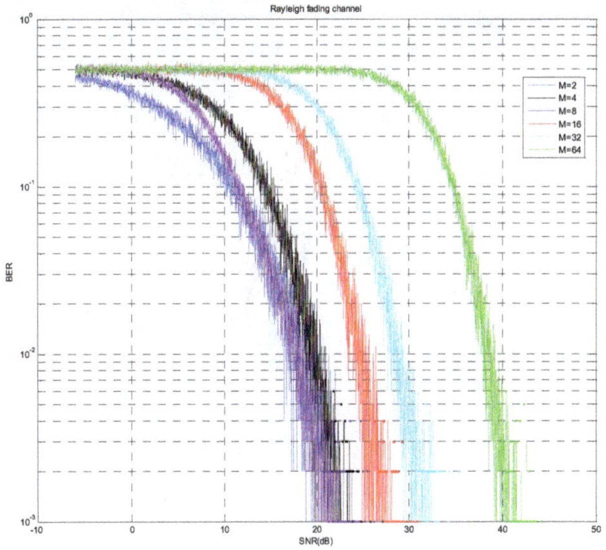

Figure 3. *Bit Error Rate vs. SNR over Rayleigh channel for 2, 4, 8, 16, 32, 64 DPSK modulations.*

5.3. Transmission through Rician Channel

Figure 4 shows Bit Error Rate vs. SNR over Rician channel for 2, 4, 8, 16, 32, 64 DPSK.

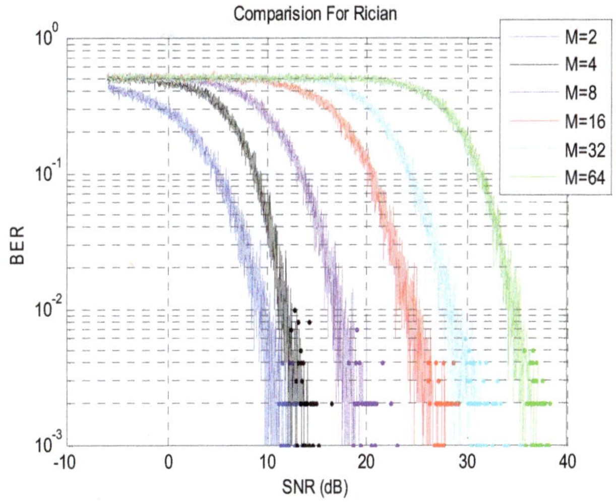

Figure 4. *Bit Error Rate vs. SNR over Rician channel for 2, 4, 8, 16, 32, 64 DPSK modulations.*

We can understand from the above figure 2, 3 and 4 that, the value of SNR is rising with the higher value of M for the constant value of BER which is 10^{-3}. If the value of M is increased, the SNR will be increased and for that reason the transmitted signal will be distorted and it will be time consuming and costly to recover the original signal.

5.4. Comparison Transmission between AWGN, Rayleigh and Rician Channel

Figure 5 shows Bit Error Rate vs. SNR over AWGN, Rayleigh and Rician channel for 4 DPSK modulations.

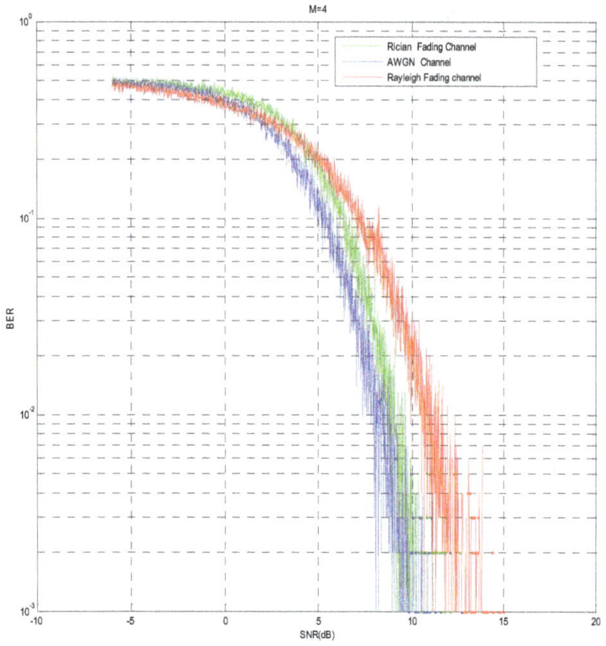

Figure 5. *Bit Error Rate vs. SNR over AWGN, Rayleigh and Rician channel for 4 DPSK modulations.*

Figure 6 shows Bit Error Rate vs. SNR over AWGN, Rayleigh and Rician channel for 16 DPSK modulations.

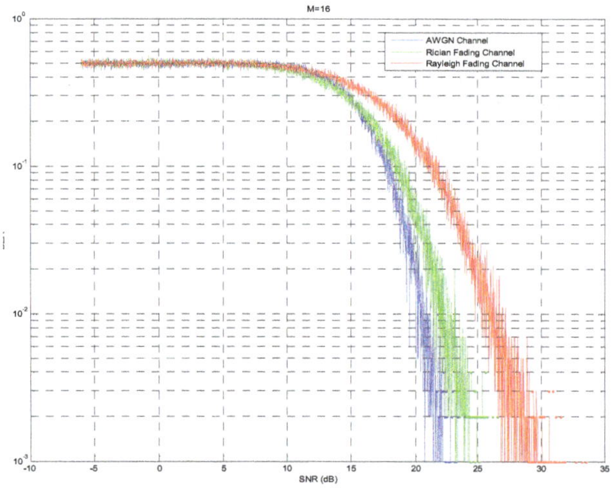

Figure 6. *Bit Error Rate vs. SNR over AWGN, Rayleigh and Rician channel for 16 DPSK modulations.*

Figure 7 shows Bit Error Rate vs. SNR over AWGN, Rayleigh and Rician channel for 64 DPSK modulations.

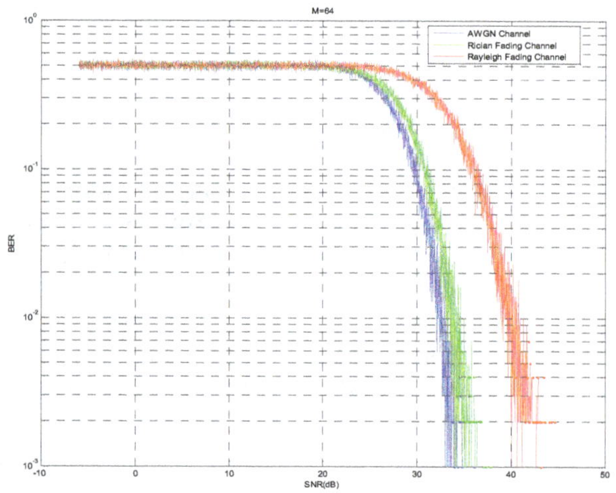

Figure 7. *Bit Error Rate vs. SNR over AWGN, Rayleigh and Rician channel for 64 DPSK modulations.*

It has seen from the figure 5, 6 and 7 that, for M=4, 16 and 64 the demodulated signals of different AWGN, Rayleigh and Rician Fading Channel are almost like waterfall with a different SNR value. AWGN channel is not a fading channel, so the BER vs. SNR rate will be compared with Rayleigh and Rician Fading channel. From those channels Rician Channel has both LOS (Line of Sight) and multipath, so the SNR value for Rician will definitely lower than the Rayleigh fading channel for the direct transmission path. The Rician Channel is much better for signal transmission compared with the Rayleigh fading channel.

Table 1. *BER vs. SNR Comparison for AWGN, Rayleigh and Rician Channel.*

BER	Values of M	SNR(dB)		
		AWGN	Rayleigh	Rician
10^{-3}	4	10	14	12
10^{-3}	16	22	28	24
10^{-3}	64	32	42	36

For DPSK, various values of SNR with constant BER 10^{-3} are obtained using different values of M shown in the table 1. When the value of M is 4, SNR is 10dB for AWGN, 14 dB for Rayleigh, 12 dB for Rician. When the value of M is 16, SNR is 22dB for AWGN, 28 dB for Rayleigh, 24 dB for Rician. For the value of M equal to 64, SNR is 32dB for AWGN, 42 dB for Rayleigh, 36 dB for Rician.

Rayleigh fading channel has comparatively worst performance with compared to AWGN and Rician fading channel. This is because Rayleigh fading has no LOS from transmitter to receiver.

6. Conclusion

In Wireless communication system, the needs of data rate are increasing day by day which requires more bit rate in same channel bandwidth. Multi level digital modulation schemes are the one of such techniques by which it is possible to transmit more bit rate in same channel. In such modulation technique, for *k*-bits of information, one of the $M=2^k$ possible symbols is used to modulate a carrier signal, which results of *k*-times increment of bit rate in same bandwidth.

From simulation results, it is seen that for same values of M the Rician channel exhibits better performance than AWGN and Rayleigh in terms of SNR with constant BER.

References

[1] Swamy M. Katta, Deepthi M., et.al, "Performance Analysis of DSSS and FHSS Techniques over AWGN Channel", International Journal of Advancements in Technology, Volume 4, No. 1, March 2013.

[2] Fating Pooja P., Ashtankar Pankaj S., "Comparative Study of Different Modulation Techniques for Multipath Communication Channel", Proceedings of 4th SARC International Conference, Nagpur, India, ISBN: 978-93-82702-70-2 March 30th, 2014.

[3] Mohammaed Slim Alouini and Andrea J. Goldsmith "Capacity of Rayleigh fading channels under different Adaptive Transmission and Diversity combining Techniques", IEEE Transactions on Vehicular Technology, Vol. 48, No. 4, July 1999.

[4] Gary Breed, High Frequency Electronics, 2003 Summit, Technical Media LLC "Bit Error Rate: Fundamental Concepts and measurement issues".

[5] Fumiyaki Adachi, "error Rate Analysis of Differentially Encoded and detected 16-APSK under Rician fading", IEEE Transactions on Vehicular Technology, Vol. 45, No. 1, February 1996.

[6] Mohammaed Slim Alouini and Andrea J. Goldsmith " Capacity of Rayleigh fading channels under different Adaptive Transmission and Diversity combining Techniques", IEEE Transactions on Vehicular Technology, Vol. 48, No. 4, July 1999.

[7] Gupta, Akhil. "Improving Channel Estimation in OFDM System Using Time Domain Channel Estimation for Time Correlated Rayleigh Fading Channel Model." International Journal of Engineering and Science Invention, vol. 2, issue 8, pp. 45-51, August. 2013

[8] "Quadrature Amplitude Modulation", digital Modulation Techniques" www.digitalmodulation.net/qam.html

[9] Kaur Harjot & Verma Amit, "BER Performance Analysis of Mary DPSK Techniques Using Simulation Modelling", International Journal of Electrical and Electronics Engineering Research (IJEEER), ISSN 2250-155X Volume 3, Issue 2, Jun 2013, pp. 93-100.

[10] James E. Gilpy, Transcript International Inc., August 2003, "Bit Error Rate Simulation using Matlab".

[11] J. P. E. Biglieri and S. Shamai, "Fading Channels: Information Theoretic and Communications Aspects," IEEE Trans. Inform. Theory, vol. 44, no. 6, pp. 2619–2692, Oct. 1998

[12] Pandey Ashish, Singh Sarala et.al, "Comparative Study of Different Modulation Technique in Chaotic Communication", International Journal of Scientific Research Engineering & Technology (IJSRET), Volume 2, Issue 11, February 2014, pp 738-743.

[13] Dixit Dutt Bohra, Avnish Bora," Bit Error Rate Analysis in Simulation of Digital Communication Systems with Different Modulation Schemes", IJISET - International Journal of Innovative Science, Engineering & Technology, Vol. 1 Issue 3, May 2014.

[14] A. Sudhir Babu and Dr. K.V Sambasiva Rao," Evaluation of BER for AWGN, Rayleigh and Rician Fading Channels under Various Modulation Schemes", International Journal of Computer Applications (0975 – 8887) Volume 26– No.9, July 2011.

Minimizing power consumption through swinging power mode in wireless sensor networks

Harish Kumar[1], Prashant Singh[2], Jai Prakash Gupta[1]

[1]Department of Computer Science Sharda University, Greater Noida, Uttar Pradesh, India
[2]Department of Information Technology, Northern India Engineering College, Delhi

Email address:
harish.kumar.phd@gmail.com (H. Kumar), prashant.ert@gmail.com (P. Singh), jaip.gupta@gmail.com (J. P. Gupta)

Abstract: Wireless Sensor Networks have predefined objectives which require them to connect and communicate with other nodes. During communication these nodes transmit and receive data and other control packets. This paper proposes Swing-MAC, a Medium Access Control protocol for energy efficiency designed for wireless sensor networks. It adjusts transmission power level in a swinging mode where transmission power level keeps swinging between minimum and maximum requirement of the network. Minimum and maximum transmission power level is based on the expected coverage between nodes. As transmission power keeps swinging between two intervals on cyclic basis this leads to energy conservation resulting in a longer battery life. Simulation results show that Swing-MAC obtains significant energy savings compared with 802.11 MAC without sleeping.

Keywords: Transmission Power Control, Wireless Sensor Networks, Medium Access Control, MAC, Energy Conservation, Minimizing Transmission Power, WSN, TPC, Energy Efficiency

1. Introduction

There are continuous advancements in electronics and wireless technology which is creating enough opportunities to create low cost, low power multifunctional wireless sensor nodes that are tiny in size and can communicate in a local area. Wireless Sensor nodes are capable of sensing, measure and gather information from the environment, and as per the network requirements they can send this information to the desired location.

There are some challenges [3][8] of WSNs as applications, control, data, nodes and run-time. Properties of wireless nodes are as they are huge in numbers with low complexity and limited battery life. There are many protocols which are developed to minimize the energy consumption [1][2][3][4][6][12][16]. These protocols focus on the problem of idle listening

[5] and overhearing. These put WSN nodes to sleep [7][10][17] periodically which saves energy consumption. During sleep other nodes are expected to wait for their cycle to be completed so they can become active and start communication. Energy efficient MAC [6] protocol must not only conserve energy but should promise minimization of delay [11] and optimal information extraction [13] [14].

Our Medium Access Control [6] protocol proposes that transmission power keeps swinging between a set of two values which are selected on the basis of minimum and maximum criteria of distance between nodes. As energy keeps swinging between 2 values this protocol is named as Swing-MAC. Process of energy swinging saves energy and throughput remains unchanged.

2. Related Works

Paulo Sérgio Sausen, José Renato de Brito Sousa, Marco Aurélio Spohn, Angelo Perkusich and Antônio Marcus Nogueira Lima [10] proposes that power saving techniques may be static / dynamic in nature. Static techniques maintains network parameters unchanged throughout network lifetime. Dynamic techniques are based on network parameter sensing which helps them to adjust some parameters to achieve more power saving. Wireless sensor nodes consume energy during transmission and reception. These techniques deal with making these nodes partially or completely turn-off so energy can be saved and network lifetime can be improved. When a node changes its state between available modes (i.e., transmission, reception and sleep) node consumes power during switching process also. Switching energy is assumed to be negligible many times. Switching energy is dependent on the technology used in wireless sensor nodes. Radio used in sensor nodes can be used to save energy between switching cycles. Battery model used in network simulators use linear discharge model which leads to wrong conclusions as switching states depends on battery capacity discharge. Only 6% error in energy consumption of WSN may lead to a shorter network lifetime up to 2 months over the expected network life of 3 years. This research gives a power management technique called Dynamic Power Management with Scheduled Switching Modes (DPM-SSM) [10].in this technique nodes are scheduled to sleep after transmission activity. Battery capacity recovery was analyzed using Differential Hybrid Petri Nets (DHPNs) [10] approach for switching energy of sensor nodes and non linear behavior of battery discharge. This model uses Rakhmatov–Vrudhula battery model and Differential Hybrid Petri Nets (DHPNs) [10] approach for simulating the technique.

Wen-Hwa Liao, Kuei-Ping Shih and Yu-Chee Lee [9] states that in WSN nodes are scattered over an area and they make a network by their own connecting in some fashion. Adjusting energy levels for energy conservation is important for the lifetime of WSN. This research does the estimation of transmission energy by the use of beacon packets [9]. Each node sends some beacon packets at different power level and receiving node estimates the distance. Sensor node estimates the distance through three landmarks. Localization is computed by triangulation method. Based on the availability there can be various types of nodes with different capabilities as low power, low cost, multi-functional nodes. They may have capacity to communicate in long / short distances. Estimation of location has advantages for coverage, deployment, routing, location service, target tracking and rescue.

Suan Khai Chong, Mohamed Medhat Gaber, Shonali Krishnaswamy and SengWai Loke [1] has given a technique that challenges to maximize sensor network lifetime by informing the node about its future operations so future readings can be inferred from previous readings. In this technique some of the sensors are chosen to be switched off for specific time to save battery. Techniques developed in this research are termed as CASE (Context Awareness for Sensing Environments) [1]. The CASE framework has components for building, learning and triggering components for energy conservation in both centralized and in-network WSN configurations. This research has developed a rule-learning algorithm Highly Correlated Rules for Energy Conservation (HiCoRE) [1] algorithm that can learn and discover rules to regulate sensing operations. For CASE Compact this research has evaluated two applications Physical clustering and Query Physical clustering is used to save energy and efficient clustering. Query processing deals with effective and efficient data transmission.

Nikolaos A. Pantazis, Dimitrios J. Vergados , Dimitrios D. Vergados and Christos Douligeris [7] proposes that WSN nodes must listen to idle channel for the prevention of network partitioning. If there is no acknowledgement then nodes are considered as dead nodes and incoming nodes create their own network. Wireless transceiver listening to idle channel consumes energy equivalent to energy consumption in transmission or reception. It is very large in comparison to a node in sleeping condition. A synchronization schemes that keeps nodes in sleeping mode as long as possible and retain network connectivity would solve the purpose and save energy. Sleep mode synchronization techniques (like S-MAC [17]) create sleep mode related delays. During sleep mode nodes are not able to communicate and this creates delay but saves power. This controls overhearing but introduces sleep-mode delay. This research uses TDMA based scheduling for creating appropriate transmission schedule which saves power and minimizes end to end transmission time between sensors to gateway. Network connectivity is retained through the scheduling of TDMA based wakeup intervals. During these intervals WakeUp [7] messages are transmitted before actual data transmissions. Due to WakeUp scheduling data packets can be delayed for one sleep interval for end-to-end transmission from sensors to appropriate gateway. This technique is advantageous over S-MAC when situation expects high power conservation and low delay simultaneously in network static in nature and requires limited traffic. Disaster detection WSNs expects this situation to keep detecting a rare event for a long time.

3. Proposed Protocol

Every WSN has some defined limits for connectivity. It is assumed that nodes requiring connection would remain within these limits. Transmission power is set to sing within these 2 levels. For smoothening of the power switching level it follows the half circular path described in the algorithm. The reason behind the swinging power levels is that nodes in the WSN are at different distance and require different power levels. Transmitter saves energy by setting different power levels at the variable of time. Bottom energy level is slashed by the minimum power requirement of the network. This helps to keep the nodes active in certain conditions where nodes are at minimum distance specified by the network.

As per our protocol Swing-MAC follows swinging path between two set of values. These values are taken by reference to provide connectivity between 100 meters to 250 meters. Transmission power follows the path described by Swing-MAC Algorithm. TPower_Max=0.2818 and TPower_Min=7.214e-3 are opted for the simulation. Pattern followed by transmission is shown in Fig.1.

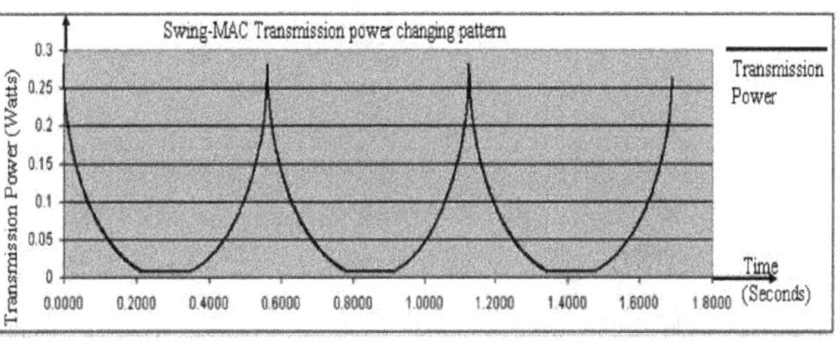

Fig. 1. *Swing-MAC Transmission Power Changing Pattern.*

3.1. Swing-MAC Algorithm for Changing Energy Pattern

```
Start
Set SimulationTime=DesiredTime
Time=0
/* Time is auto incremented*/
Repeat

Set TPower_Max=UpValue
Set TPower_Min=DownValue

X=0
While(X<=2* TPower_Max){
AT Time=X{
Y= TPower_Max – Sqrt(Sqr(TPower_Max) –Sqr(X- TPower_Max))
If(Y< TPower_Min) Y= TPower_Min
}
Increment X by 0.001
}
Until Time <=SimulationTime
End


/* TPower_Max, TPower_Min are transmission power in Watt */
```

4. Simulation Results

Simulation was done in NS 2.35 with following transmission power levels.

Pt_ = 8.5872e-4 Watt
// 40 m transmission range
Pt_ =7.214e-3 Watt
// 100 m transmission range
Pt_ = 0.2818 Watt
// 250m transmission range

4.1. Simulation with 2 WSN Nodes

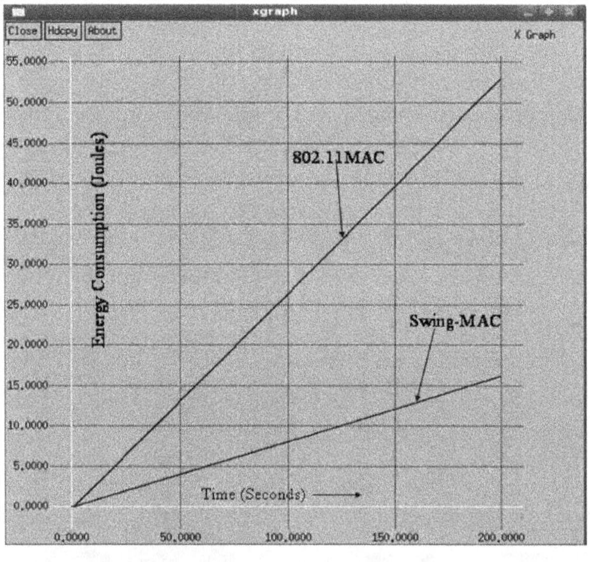

Fig. 2. Comparison of Energy Consumption with 2 WSN nodes.

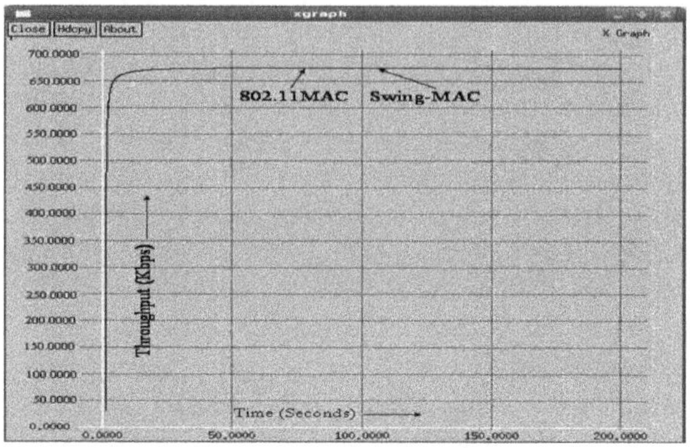

Fig. 3. Network Throughput comparison using 2 WSN nodes.

Graphical representation of energy consumption between 802.11 MAC and Swing-MAC is shown in Fig.2. Graph produced by XGRAPH shows energy advantages. Comparison between 802.11 MAC and Swing-MAC protocol shows 69.48507278% energy savings. Throughput between these protocols is shown in Fig.3. Graph produced by XGRAPH shows no changes in throughput compared to 802.11 MAC. Two lines exactly overlap each other showing no change in throughput. Comparison showing energy consumption and throughput is depicted in Table 1.

Table 1. Energy consumption and network throughput with 2 Nodes.

S.N.	Protocol	Energy Consumption (Joules)	Throughput Achieved
1.	Swing-MAC	16.170000	675.25
2.	802.11 MAC	52.990459	675.25

4.2. Simulation with 6 WSN Nodes

Graphical representation of energy consumption between 802.11 MAC and Swing-MAC is shown in Fig.4. Graph produced by XGRAPH shows energy advantages. Comparison between 802.11 MAC and Swing-MAC protocol shows 12.9733317% energy savings. Throughput between these protocols is shown in Fig.5. Graph produced by XGRAPH shows no changes in throughput compared to 802.11 MAC. Two lines exactly overlap each other showing no change in throughput. Comparison showing energy consumption and throughput is depicted in Table 2.

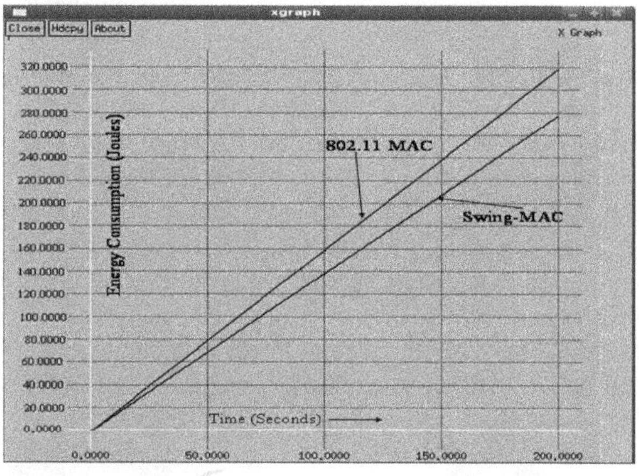

Fig. 4. Comparison of Energy Consumption with 6 WSN nodes.

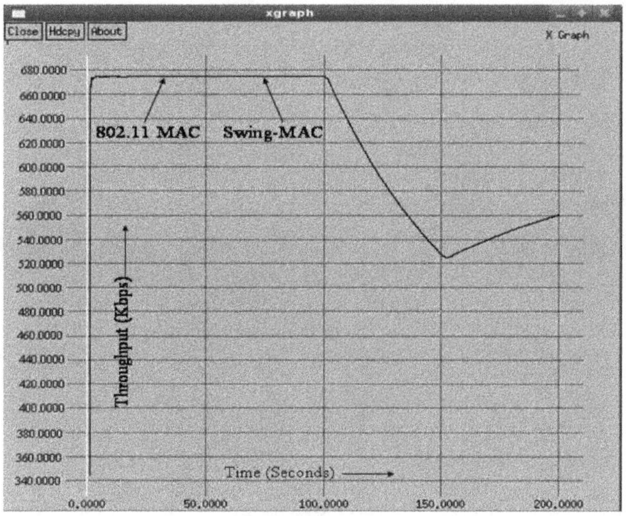

Fig. 5. Network Throughput comparison using 6 WSN nodes.

Table 2. Energy consumption and network throughput with 6 Nodes.

S.N.	Protocol	Energy Consumption	Throughput Achieved
1.	Swing-MAC	276.931299	560.385
2.	802.11 MAC	318.214295	560.385

4.3. Simulation with 50 WSN Nodes

Graphical representation of energy consumption between 802.11 MAC and Swing-MAC is shown in Fig.6. Graph produced by XGRAPH shows energy advantages. Comparison between 802.11 MAC and Swing-MAC protocol shows 4.757737803% energy savings. Throughput between these protocols is shown in Fig.7. Graph produced by XGRAPH shows no changes in throughput compared to 802.11 MAC. Two lines exactly overlap each other showing no change in throughput. Comparison showing energy consumption and throughput is depicted in Table 3.

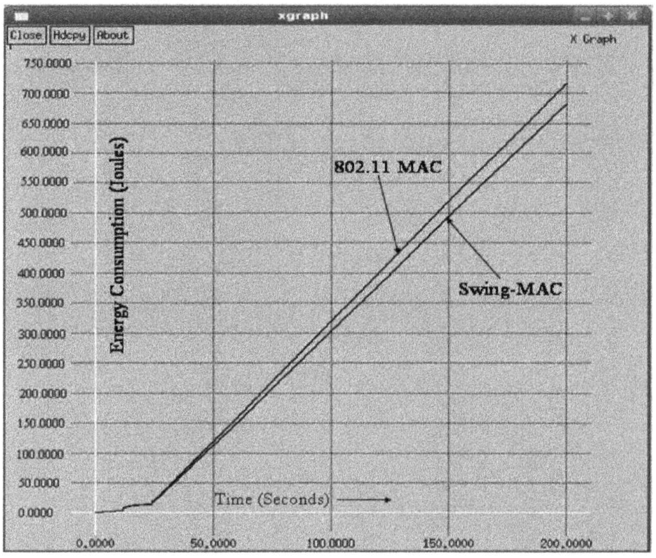

Fig. 6. Comparison of Energy Consumption with 50 WSN nodes.

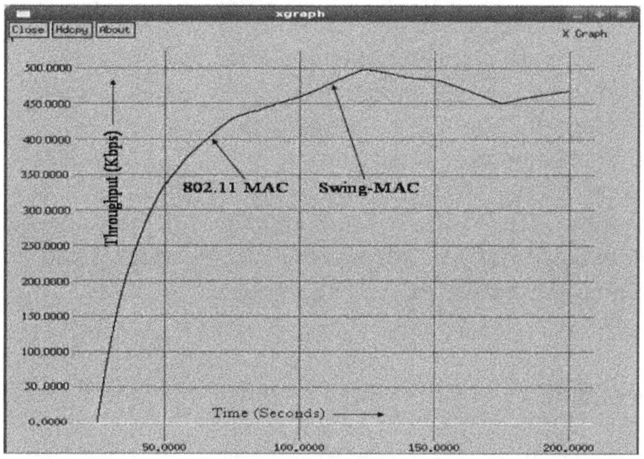

Fig. 7. Network Throughput comparison using 50 WSN nodes.

Table 3. Energy consumption and network throughput with 50 Nodes.

S.N.	Protocol	Energy Consumption	Throughput Achieved
1.	Swing-MAC	682.643175	468.237
2.	802.11 MAC	716.743974	468.237

4.4. Summary of Observations

Table 4. Comparison of Energy Consumption and Throughput.

S. N.	Protocol	Number of Nodess	Energy Consumption(N/W)	Energy Saved	Throughput Achieved (N/W)	Throughput Change
1	Swing-MAC	2	16.17	69.49%	675.13	NIL
2	802.11-MAC		52.99		675.13	
3	Swing-MAC	6	276.93	12.97%	560.39	NIL
4	802.11-MAC		318.21		560.39	
5	Swing-MAC	50	682.64	4.76%	468.24	NIL
6	802.11-MAC		716.74		468.24	

Table 4 shows summary of energy consumption with respect to percentage of energy saved using various sets of nodes comparing Swing-MAC against 802.11 MAC. Table 4 also shows throughput consistency for performance.

5. Conclusions

Energy is the important criteria in WSN networks. Some protocols put WSN nodes to sleep to save energy. In this protocol nodes remain active throughout their battery lifetime. Energy in the nodes keeps swinging between two values decided by the network range. This protocol promises savings up to 69.49% (approximately) in energy and no compromises on throughput. This protocol does not put nodes to sleep and is suitable to critical situations where there is a need to transfer continuous data on very small interval gaps. Swing-MAC has shown energy savings for static WSN nodes. Simulation and research is yet to be done for nodes under certain motion criteria. Comparison of this protocol to other protocols can also be done to prove its efficiency. Impact to other network parameters like latency, packet loss, congestion, etc. are also for futuristic research.

References

[1] Suan Khai Chong, Mohamed Medhat Gaber, Shonali Krishnaswamy, SengWai Loke, Energy conservation in wireless sensor networks: a rule-based approach, Knowledge Information Systems, DOI 10.1007/s10115-011-0380-x, Springer-Verlag London Limited 2011

[2] Moshaddique Al Ameen, S.M. Riazul Islam, and Kyungsup Kwak, Energy Saving Mechanisms for MAC Protocols in Wireless Sensor Networks, Hindawi Publishing Corporation International Journal of Distributed Sensor Networks Volume 2010

[3] Jorge M. Soares and Bruno J. Gonçalves, Rui M. Rocha, Practical issues in the development of a minimalistic power management solution for WSNs, Int. J. Sensor Networks, Vol. 8, Nos. 3/4, 2010, Inderscience Enterprises Ltd.

[4] Tao Shu, Marwan Krunz, Energy-efficient power/rate control and scheduling in hybrid TDMA/CDMA wireless sensor networks, Computer Networks 53 (2009) 1395–1408, Elsevier

[5] LI Cheng , WANG Kui-ru, ZHANG Jin-long, ZHAO De-xin, LI Wang, Optimization of listening time of S-MAC for wireless sensor networks, The Journal of China Universities of Posts and elecommunications, Issue 5,2009 Elsevier

[6] Bashir Yahya and Jalel Ben-Othman, Towards a classification of energy aware MAC protocols for wireless sensor networks, Wireless Communications And Mobile Computing, 2009 John Wiley & Sons, Ltd.

[7] Nikolaos A. Pantazis, Dimitrios J. Vergados , Dimitrios D. Vergados , Christos Douligeris, Energy efficiency in wireless sensor networks using sleep mode TDMA scheduling, Ad Hoc Networks 7 (2009) 322–343, Elsevier

[8] Jennifer Yick, Biswanath Mukherjee, Dipak Ghoshal, Wireless sensor network survey,Computer Networks, 52 (2008) 2292-2330,Elsevier

[9] Wen-Hwa Liao, Kuei-Ping Shih, Yu-Chee Lee, A localization protocol with adaptive power control in wireless sensor networks, Computer Communications 31 (2008) 2496–2504, Elsevier

[10] Paulo Sérgio Sausen, José Renato de Brito Sousa, Marco Aurélio Spohn , Angelo Perkusich, Antônio Marcus Nogueira Lima, Dynamic Power Management with Scheduled Switching Modes, Computer Communications 31 (2008) 3625–3637,Elsevier

[11] Celal Ceken, An energy efficient and delay sensitive centralized MAC protocol for wireless sensor networks, Computer Standards & Interfaces 30 (2008) 20–31, Elsevier

[12] Luiz H.A. Correia, Daniel F. Macedo, Aldri L. dos Santos, Antonio A.F. Loureiro, Jose´ Marcos S. Nogueira, Transmission power control techniques for wireless sensor networks, Computer Networks 51 (2007) 4765–4779, Elsevier

[13] Fernando Ordóñez, Bhaskar Krishnamachari, Optimal Information Extraction in Energy-Limited Wireless Sensor Networks, IEEE Journal on Selected Areas In Communications, Vol. 22, No. 6, August 2004, IEEE

[14] Abdelmalik Bachir, Mischa Dohler, Thomas Watteyne, Kin K. Leung, MAC Essentials for Wireless Sensor Networks, IEEE communications surveys & tutorials, vol. 12, no. 2, second quarter 2010

[15] Issa M. Khalil, ELMO: Energy Aware Local Monitoring in Sensor Networks, IEEE Transactions on Dependable and Secure Computing, Vol. 8, No. 4, July/August 2011

[16] Antonio G. Marques, Xin Wang, Georgios B. Giannakis, Minimizing Transmit Power for Coherent Communications in Wireless Sensor Networks With Finite-Rate Feedback, IEEE Transactions On Signal Processing, Vol. 56, No. 9, September 2008

[17] Wei Ye and John Heidemann, Medium Access Control in Wireless Sensor Networks, USC/ISI TECHNICAL REPORT ISI-TR-580, OCTOBER 2003

[18] http://www.isi.edu/nsnam/ns (official website of network simulator)

[19] http://www.dictionary.com

5

Design and Development of a BlueGS Gateway for Bluetooth and GSM Protocols

Edgar Manuel Cano Cruz, Juan Gabriel Ruiz Ruiz, Luis Alberto Hernández Montiel

Computer Science Department, University of the Istmo Region, Ixtepec, Mexico

Email address:

ie.edgarcano@gmail.com (E. M. C. Cruz)

Abstract: The new generation of mobile devices with embedded applications is one of the most rapidly growing technologies in the Wireless Networks. In this paper, we propose to combine the functions and capabilities of the maturity wireless protocols Bluetooth and GSM technologies, to design a powerful gateway tool for embedded systems: the BlueGS system. In addition, also included, is the construction of the BlueGS node, providing a flexible platform with the possibility of expanding the functionality. The system incorporates a Bluetooth and GSM module, thus achieving the development of a low cost system that does not require complex infrastructure to operate and is easily accessible to the general public.

Keywords: Security System, Embedded System, Motion Sensor, Bluetooth, GSM

1. Introduction

In the last two decade, many users have an interest in both Bluetooth and GSM technologies, because of their advantages such as, low-cost implementation and low energy consumption, which offer different advantages in the delivery of embedded systems, various types of devices have adopted wireless technologies such as, Wi-Fi, Bluetooth, and ZigBee, and have opened up new opportunities for new and innovative means of embedded systems delivery [1, 2]. Consequently, these capabilities make these protocols a suitable wireless standard to build a platform with different levels of scalability, flexibility and easy to use adaptation on several projects.

The implementation of an embedded gateway system using wireless technologies, will in time, exploit the recent advances in WPAN, ubiquitous computing, wireless sensors and others areas [3, 4, 5]. The interoperability between Short-Range Wireless Protocols on embedded systems devices, can give to available to the user several functions that maximize the level of usability.

Bluetooth is clearly planned for short-range cable replacement for medium bandwidth device to device connections, and its most likely uses are ad hoc communication between mobile computing devices and fixed equipment.

Based on the extensive increase of applications on smart phones with Bluetooth, and in the analysis of the development of smart appliance devices with GSM technology in the industry [1], we propose combining both capabilities of these protocols for diagnostics, data transfer or configuration systems. Bluetooth and GSM are different by design and are optimized for different applications. In this way, the real industrial wireless networks will inevitably be hybrids, including Bluetooth and GSM, in complementary roles that suit the characteristics of each. The key to success will be in deploying the right wireless technologies for the requirements of the application and avoiding the temptation of trying to make one technology meet all needs.

This paper shows the design and implementation of a wireless embedded system gateway tool based on Bluetooth and GSM protocols. The rest of this paper is organized as follows: Section II introduces a brief related work; Section III of Short-Range Wireless Networks and presents the basics of Bluetooth and GSM wireless protocols; Section IV describes the design and implementation of the BlueGS system. In Section V, we present the preliminary results of the BlueGS system with a one basic application on a move sensor node. Finally, Section VI concludes this paper.

2. Related Work

In related work there are some emergent hybrid platforms to communicate devices between different wireless protocols. In the work [5], Galinina and colleagues envision a scenario where many in-home sensors are communicating with a smart gateway over the Bluetooth Low Energy protocol, while at the same time harvesting RF energy transmitted from the gateway wirelessly via a dedicated radio interface. The authors thoroughly investigate performance limitations of such wireless energy transfer interface (WETI) with dynamic analytical model and with important practical considerations. Their methodology delivers the upper bound on WETI operation coupled with BLE-based communication, which characterizes ultimate system performance over the class of practical radio and energy resource management algorithms.

Ruta et al. [6], presented a hybrid ZigBee/Bluetooth grid infrastructure to interconnect the user with interface nodes and to perform an advanced resource discovery. Li et al. [7], introduced a new method of conversion to achieve the ZigBee communication with CAN bus, realizing the link of CAN bus and the wireless network in order to conquer the conflictions brought from bus configuration and the protocol of controllers. Finally, Ren Xiaoghon et al. [8], gives a kind of design of CAN bus network based on Bluetooth technology, they indicated that the system could operate reliably and steadily and CAN-Bluetooth nodes could deliver the data correctly transmitted.

At work [8], Laine, Chaewoo and Haejung present a technical design of a Bluetooth-based mobile gateway that bridges the connection between a ZigBee sensor network and the Internet. Their system enables ubiquitous health care experience while providing a platform for additional services such as alarms, notifications and analysis of medical data. Controlling a sensor network from the mobile gateway is also possible. The flexible design of the system does not restrict its usage only to health care services - the gateway can be configured to work with any kind of sensor network having a sink node with Bluetooth capability.

3. GSM and Bluetooth Network: A Brief Review

3.1. GSM

According with [9], a GSM network has defined, the three subsystems involved are the mobile station subsystem, the base station subsystem, and the home subsystem (see Figure 1).

- The mobile station subsystem consists of the mobile equipment (ME) and a smart card called the Subscriber Identity Module (SIM). The mobile equipment is uniquely identified by the International Mobile Equipment Identity (IMEI). The SIM card contains the International Mobile Subscriber Identity (IMSI) used to identify the subscriber to the system, a secret key for authentication, and other information.
- The base station subsystem consists of the Base Transceiver Station (BTS) and the Base Station Controller (BSC). These are the connections between the mobile stations and the Mobile Switching Center (MSC).
- The home subsystem is composed of five parts, the Mobile Switching Center (MSC), the Home Location Register (HLR), the Visitor Location Register (VLR), the Authentication Center (AuC), and the Equipment Identity Register (EIR).
- The HLR is a database that stores complete local customer information. It is the main database. Your carrier puts your information on its nearest HRL, or the one assigned to your area. That info includes your IMEI, your directory number, and the class of service you have. It also includes your current city and your last known "location area" the place you last used your mobile.
- The VLR contains roamer information. Once the visited system detects your mobile, its VLR queries your assigned HLR. The VLR makes sure you are a valid subscriber, then retrieves just enough information from the now distant HLR to manage your call. It temporarily stores your last known location area, the power your mobile uses, special services you subscribe to and so on.
- The AuC stores a copy of the secret key kept in each subscriber's SIM card and generates authentication parameters for the authentication protocol on the request of HLRs.
- The EIR is a database that contains a list of all the valid mobile devices on the network, where each mobile station is identified by its International Mobile Equipment Identity (IMEI). The mobile stations communicate through radio links with the base stations, which are in turn connected to the MSC.
- The MSC is responsible for transiting signals between radio links and wire-lined networks.

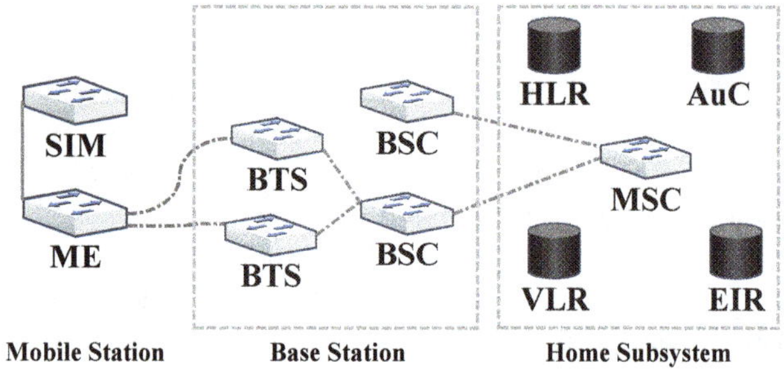

Figure 1. *GSM network.*

3.2. Bluetooth

Meanwhile, Bluetooth, also known as the IEEE 802.15.1 standard, is based on a wireless radio system designed for short-range and has cheap devices to replace cables [10]. The Figure 2 shown the Bluetooth topology that provides a point-to-point connection as well as a point-to-multipoint connection.

There are 79 Bluetooth channels in the 2.4GHz ISM band, each Bluetooth channel is divided into time slots and the duration of each time slot is 625µs. Bluetooth devices can communicate with each other using the frequency hopping spread spectrum (FHSS). Bluetooth devices are divided into three power classes, where the maximum output power levels of classes 1, 2 and 3 are set to 100 mW, 2.5 mW, and 1 mW respectively [11].

The Bluetooth topology defines the basic network called piconet, and can form point-to-point (see Figure 2a) or point-to-multipoint links (see Figure 2b). The piconet is a set of two to eight Bluetooth devices. One device manages the transmission in each piconet, this device is called master, while the others devices are called slaves.

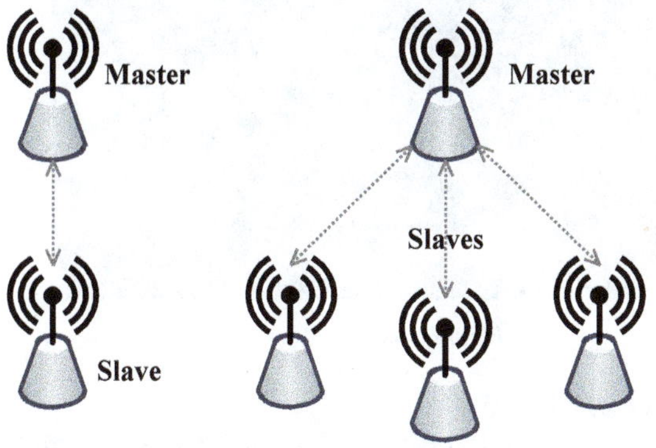

(a) Point-to-point (b) Point-to-multipoint

Figure 2. *Bluetooth piconet.*

4. Design of the BlueGS System

BlueGSM system implementation was done with the support of the methodology for the development of embedded systems SPIES [11], which is a specific methodology for modeling relationships between functionality, components, equipment and materials of an embedded system. SPIES consist of five phases describe at the follow (see Figure 3).

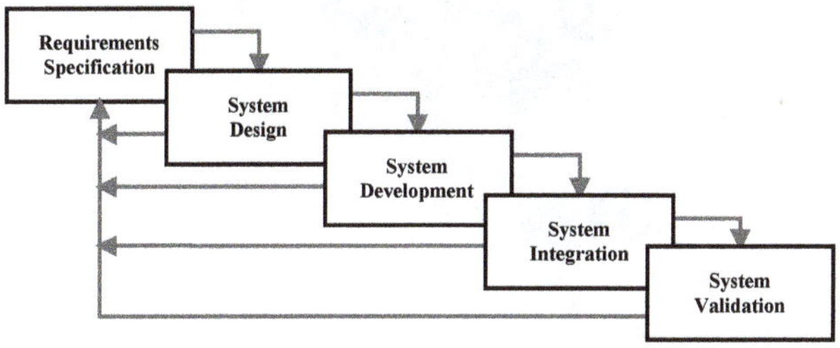

Figure 3. *SPIES methodology.*

4.1. Requirements Specification

The BlueGSM system is a gateway tool that converts data from Bluetooth devices to GSM data, and sends information through GSM network. The BlueGSM node (see Figure 4) consists of the follow components:

Arduino Uno is a microcontroller board based on the ATmega328 microcontroller [14]. This board has 14 digital pins of input / output (of which 6 can be used as PWM outputs), 6 analog inputs, a ceramic resonator of 16 MHz, a USB connection, a power jack an ICSP header, and a button reset. It contains everything needed to support the microcontroller.

Through the DTE module the user configures the node and displays the messages that are sending between devices. Within the DTE module will find the following components.

The DCE modules are responsible for establishing the different connections according to the protocol used and for transmitting or receiving information in serial format.

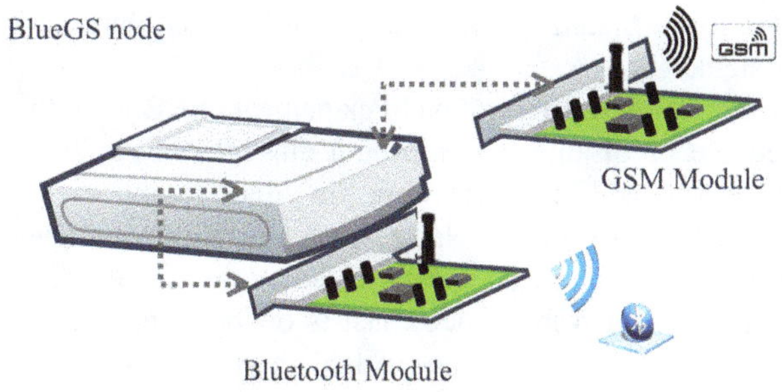

Figure 4. *Gateway System.*

The DCE module consists of follow components:

Bluetooth Hc-05The HC-05 module is a Bluetooth device that is configured at the factory to work as master or slave. In the master mode you can connect with other Bluetooth modules, while in the slave mode is listening to connection requests. This module allows remote control from a cell phone or a laptop all the functionality desired in a system.

The BlueGS node operates in two main modes: Configuration Mode and Data Mode (see Figure 5). These operating main modes are managed by the Operation Management subsystem. The BlueGS node starts up in Data Mode and the user can be requested to move to Configuration Mode by sending an escape sequence through the module DTE. Although in some situations it is necessary to restore the BlueGS node settings to their default values.

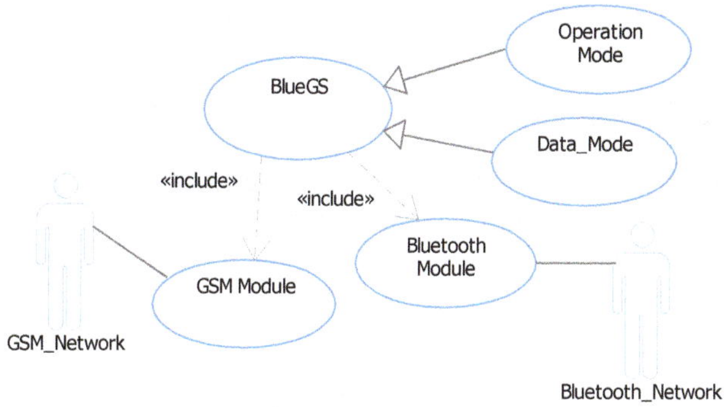

Figure 5. *Use case diagram for BlueGS node.*

4.2. System Design

In the BlueGS Configuration Mode the user set up both GSM and Bluetooth modules, through the Bluetooth Configuration Management and GSM Configuration Management subsystems. A Bluetooth module can be in several different operation modes, which determines whether or not the BlueGS node can be connected to and whether or not other Bluetooth devices performing searches can discover a BlueGS node. Serial connections are Bluetooth connections based on the Serial Port Profile (SPP). In this sense, it is important that devices, the BlueGS node and Bluetooth device have the same profile to establish the link and can communicating between them.

Into the functionality of the BlueGS system, there exist tow subsystem to manage all operation of the gateway: Bluetooth Configuration Management and GSM Configuration Management.

In the Bluetooth Configuration Management, the Bluetooth Mode enable or disable the Bluetooth characteristic of the BlueGS node and if can work in discoverable and non-discoverable mode. The Name Mode is useful to change the friendly name of the Bluetooth Module into the BlueGS node. Finally, the Pairing Mode consists of four methods to search a Bluetooth device and establish a security link connection with the BlueGS node.

In the GSM Configuration Management, to enable the GSM module and establish a real cell phone number where the data information will receive, the GSM module is configured.

In the BlueGS Data Mode, the system can manage both Bluetooth and GSM, and convert Bluetooth data in GSM data, but not in vice versa mode. Within the Data Management subsystem, the BlueGS can send and receive Bluetooth and GSM data, but only one packet at time, also is necessary that in a previously steps, the user active and configure the Bluetooth and GSM modules and change the operation mode of the BlueGS node to data mode to establish connection with the respective devices.

Once that the modules are configured, is possible that the BlueGS convert GSM packets to Bluetooth packets. When the BlueGS node is in the Converting Data Mode subsystem, the BlueGS node can listen and receive packets from GSM network and convert to Bluetooth data, and then send data to other Bluetooth device (both devices have been previously paired).

4.3. System Development

To prepare the project, we have considered a maximum budget of $ 50.00 USD, considering the features that are contemplated in the system, the table 1 shows the tools currently in the system.

Table 1. *Costs of the tools [12].*

Tool	Cost (USD)
1 Arduino Mega 2560	$ 24.95
1 Bluetooth Hc-05	$ 3.90
1 GPRS SIM800L	$13.83
Arduino IDE	$ 0.00
Android Studio	$ 0.00
Total	$42.68

4.3.1. Arduino Uno

Arduino Uno is a microcontroller board based on the ATmega328 microcontroller [13]. This board has 14 digital pins of input / output (of which 6 can be used as PWM outputs), 6 analog inputs, a ceramic resonator of 16 MHz, a USB connection, a power jack an ICSP header, and a button reset (see Figure 6). It contains everything needed to support the microcontroller.

Figure 6. *Arduino Uno.*

4.3.2. Bluetooth Hc-05

The HC-05 module is a Bluetooth device that is configured at the factory to work as master or slave (see Figure 7). In the master mode you can connect with other Bluetooth modules, while in the slave mode is listening to connection requests. This module allows remote control from a cell phone or a laptop all the functionality desired in a system.

***Figure** 7. Bluetooth Hc-05 wireless module.*

4.3.3. GPRS SIM800L

The GPRS module based on SIM800L, supports quad-band GSM/GPRS network, available for GPRS and SMS message data remote transmission. The GSM module can be used with a direct link to the microcontroller through its TTL serial port. No need MAX232. Power on the module automatically boot automatically search network. Onboard signal lights (with signal flash slowly, no signal flash quickly).

5. Results

A partial integration of BlueGS was performed, and test the operation of a move sensor. In the Figure 8 the plan implemented between the move sensor and the Arduino board is shown.

The system was validated through a monitoring application with the move sensor running on the Arduino board and an application on a smartphone with Android operating system.

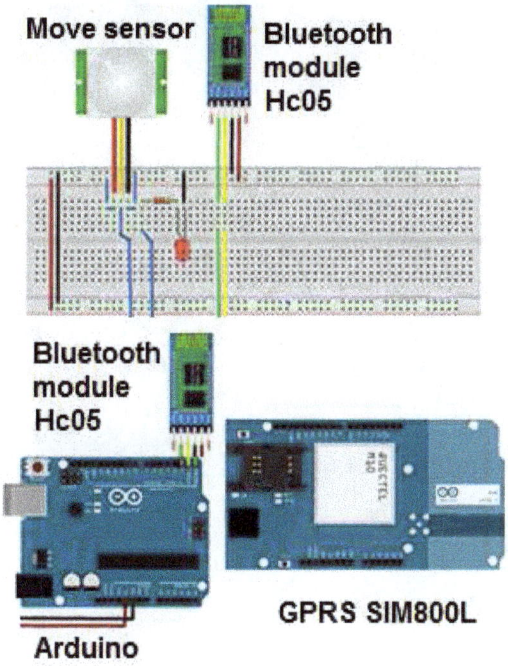

Figure. 8. Scheme system for validation.

System operation is very simple, the move sensor is connected at Arduino board through Bluetooth connection (see Figure 9), if the move sensor detects movement, and then the android device will process the data and send a text message to the number that was established.

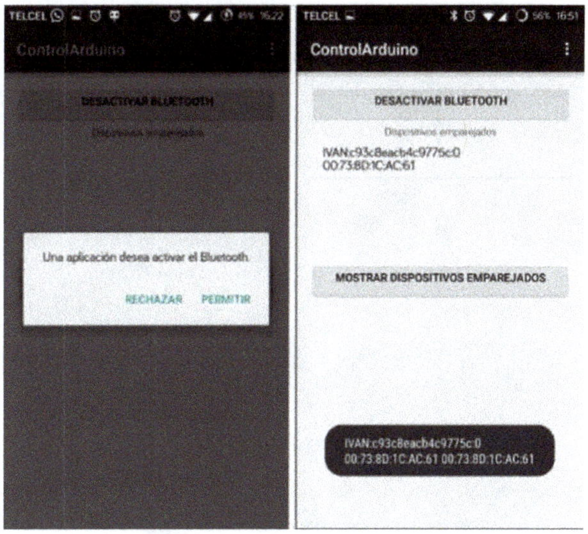

Figure 9. Management of the Bluetooth connection.

6. Conclusions

The development of a Bluetooth and GSM gateway system is proposed in this paper. The system validation consisted of the move sensor that is detect measures variables whereas Bluetooth protocol establishes the communication between sensors and the Arduino system where this parameters trigger a gsm module to send a sms text to the number that was previously established.

In future research, we hope to develop the weather system by using the ARDUINO Mega card to replace the current card and provide in this form more functionalities to the BlueGS gateway system. The fusion of both wireless technologies Bluetooth and GSM, made it achievable to delivery sensor networks, industrial control systems, location and position throughout mobile users environments, in order to capture different types of data information. Also, using the capabilities of these wireless protocols, it gives different solutions such as low-cost and low energy consumption to delivery and build embedded systems.

References

[1] Wang, Q., Rackers, J. A., He, C., Qi, R., Narth, C., Lagardere, L., & Ren, P. (2015). A General Model for Treating Short-Range Electrostatic Penetration in a Molecular Mechanics Force Field. *Journal of Chemical Theory and Computation.*

[2] Bazydło, P., Dąbrowski, S., & Szewczyk, R. (2015). Wireless temperature measurement system based on the IQRF platform. In *Mechatronics-Ideas for Industrial Application* (pp. 281-288). Springer International Publishing.

[3] Kaur Kapoor, N., Majumdar, S., & Nandy, B. (2015). Techniques for Allocation of Sensors in Shared Wireless Sensor Networks. *Journal of Networks*, 10(01), 15-28.

[4] De Souza, R. H., Savazzi, S., & Becker, L. B. (2015). Network design and planning of wireless embedded systems for industrial automation. *Design Automation for Embedded Systems*, 1-22.

[5] Galinina, O., Mikhaylov, K., Andreev, S., Turlikov, A., & Koucheryavy, Y. (2015). Smart home gateway system over Bluetooth low energy with wireless energy transfer capability. In *EURASIP Journal on Wireless Communications and Networking*. Springer International Publishing.

[6] Ruta, M., Scioscia, F. Di Noia, T., & Di Sciascio, E. A hybrid ZigBee/Bluetooth approach to mobile semantic grids, *International Journal of Computer Systems Science and Engineering* (IJCSSE) - Special issue on Mobile Data Management: Models, Methodologies and Services, 2009.

[7] Y. Li, L. Yu, J. Yan, and H. Li, "The design of Zig Bee communication convertor based on CAN," International Conference on Computer Application and System Modeling (ICCASM), IEEE Computer Society, 245-249, 2010.

[8] R. Xiaohong, F. Chenghua, W. Tianwen, and J. Shuxiang, "CAN Bus Network Design Based on Bluetooth Technology," Electrical and Control Engineering (ICECE), 2010 International Conference on, vol., no., pp.560-564, 25-27 June 2010.

[9] GSMA, «GSMA,» 02 03 2015. [En línea]. Available: http://www.gsma.com/aboutus/gsm-technology/gsm.

[10] Bluetooth SIG, "Bluetooth Specification v4.0", 2009.

[11] Garcia. I, and Cano. E. Designing and implementing a constructionist approach for improving the teaching–learning process in the embedded systems and wireless communications areas, Computer Applications in Engineering Education: Wiley Periodicals, Inc. 2011.

[12] Arduino, «The community of Arduino,» 2015. [En línea]. Available: http://www.arduino.cc/. [Accessed May 2015].

[13] Electrodragon, "Electrodragon," 2015. [Online]. Available: http://www.electrodragon.com/. [Accessed 2015].

6

Assessment of Secured Voice Frequency Signal Transmission in Dual Polarized DWT Aided MIMO SC-FDMA Wireless Communication System

Shammi Farhana Islam[1], Mahmudul Haque Kafi[2], Sk. Sifatul Islam[2]

[1]Department of Material Science and Engineering, Rajshahi University, Rajshahi, Bangladesh
[2]Department of Applied Physics and Electronic Engineering, Rajshahi, Bangladesh

Email address:
shammi.farhana@gmail.com (S. F. Islam), mahmudkafi49@gmail.com (M. H. Kafi), chisty56@gmail.com (Sk. S. Islam)

Abstract: In this paper, we have emphasized the BER performance of dual polarized DWT aided MIMO SC-FDMA wireless communication system. The simulated system under investigation with 4 × 4 antenna configuration implements various types of channel coding(LDPC& Turbo) and signal detection (MMSE-SIC, BLUE, ZF & OSIC) techniques. On considering transmission of secured voice frequency signal in a hostile fading channel under MATLAB based simulative study, it is observable that the simulated system shows satisfactory performance in retrieving transmitted audio signal under scenario of implementing Turbo channel coding, MMSE-SIC signal detection and 8-QAM digital modulation schemes.

Keywords: SC-FDMA, Dual Polarized Antennas, DWT, Signal to Noise Ratio, Channel Coding and Signal Detection

1. Introduction

Multiple input multiple output (MIMO) techniques utilize multiple antenna elements at the transmitter and the receiver to improve communication link quality and/or communication capacity. A MIMO system can provide two types of gain such as spatial diversity gain and spatial multiplexing gain. The spatial diversity improves the reliability of communication in fading channels and the spatial multiplexing increases the capacity by sending multiple streams of data in parallel through multiple spatial channels.

Single Carrier Frequency Division Multiple Access (SC-FDMA) is a novel method of radio transmission under consideration for deployment in cellular systems[1].It has been known from literature review that over the past time, several cellular technologies have been surfaced with commercial deployment of the long term evolution (LTE) and its successor of LTE-advanced (LTE-A) networks. The LTE-A networks use MIMO SC-FDMA for uplink transmissions. The SC-FDMA signals have inherently lower peak-to-average power ratio (PAPR) than the OFDMA signals. In comparison to OFDMA, the SC-FDMA significantly reduces the envelope fluctuations in the transmitted waveform [2].In 2012, Umaria and et. al., made performance comparison study for FFT based OFDM system and DWT based OFDM system using different wavelet families and found that the DWT based OFDM system is better than FFT based OFDM system with regards to the bit error rate (BER) performance[3]. In Fifth generation (5G), millimetre wave (mmWave) multiple-input multiple output (MIMO) wireless communication systems are being preferred to provide the throughput enhancements needed to meet up the expected demands for mobile data. The dual-polarized antenna systems are expected to be incorporated in mmWave systems [4].The present study represents SC FDMA system performance on secured audio signal transmission under implementation of dual polarization antenna configuration, Haar's based discrete wave transformation (DWT) schemes and various channel coding and signal detection schemes.

2. Signal Processing Techniques

In our present study, various signal processing schemes have been used. A brief overview of these schemes is given below:

2.1. Turbo Coding

Turbo code is systematic code with its coding rate is of ⅓ formed by concatenating in parallel two recursive systematic convolutional (RSC) codes separated by an interleaver. In such coding scheme, the encoder produces three code bits. One is the message bit treated as systematic bit and the other two are the parity bits generated by the two RSC encoders. The code may also be punctured to obtain a coding rate of 1/2. Puncturing operates only on the parity sequences; the systematic bits are not touched. In maximum a posteriori (MAP) turbo decoding, the transmitted message bits are retrieved iteratively through computation of their log likelihood ratio (LLR). Let $\overline{C} = C_0, C_1, C_2 \ldots\ldots C_{N-1}$ be a coded sequence produced by the rate ½ RSC encoder and $\bar{r} = r_0, r_1, r_2 \ldots\ldots r_{N-1}$ be the noisy received sequence where the codeword is $c_k = (c_k^{(1)} \quad c_k^{(2)})$ with the first bit being the message bit and the second bit being the punctured parity bit. The corresponding received word is $r_k = (r_k^{(1)} \quad r_k^{(2)})$

The coded bit in 0/1 format is converted to a value of +1/-1. The maximum a posteriori(MAP) decodingis carried out as:

$$c_k^{(1)} = \{\frac{+1, \text{if } P(c_k^{(1)} = +1|\bar{r}) \geq P(c_k^{(1)} = -1|\bar{r})}{-1, \text{if } P(c_k^{(1)} = +1|\bar{r}) < P(c_k^{(1)} = -1|\bar{r})}\} \; (i = 0,1,2,3, N-1) \quad (1)$$

A posteriori log likehood ratio(LLR) of $c_k^{(1)}$ is given by

$$L(c_k^{(1)}) = \ln\left[\frac{P(c_k^{(1)} = +1|\bar{r})}{P(c_k^{(1)} = -1|\bar{r})}\right] \quad (2)$$

The MAP decoding rule in Equation (1) can be written alternatively as:

$$c_k^{(1)} = \text{sign}\left[L(c_k^{(1)}|\bar{r})\right] \quad (3)$$

The magnitude LLR, $\left[L(c_k^{(1)}|\bar{r})\right]$ measures the like hood of $c_k^{(1)}$ =+1 or $c_k^{(1)}$ =-1. The LLR can be expressed as a function of the probability $P(c_k^{(1)}|\bar{r})$ as[5]:

$$L(c_k^{(1)}) = \ln\left[\frac{P(c_k^{(1)} = +1|\bar{r})}{P(c_k^{(1)} = -1|\bar{r})}\right] = \left[\frac{P(c_k^{(1)} = +1|\bar{r})}{1 - P(c_k^{(1)} = +1|\bar{r})}\right] \quad (4)$$

2.2. LDPC Coding

In ½.-rated irregular LDPC coding, a code length of 1024 bits is used. Its parity-check matrix [H_{parity}] is a sparse matrix with a dimension of 512×1024 and contains only three 1's in each column and six 1's in each row. The parity-check matrix [H_{parity}] is formed from a concatenation of two matrices [A] and [P]([H_{parity}]=[A]|[P]), each has a dimension of 512 × 512). The columns of the parity-check matrix [H_{parity}] is rearranged to produce a new parity-check matrix [newH]. With rearranged matrix elements, the matrix [A] becomes non-singular and it is further processed to undergo LU decomposition. The parity bits sequence [p] is considered to have been produced from a block based input binary data sequence [u]=[u1u2u3u4…….u512]^{T}and three matrices [P](of [newH]),[L] and [U]using the following MATLAB notation :

p = mod(U\(L\z), 2);where, z = mod(P*u, 2); The LDPC encoded 1024× 1 sized block based binary data sequence [c] is formulated from concatenation of parity check bit p and information bit u as :

[c]=[p;u] (5)

The first 512 bits of the codeword matrix [c] are the parity bits and the last 512 bits are the information bits. In iterative Log Domain Sum-Product LDPC decoding Algorithm, the transmitted bits are retrieved [6,7].

2.3. Dual Polarized MIMO Channel

A 4× 4 dual polarized MIMO channel $H_\chi \in \mathbb{C}^{4x4}$ is parameterized by a single parameter and can be modeled as:

$$H\chi = X \odot H \quad (6)$$

where, $H_w \in \mathbb{C}^{4x4}$ denotes a single polarized MIMO channel having i.i.d. entries with $\mathcal{C}(0, 1)$,X $\in \mathbb{C}^{4x4}$is a matrix describing the power imbalance between the orthogonal polarizations. It is modeled as:

$$X = \begin{bmatrix} 1 & \sqrt{\chi} \\ \sqrt{\chi} & 1 \end{bmatrix} \otimes 1_{2\times 2} \quad (7)$$

The parameter $0 \leq \chi \leq 1$ stands for the inverse of the cross-polar discrimination (XPD), where $1 \leq XPD \leq \infty$. The XPD refers to the physical ability of the antennas to distinguish the orthogonal polarization. In Equation 1, ⊙is theHadamard product of X and $H_{w.}$ Equation 6 can be written in s a block matrix representation as: [8].

$$H_\chi = \begin{bmatrix} H_{w,11} & \sqrt{\chi}\, H_{w,12} \\ \sqrt{\chi}\, H_{w,21} & H_{w,22} \end{bmatrix} \quad (8)$$

2.4. Best Linear Unbiased Estimation (BLUE)

In BLUE based signal detection scheme, it is assumed that the channel matrix.His deterministic and the covariance matrix Ree (=E{NNT}) of the contaminated noise N is positive definite and its inversion matrix $R^{-1}{}_{ee}$ is known or can be estimated. The noise covariance matrix Ree is of dimension 4 × 4. The estimated transmitted signal X_{BLUE} using such scheme can be written in terms of Y(Received signal),HχandRee, as[9]:

$$X_{BLUE} = (H\chi T\ R\text{-}1ee\ H\chi)\text{-}1H\chi T\ R\text{-}1eeY \quad (9)$$

2.5. Haar Wavelet Transform

In Wavelet decomposition technique, a discrete signal X(z) is decomposed into coarse approximation a(m) and detail d(m) components using four sets of wavelet filters H_0, H_1, G_0, and G_1. An important property of the wavelet transform is the perfect reconstruction which is the process of rebuilding a decomposed signal into its original transmitted form without deterioration. In Haar wavelet transform, the discrete signalX(z) is decomposed into two components of half the length of original signal. At each decomposition level, the high-pass filter produces the detail component and the low pass filter produces the coarse approximation component. The filtering and decimation process continues until the desired decomposition level is reached. The maximum number of levels depends on the length of the signal. In Haar wavelet transform, the polynomial, P (z) is given by,

$$P(z) = \frac{1}{2}(z+2+z\text{-}1) = \frac{1}{2}(z+1)(1+z\text{-}1) = G0\ (z)\ H0\ (z) \quad (10)$$

the filter $H_0(z)$ and $G_0(z)$ are estimated using the following relation:

$$H0\ (z) = \frac{1}{2}(1+z\text{-}1) \quad (11)$$

$$G0\ (z) = (z+1) \quad (12)$$

The other two filters $H_1(z)$ and $G_1(z)$ are estimated using the following relation:

$$G1(z) = zH0\ (\text{-}z) = \frac{1}{2}z(1\text{-}z\text{-}1) = \frac{1}{2}(z\text{-}1) \quad (13)$$

$$H1(z) = z\text{-}1\ G0\ (\text{-}z) = z\text{-}1(\text{-}z+1) = (z\text{-}1\text{-}1) \quad (14)$$

The approximation and detail coefficients can be expressed as follows [10]:

$$a(m) = \sum_{k=-\infty}^{\infty} x(k)H_0\,(2m-k) \quad (15)$$

$$d(m) = \sum_{k=-\infty}^{\infty} x(k)H_1\,(2m-k) \quad (16)$$

where, m ranges from 1,2,3……. 32 as the total number of samples used in a single block wise processing is 64

2.6. Zero Forcing (ZF)

In Zero-Forcing (ZF) signal detection scheme, the ZF weight matrix is given by

$$W_{ZF} = (H_{\chi}{}^{H} H_{\chi})^{-1} H_{\chi}{}^{H} \quad (17)$$

and the detected desired signal $\tilde{X}_{ZF}$ from the transmitting antenna in terms of ZF weight matrix and received signal Y is given by

$$\tilde{X}_{ZF} = W_{ZF} Y \quad (18)$$

2.7. MMSE-SIC

In Minimum mean square error successive interference cancellation (MMSE-SIC) scheme,

The extended channel matrix $\hat{H}$ and the extended received signal $\hat{Y}$ in terms of identity and null matrices are given by

$$\hat{H} = \begin{bmatrix} H_{\chi} \\ (\sqrt{\sigma^2{}_n})I_{4x4} \end{bmatrix} \quad (19)$$

$$\hat{Y} = \begin{bmatrix} Y \\ \mathbf{0}_{4\times146445} \end{bmatrix} \quad (20)$$

On QR decomposition of $\hat{H}$, a 8×8 orthogonal matrix $\hat{Q}$ and a 8×4upper triangular matrix $\hat{R}$ are produced. Equation (20) is multiplied with $\hat{Q}^T$ to provide a modified form of received signal with neglected noise component

$$\hat{\hat{Y}} = \hat{Q}^T\,\hat{Y} = \hat{Q}^T\hat{H}X = \hat{\hat{R}}X \quad (21)$$

Considering a single time slot, the transmitted four signals $\hat{\hat{X}}_1, \hat{\hat{X}}_2, \hat{\hat{X}}_3$ and $\hat{\hat{X}}_4$ in terms of four received signals $\hat{\hat{Y}}_1$, $\hat{\hat{Y}}_2$, $\hat{\hat{Y}}_3$ and $\hat{\hat{Y}}_4$ (First through Fourth rows of $\hat{\hat{Y}}$ and neglecting other row data) and the components of matrix $\hat{\hat{R}}$ in first through fourth row) can be obtained from a matrix equation as:

$$\hat{\hat{Y}}[(:,1)] = \begin{bmatrix} \hat{\hat{Y}}_1 \\ \hat{\hat{Y}}_2 \\ \hat{\hat{Y}}_3 \\ \hat{\hat{Y}}_4 \\ \hat{\hat{Y}}_5 \\ \hat{\hat{Y}}_6 \\ \hat{\hat{Y}}_7 \\ \hat{\hat{Y}}_8 \end{bmatrix} = \begin{bmatrix} \hat{\hat{R}}_{1,1} & \hat{\hat{R}}_{1,2} & \hat{\hat{R}}_{1,3} & \hat{\hat{R}}_{1,4} \\ 0 & \hat{\hat{R}}_{2,2} & \hat{\hat{R}}_{2,3} & \hat{\hat{R}}_{2,4} \\ 0 & 0 & \hat{\hat{R}}_{3,3} & \hat{\hat{R}}_{3,4} \\ 0 & 0 & 0 & \hat{\hat{R}}_{4,4} \\ 0 & 0 & 0 & 0 \\ 0 & 0 & 0 & 0 \\ 0 & 0 & 0 & 0 \\ 0 & 0 & 0 & 0 \end{bmatrix} \begin{bmatrix} \hat{\hat{X}}_1 \\ \hat{\hat{X}}_2 \\ \hat{\hat{X}}_3 \\ \hat{\hat{X}}_4 \end{bmatrix} \quad (22)$$

the transmitted signals are detected from equation 22

2.8. Ordered Successive Interference Cancellation (OSIC)

In Ordered successive interference cancellation (OSIC) signal detection scheme, its implementation is performed in four steps. In first step, the first detected signal/data stream $\tilde{X}_{OSIC-1}$ and modified form of received signal $\tilde{Y}_{OSIC-1}$ can be written as:

$$\tilde{X}_{OSIC-1} = W_{(MMSE(1,:)}Y \quad (23)$$

$$\tilde{Y}_{OSIC-1} = Y - H_{\chi(:,1)}\tilde{X}_{OSIC-1}$$

In second step, the second detected signal/data stream $\tilde{X}_{OSIC-2}$ and modified form of received signal $\tilde{Y}_{OSIC-2}$ can be written as:

$$\tilde{X}_{OSIC-2} = W_{(MMSE(2,:)}\tilde{Y}_{OSIC-1} \quad (24)$$

$$\tilde{Y}_{OSIC-2} = \tilde{Y}_{OSIC-1} - H_{\chi(:,2)}\tilde{X}_{OSIC-2}$$

In third step, the third detected signal/data stream $\tilde{X}_{OSIC-3}$

and modified form of received signal $\tilde{Y}_{OSIC-3}$ can be written as:

$$\tilde{X}_{OSIC-3} = W_{(MMSE(3,:)}\tilde{Y}_{OSIC-2} \tag{25}$$

$$\tilde{Y}_{OSIC-3} = \tilde{Y}_{OSIC-2} - H_{\chi(:,3)}\tilde{X}_{OSIC-3}$$

In fourth step, the fourth detected signal/data stream $\tilde{X}_{OSIC-4}$ and modified form of received signal $\tilde{Y}_{OSIC-4}$ can be written as:

$$\tilde{X}_{OSIC-4} = W_{(MMSE(4,:)}\tilde{Y}_{OSIC-3} \tag{26}$$

$$\tilde{Y}_{OSIC-4} = \tilde{Y}_{OSIC-3} - H_{\chi(:,4)}\tilde{X}_{OSIC-4}$$

where, W(MMSE(1,:),W(MMSE(2,:), W(MMSE(3,:)and W(MMSE(4,:)are the first, second, third and fourth rows of MMSE weight matrix and $H\chi(:,1)$, $H\chi(:,2)$,$H\chi(:,3)$ and $H\chi(:,4)$are the first, second, third and fourth columns of the dual polarized channel matrix respectively. The detected desired signal $\tilde{X}_{OSIC}$ from the transmitting antenna is given by[11-13]

$$\tilde{X}_{OSIC} = \begin{bmatrix} \tilde{X}_{OSIC-1} \\ \tilde{X}_{OSIC-2} \\ \tilde{X}_{OSIC-3} \\ \tilde{X}_{OSIC-4} \end{bmatrix} \tag{27}$$

3. System Description

The simulateddual polarized DWT aided MIMO SC-FDMA wireless communication system is depicted in Figure 1.In such system, a segment of audio signal is processed primarily for generating binary bit stream and subsequently encrypted[14]. The encrypted binary data are channel coded and.digitally modulated and are spatially demultiplexed to produce four data stream. Each data stream is rearranged into blocks with each block consisting of 64 symbols. A 64 point discrete wavelet transformation (DWT) algorithm is applied to each block to produce details and approximate coefficients. These coefficients are concatenated block wise, spatially mapped into a block of data symbols, serial to parallel converted with 2048 parallel data symbols and are transformed with inverse discrete wavelet transformation(IDWT), cyclically prefixed,parallel to serially converted and transmitted from four dual polarized antennas. In receiving end, the transmitted signals are detected using various signal detection techniques. The detected signals are processed with subsequent cyclic prefix removing, 64 point DWT transformed with its output coefficients concatenation, inverse discrete wavelet transformed(IDWT), spatially multiplexed, digitally demodulated, deinterleaved, channel decoded and decrypted to recover transmitted audio signal.

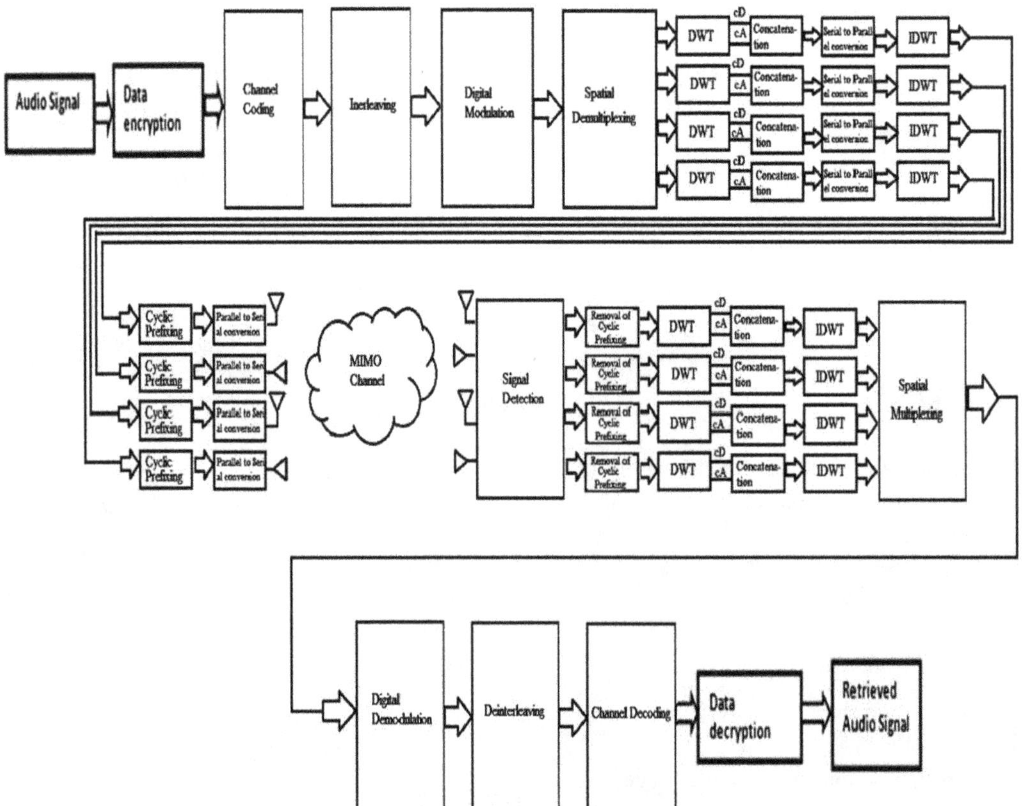

Figure 1. *Block diagram of dual polarized DWT aided MIMO SC-FDMAwireless communication system.*

4. Results and Discussion

In this section, we present a series of simulation results using MATLAB R2014ato illustrate the significant impact of various channel and signal detection techniques on dual polarized DWT aided MIMO SC-FDMA system performance in terms of BER based on the parameters given in Table 1. It is assumed that the channel state information (CSI) of the dual polarized MIMO fading channel is available at the receiver and the fading process is approximately constant during the whole audio signal transmission.

Table 1. *Summary of the Simulated Model Parameters.*

Parameters	Types
Data type	Audio signal
No. of samples	32768
Sampling Frequency(Hz)	12000
No of binary bits for a single sample	16
Total number of binary bits from audio samples	524288
Digital modulation	8PSK and 8QAM
Signal detection	BLUE, MMSE-SIC,ZF and OSIC
Channel coding	Turbo and LDPC
Antenna configuration	4 × 4
Value of inverse of the cross-polar discrimination (XPD)	0.85
SNR	0-10 dB
Channel	AWGN and Rayleigh

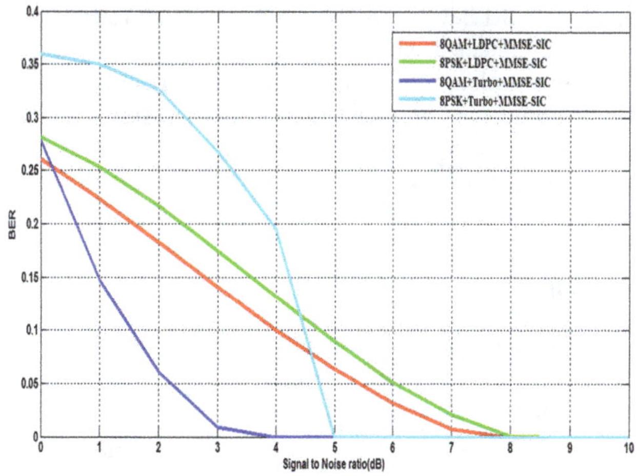

Figure 2. *BER performance of Dual Polarized DWT aided MIMO SC-FDMA Wireless Communication systemusing MMSE-SIC signal detection technique.*

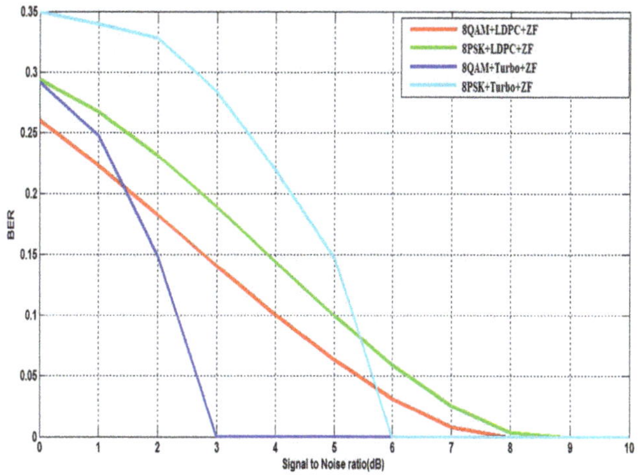

Figure 3. *BER performance of Dual Polarized DWT aided MIMO SC-FDMA Wireless Communication systemusing ZF signal detection technique.*

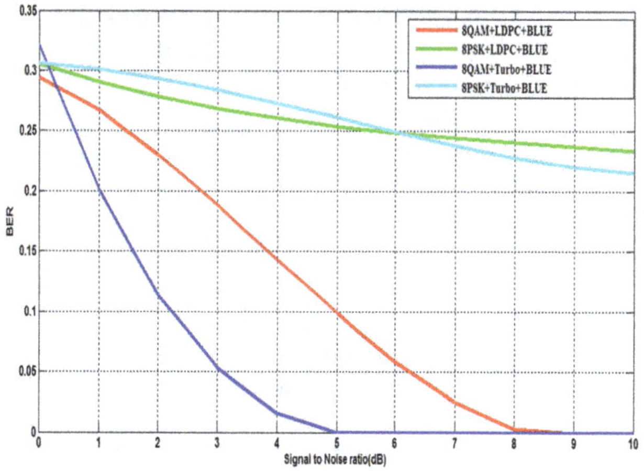

Figure 4. *BER performance of Dual Polarized DWT aided MIMO SC-FDMA Wireless Communication systemusing BLUE signal detection technique.*

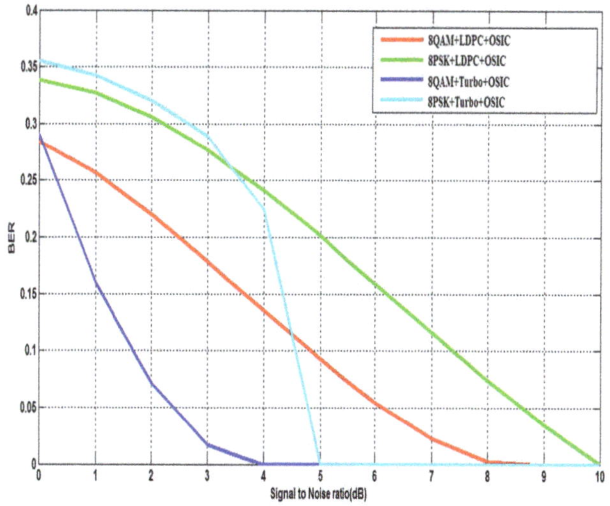

Figure 5. *BER performance of Dual Polarized DWT aided MIMO SC-FDMA Wireless Communication systemusing OSIC signal detection technique.*

The graphical illustrations presented in Figure 2 through Figure 5 show system performance comparison with implementation of MMSE-SIC, ZF, BLUE and OSIC based signal detection schemes under various digital modulations and channel coding schemes. In all cases, the system outperforms in 8QAM and shows worst performance in 8PSK digital modulations. The BER performance difference is quite obvious in lower SNR areas and the system's BER declines with increase in SNR values. In Figure 2, it is noticeable that for a typically assumed SNR value of 2 dB and MMSE-SIC signal detection, the estimated BER values are 0.0128 and 0.2263 in case of Turbo channel coding with 8-QAM and Turbo coding with 8-PSK which is indicative of a system performance improvement of 7.27dB.

In Figure 3, it is observable that for a SNR value of 2 dB and ZF signal detection, the estimated BER values are 0.1495 and 0.3280 in case of Turbo channel coding with 8-QAM and Turbo coding with 8-PSK which is indicative of a system performance improvement of 3.41dB. At 5% BER, SNR gains of 2.7,2.9 and 3.6 dB are achieved in case of 8-QAM with Turbo in comparison with 8-QAM with LDPC, 8-PSK with Turbo and 8-PSK with LDPC respectively.

In Figure 4, it is noticeable that for a SNR value of 2 dB and BLUE signal detection, the estimated BER values are 0.1139 and 0.2939 in case of Turbo channel coding with 8-QAM and Turbo coding with 8-PSK which is indicative of a system performance improvement of 4.11dB.

In Figure 5, it is observable that for a SNR value of 2 dB and OSIC signal detection, the estimated BER values are 0.0717 and 0.3207 in case of Turbo channel coding with 8-QAM and Turbo coding with 8-PSK which is indicative of a system performance improvement of 6.51dB. At 5% BER, SNR gains of 2.4, 3.7 and 6.3 dB are achieved in case of 8-QAM with Turbo in comparison with 8-PSK with Turbo, 8-QAM with LDPC and 8-PSK with LDPC respectively.

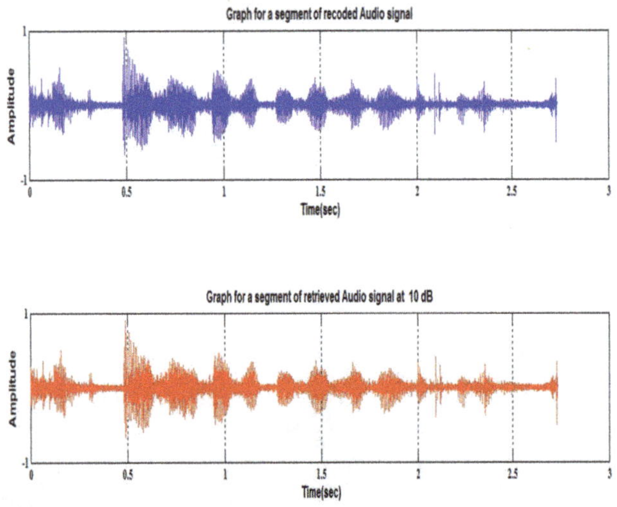

Figure 6. *Transmitted and Retrieved audio signals at SNR value of 8 dB under implementation of Turbo channel coding, 8-QAM digital modulation and MMSE-SIC signal detection technique.*

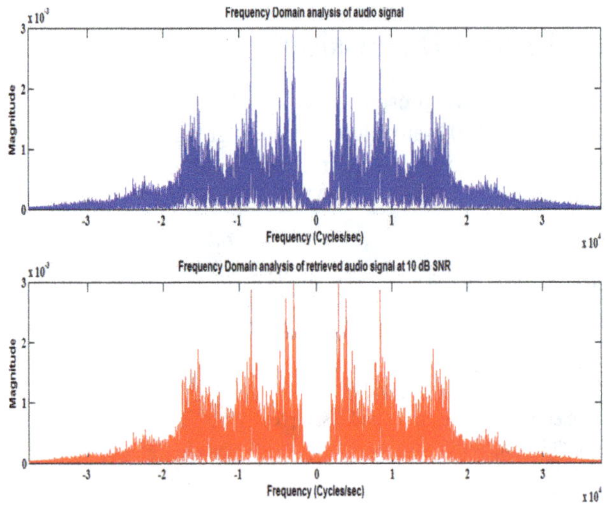

Figure 7. *Spectral analysis of Transmitted and Retrieved audio signals at SNR value of 8 dB under implementation of Turbo channel coding, 8-QAM digital modulation and MMSE-SIC signal detection technique.*

In Figure 6, the transmitted and retrieved audio signals at SNR value of 8 dB have been presented which are found to have great resemblance with each other. In perspective of spectral representation, the amplitude of different significant frequency components fortransmitted and retrieved audio signals are shown in Figure 7. It is noticeable that a unique spectral response is achieved in case of both transmitted and received audio signals.

5. Conclusions

In this paper, the performance of dual polarized DWT aided MIMO SC FDMA wireless communication system has been investigated on secured audio signal transmission using 8QAM and 8PSK digital modulations, various types of channel coding and signal detection techniques. The simulation results show that the implementation of Turbo channel coding, MMSE-SIC signal detection and 8QAM digital modulation schemes ratifies the robustness of dual polarized DWT aided MIMO SC-FDMA wireless communication system in retrieving audio signal transmitted over noisy and Rayleigh fading channels.

References

[1] Hyung G. Myung, David J. Goodman, 2008: Single carrier FDMA A new air interface for Long Term Evolution, John Wiley and Sons, Ltd, Publication, United Kingdom

[2] Syed Tariq Shah, Jaheon Gu, Syed Faraz Hasan, and Min Young Chung. (2015). SC-FDMA-based resource allocation and power control scheme for D2D communication using LTE-A uplink resource, EURASIP Journal on Wireless Communications and Networking, 137, 1-15.

[3] Umaria, Krupali N.; SCET, Surat and Joshi, Ketki, (2012). Comparative analysis of BER performance of DWT based OFDM system with conventional FFT based OFDM system, International Conference on Emerging Technology Trends in Electronics, Communication and Networking (ET2ECN).

[4] Jiho Song, Junil Choi, Stephen G. Larew, David J. Love, Timothy A. Thomas, and Amitava Ghosh.(2015). Adaptive Millimeter Wave Beam Alignment for Dual-Polarized MIMO Systems, http://arxiv.org/abs/1408.2098

[5] Yuan Jiang, 2010: A Pratical Guide to Error - Control Coding Using MATLAB, Jiang Artech House, Boston, USA

[6] Christian B. Schlegel and Lance C. Perez,2004:Trellis and turbo coding, John Wiley and Sons, Ltd, Publication, United Kingdom

[7] Bagawan Sewu Nugroho, https://sites.google.com/site/bsnugroho/ldpc

[8] Taejoon Kim, Bruno Clerckx, David J. Loveand Sung Jin Kim. 2010: Limited Feedback Beamforming Systems in Dual-Polarized MIMO Channel. EURASIP Journal on Wireless Communications and Networking, New york, USA 9,3425-3439.

[9] Andrzej CICHOCKI and Shun-ichi AMARI,2002. Adaptive Blind Signal and Image Processing Learning Algorithms and Applications. John Wiley and Sons Inc., New York, USA.

[10] Fathi E. Abd El-Samie, Faisal S. Al-kamali, Azzam Y. Al-nahari and Moawad I. Dessouky, 2013:.SC-FDMA for Mobile Communications, CRC Press, Taylor and Francis Group, Florida, USA

[11] Yong Soo Cho, Jackson Kim, Won Young Yang, Chung G. Kang, 2010: MIMO-OFDM Wireless Communications with MATLAB, John Wiley and Sons (Asia) PTE Limited, Singapore

[12] Md. Sarwar Hosain, Mousumi Haque and Shaikh Enayet Ullah, 2015:Performance Assessment of Mimo Mccdma System on Video Signal Transmission with Implementation of Various Digital Signal Processing Techniques, International Journal of Wireless Communications and Mobile Computing (WCMC), vol.3, no.2,pp.18-26, Science Publishing Group, New York, USA

[13] Lin Bai and Jinho Choi, 2012:Low Complexity MIMO Detection, Springer Science and Business Media, LLC, New York, USA

[14] William Stalling, 2005: Cryptography and Network Security Principles and Practices, Prentice Hall, USA.

7

Performance Assessment of Mimo Mccdma System on Video Signal Transmission with Implementation of Various Digital Signal Processing Techniques

Md. Sarwar Hosain[1], Mousumi Haque[2], Shaikh Enayet Ullah[3]

[1]Department of Information and Communication Engineering (ICE), Pabna University of Science and Technology, Pabna, Bangladesh
[2]Department of Information and Communication Engineering (ICE), University of Rajshahi, Rajshahi, Bangladesh
[3]Department of Applied Physics and Electronic Engineering (APEE), University of Rajshahi, Rajshahi, Bangladesh

Email address:
sarwar@ice.pust.ac.bd (M. S. Hosain), Mishiape@yahoo.com (M. Haque), enayet_apee@ru.ac.bd (S. E. Ullah)

Abstract: In this paper, we made a comprehensive study to evaluate the performance of MIMO MCCDMA wireless communication system on video signal transmission. The 4-by-4multi antenna supported MCCDMA system incorporates various digital signal processing techniques such as Minimum mean square error (MMSE) , Zero-Forcing (ZF), Ordered successive interference cancellation (OSIC) and Lattice Reduction aided MMSE for transmitted signal detection and two-dimensional nonlinear Median filtering for noise reduction. It is observable from MATLAB based simulations that the system shows quite satisfactory performance under scenario of MMSE signal detection scheme.

Keywords: Mimo, Mccdma, ZF, Mmse, Osic, Lr-Mmse, 2-D Median Filtering, Snr

1. Introduction

Multi Carrier CDMA(MC-CDMA) technique has become increasingly popular in wireless communications, mainly due to its high spectral efficiency, robustness to frequency selective fading and flexibility to support integrated applications. Various MC-CDMA schemes have been proposed and can be classified into two main categories based on two different ways of combining CDMA and Orthogonal Frequency Division Multiplexing (OFDM). One is to spread the original data sequence in the time domain and the other is to spread in frequency domain. The former scheme, which is so called MC-DS-CDMA has some advantages, such as easy tracking of the fading process over subcarriers. It, however, cannot achieve the frequency diversity gain, which is the main advantage of using multicarrier modulation technique. The latter scheme, which is usually called MCCDMA, can exploit frequency diversity. The Multi-Carrier Code-Division Multiple Access (MCCDMA) systems, which are capable of supporting the interworking of existing as well as future broadcast and personal communication systems[1,2]. In 2009, Antonis et.al., made a comparative study on the error rate performance of downlink coded multiple-input multiple-output multi-carrier code division multiple access (MIMO MC-CDMA) and coded MIMO orthogonal frequency division multiple access (MIMO OFDMA) systems under frequency selective fading channel conditions and showed that MIMO MC-CDMA system outperforms as compared to MIMO OFDMA system. In MC-CDMA multiplexing technique, multiple users are permitted to access the wireless channel simultaneously by modulating and spreading their input data signals across the frequency domain using different spreading sequences. In such scheme, the reliability of detecting individual user's transmitted signal is not highly dependent on the signal's strength [3,4].

In this present study, an effort has been made to observe the performance of MIMO MCCDMA system under a scenario of video signal transmission.

2. Signal Processing Techniques

We assumed that the captured video signal is preprocessed

in channel coding, digital modulation, interleaving, multicarrier CDMA encoding and spatial multiplexing schemes prior to transmission through a MIMO fading channel. The received signal $Y \in C^{4\times4562325}$ in terms of channel matrix $H \in C^{4\times4}$, transmitted signal $X \in C^{4\times4562325}$ and concatenated additive white Gaussian noise (AWGN) with a variance of σ_n^2 and impulsive (salt and pepper) noise $N \in C^{4\times4562325}$ can be written as

$$Y = HX + N \tag{1}$$

In Minimum mean square error (MMSE) based signal detection scheme, the MMSE weight matrix is given by

$$W_{MMSE} = (H^H H + \sigma_n^2 I)^{-1} H^H \tag{2}$$

Where $(.)^H$ denotes the Hermitian transpose operation and the detected desired signal $\tilde{X}_{MMSE} \in C^{4\times4562325}$ from the transmitting antenna is given by

$$\tilde{X}_{MMSE} = W_{MMSE} Y \tag{3}$$

In Zero-Forcing (ZF) signal detection scheme, the ZF weight matrix is given by

$$W_{ZF} = (H^H H)^{-1} H^H \tag{4}$$

and the detected desired signal $\tilde{X}_{ZF} \in C^{4\times4562325}$ from the transmitting antenna is given by

$$\tilde{X}_{ZF} = W_{ZF} Y \tag{5}$$

In Ordered successive interference cancellation (OSIC) signal detection scheme, its implementation is performed in four steps. In first step, the first detected signal/data stream $\tilde{X}_{OSIC-1}$ and modified form of received signal $\tilde{Y}_{OSIC-1}$ can be written as:

$$\begin{aligned} \tilde{X}_{OSIC-1} &= W_{(MMSE(1,:)} Y \\ \tilde{Y}_{OSIC-1} &= Y - H_{(:,1)} \tilde{X}_{OSIC-1} \end{aligned} \tag{6}$$

In second step, the second detected signal/data stream $\tilde{X}_{OSIC-2}$ and modified form of received signal $\tilde{Y}_{OSIC-2}$ can be written as:

$$\begin{aligned} \tilde{X}_{OSIC-2} &= W_{(MMSE(2,:)} \tilde{Y}_{OSIC-1} \\ \tilde{Y}_{OSIC-2} &= \tilde{Y}_{OSIC-1} - H_{(:,2)} \tilde{X}_{OSIC-2} \end{aligned} \tag{7}$$

In third step, the third detected signal/data stream $\tilde{X}_{OSIC-3}$ and modified form of received signal $\tilde{Y}_{OSIC-3}$ can be written as:

$$\begin{aligned} \tilde{X}_{OSIC-3} &= W_{(MMSE(3,:)} \tilde{Y}_{OSIC-2} \\ \tilde{Y}_{OSIC-3} &= \tilde{Y}_{OSIC-2} - H_{(:,3)} \tilde{X}_{OSIC-3} \end{aligned} \tag{8}$$

In fourth step, the fourth detected signal/data stream $\tilde{X}_{OSIC-4}$ and modified form of received signal $\tilde{Y}_{OSIC-4}$ can be written as:

$$\begin{aligned} \tilde{X}_{OSIC-4} &= W_{(MMSE(4,:)} \tilde{Y}_{OSIC-3} \\ \tilde{Y}_{OSIC-4} &= \tilde{Y}_{OSIC-3} - H_{(:,4)} \tilde{X}_{OSIC-4} \end{aligned} \tag{9}$$

where, $W_{(MMSE(1,:)}$, $W_{(MMSE(2,:)}$, $W_{(MMSE(3,:)}$ and $W_{(MMSE(4,:)}$ are the first, second, third and fourth rows of MMSE weight matrix and $H_{(:,1)}$, $H_{(:,2)}$, $H_{(:,3)}$, and $H_{(:,4),:)}$ are the first, second, third and fourth columns of the channel matrix respectively. The detected desired signal $\tilde{X}_{OSIC} \in C^{4\times4562325}$ from the transmitting antenna is given by

$$\tilde{X}_{OSIC} = \begin{bmatrix} \tilde{X}_{OSIC-1} \\ \tilde{X}_{OSIC-2} \\ \tilde{X}_{OSIC-3} \\ \tilde{X}_{OSIC-4} \end{bmatrix} \tag{10}$$

In Lattice Reduction aided MMSE signal detection technique, Lenstra-Lenstra-Lovasz (LLL) algorithm is implemented. Under such algorithm implementation scenario, the channel matrix H is transformed into a 8×8 real -valued matrix H_{real} as:

$$H_{real} = \begin{bmatrix} \Re(H) & -\Im(H) \\ \Im(H) & \Re(H) \end{bmatrix} \tag{11}$$

The equation (1) can be rewritten for real valued 8×4562325received signal Y_{real} as:

$$Y_{real} = \begin{bmatrix} \Re(Y) \\ \Im(Y) \end{bmatrix} = \begin{bmatrix} \Re(H) & -\Im(H) \\ \Im(H) & \Re(H) \end{bmatrix} \begin{bmatrix} \Re(X) \\ \Im(X) \end{bmatrix} + \begin{bmatrix} \Re(N) \\ \Im(N) \end{bmatrix} \tag{12}$$

The extended 16×8 channel matrix $\hat{H}$ and the16×4562325 extended received signal $\hat{Y}$ in terms of8×8identity and 8×4562325 null matrices are given by

$$\hat{H} = \begin{bmatrix} H_{real} \\ (\sqrt{\sigma_n^2}) I_{8\times8} \end{bmatrix} \tag{13}$$

$$\hat{Y} = \begin{bmatrix} Y_{real} \\ \mathbf{0}_{8\times4562325} \end{bmatrix} \tag{14}$$

On sorted QR decomposition of $\hat{H}$, a 8×8 permutation matrix P, 16×8 orthogonal matrix Q and a 8×8upper triangular matrix R with large condition number are produced. Using a typically chosen scaling parameter of value 0.5 in Lenstrat-Lenstra-Lovasz (LLL) algorithm, three new matrices Q_{LLL}, R_{LLL} with small condition number and 8×8 unimodular matrix T_{LLL} are produced.

The modified form of extended channel matrix $\tilde{H}$ in terms of unimodular matrix, permutation matrix and extended channel matrix can be written as:

$$\tilde{H} = \hat{H}PT_{LLL} \quad (15)$$

The real valued extended 8×4562325transmitted signal X_TEMP is given by

$$X\_TEMP = PT_{LLL}(\tilde{H}^H \tilde{H})^{-1} \tilde{H}^H \hat{Y} \quad (16)$$

In MATLAB notation using for end loop with specification of 4 iterations and real to complex conversion, the estimated transmitted signal X_LRMMSE can be written as:

for i=1:4
X_LRMMSE(i,:) = X_TEMP(i,:) + sqrt(-1) * X_TEMP(i + 4,:) end

In Equation(16), the first through fourth rows are the estimated real components and the real values presented in fifth through eight rows are multiplied by sqrt(-1) to get the estimated imaginary components. In matrix form, we can write

$$X\_LRMMSE = \begin{bmatrix} X\_TEMP(1,:) + sqrt(-1) * X\_TEMP(5,:) \\ X\_TEMP(2,:) + sqrt(-1) * X\_TEMP(6,:) \\ X\_TEMP(3,:) + sqrt(-1) * X\_TEMP(7,:) \\ X\_TEMP(4,:) + sqrt(-1) * X\_TEMP(8,:) \end{bmatrix} \quad (17)$$

However, the detected signal is filtered for salt and pepper noise reduction using 2D median filtering technique**.** In 2-D Median Filtering scheme, a 3×3 neighborhood windowing mask is used for simply sorting all the pixel values within the window and finding the median value and replacing the original pixel value with the median value[5-7].

3. System Description

A video file in mp4 format is downloaded from a website at https://www.youtube.com/watch?feature=player_detailpage&v=J5XESDvBdsY#t=3.The total number of video frame is 2910 with a frame rate of 30 RGB frames/sec. The number of video frame used in this present simulation study is 5. The resolution of each video frame is of 480 pixels (width) ×360 pixels (height) with identical 96 dpi in both vertical and horizontal dimension. The selected video frames are processed in a MIMO MCCMA system depicted in Figure 1.The captured color video frames are converted into their respective three Red, Green and Blue components with each component is of 480 pixels ×360 pixels in size.The pixel integer values[0-255] are converted into 8 bits binary form and channel coded and interleaved and digitally modulated using 16QAM. The modulated complex symbols are copied and multiplied with Walsh–Hadamard (WH) orthogonal codes. The orthogonally encoded signals are processed in Spatial multiplexing(SM) Encoding section to produce four independent data streams. Each of four data streams are serial to parallel converted, OFDM modulated, cyclically prefixed and subsequently parallel to serially converted and transmitted. In receiving section, the transmitted signals are detected and processed for serial to parallel conversion, cyclic prefix removal, and OFDM demodulation, parallel to serial conversion and decoded in Spatial multiplexing (SM) Decoding section to produce data in single channel. The retrieved data are multiplied with Walsh–Hadamard (WH) orthogonal codes and decopied, digitally demodulated, deinterleaved, channel decoded, binary to pixel integer converted and filtered to reconstruct frames for transmitted video frame retrieval [8-10].

4. Result and Discussion

Table 1. *Summary of the simulated model parameters.*

Parameters	Values
Data Type	Video Signal
Number of frames	5
Digital modulation	16 - QAM
Noise type	Impulse (Salt and pepper) and Gaussian
Signal to Noise Ratio (SNR) in dB	0 to 10 dB
Noise reduction Filter	2-D Median filtering
Channel Equalization	MMSE, ZF, OSIC and LR-aided MMSE
Orthogonal spreading code	Walsh-Hadamard
Processing gain of Walsh-Hadamard code	8
Antenna configuration	4 × 4

In this section, computer simulations using MATLAB have been performed to evaluate the BER performance of a 4x4 multi antenna supported and implemented 2D median filtering for noise reduction in MC-CDMA wireless communication system based on the parameters presented in Table 1. In Figure 2, the transmitted five video frames and their corresponding impulsive (salt and pepper) noise contaminated versions have been presented. Contamination rate is 5% viz. 8640 pixels out of 172800 pixels are contaminated with impulsive noise for each 480pixels ×360 pixels sized Red, Green and Blue components of an individual video frame. In Figure 3, histograms of captured RGB to Gray converted five video frames are presented. The histograms are indicative of pixel intensity values(0 to 255)and the absence of intensity values in the upper range: 200-255 confirms that the captured 100th video frame is not bright. In other cases except 2000th video frame , pixel intensity values are distributed over the almost whole band of pixel values.The graphical illustrations presented in Figure 4 through Figure 7 show system performance comparison(Bit error rate(BER) vs SNR values) for 1/2- rated Convolutionally encoded MIMO MC CDMA system with various signal detection, noise reduction and 16-QAM digital modulation schemes. In all cases, the impact of MMSE signal detection technique on system performance enhancement is clearly observable at low SNR value area. It is also noticeable in case of all the captured video frames that the simulated system shows identical performance over a significant part of SNR values. The estimated bit error rate at a typically assumed SNR value of 5 dB with adaptation of MMSE signal detection technique for 100th, 600th, 1100th, 2000th and 2500th video frames are 13.97%, 10.33%,12.19%, 12.36% and 16.01% respectively. The system performance is quite satisfactory for 600th video frame which is found to contain intermediate pixel values around the range: 100-170. In Figure 4 through Figure 8, the system performance improvement at identical signal and noise power(SNR=0dB) in MMSE signal detection as compared to ZF signal detection are 0.09422 dB,0.09478 dB, 0.10024 dB, 0.06654 dB and 0.07471 dB respectively. In Figure 9, transmitted 1200th video frame, its noise contaminated and retrieved video frames have been presented. The retrieved video frame has a great resemblance with the transmitted video frame and the estimated bit error rate was found to be of 0.066 viz. 3873584 bits are correctly retrieved out of 4147200 bits (480×360×24) for the captured video frame.

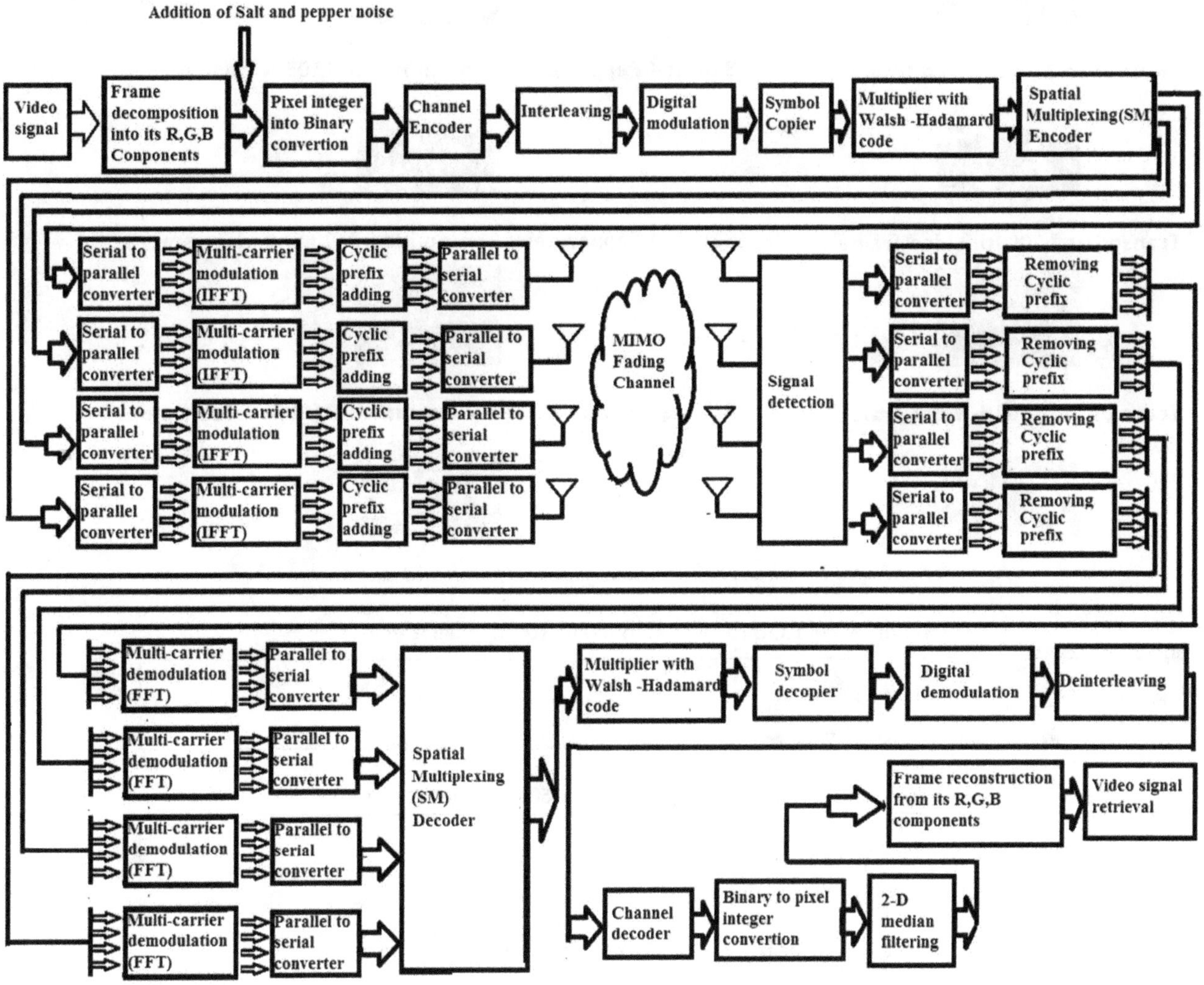

Figure 1. Block diagram of simulated MIMOMCCMA system.

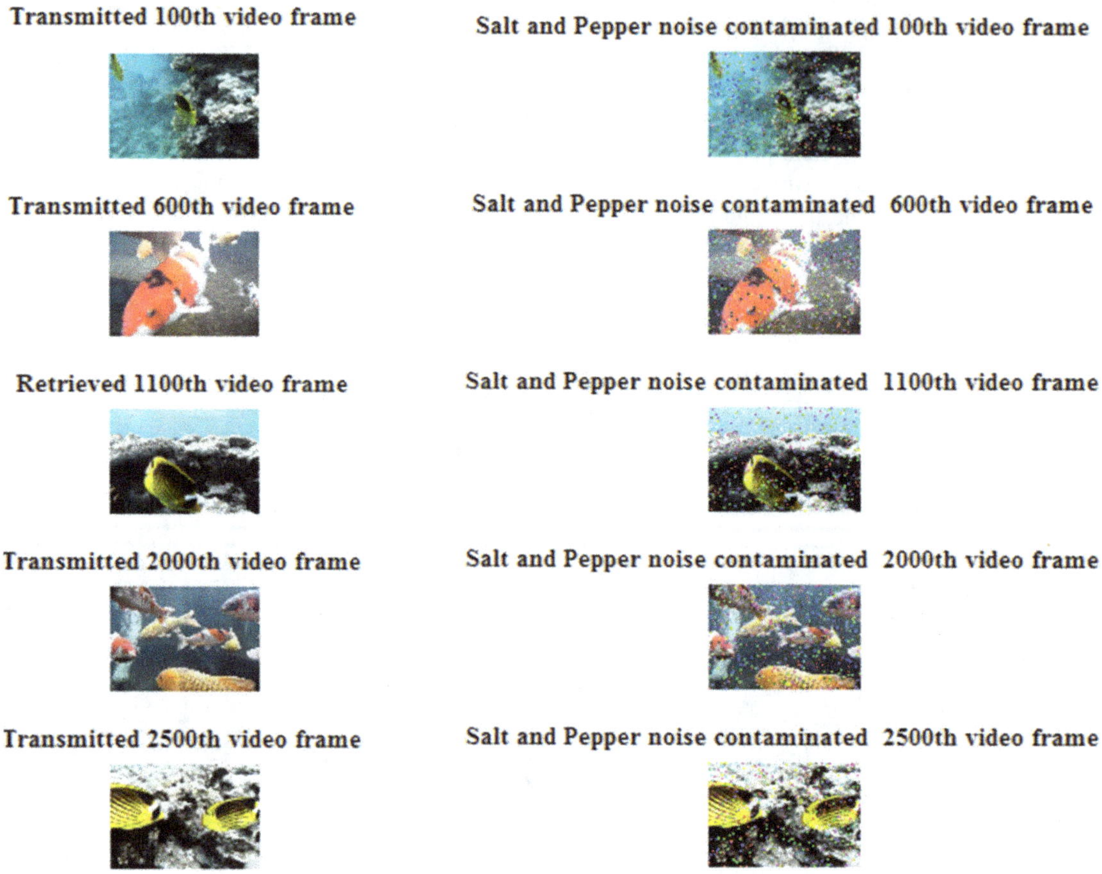

Figure 2. *Transmitted and Salt and Pepper noise contaminated five selected video frames.*

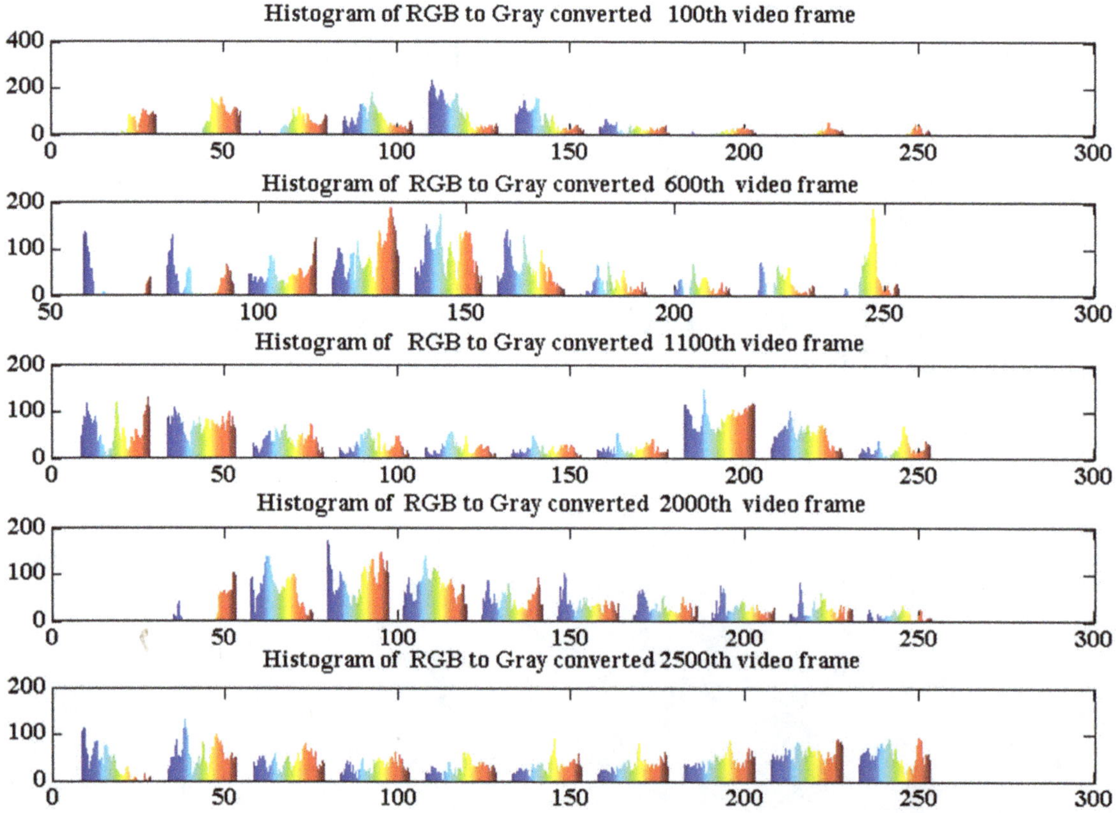

Figure 3. *Histogram of transmitted RGB to Gray converted five selected video frames.*

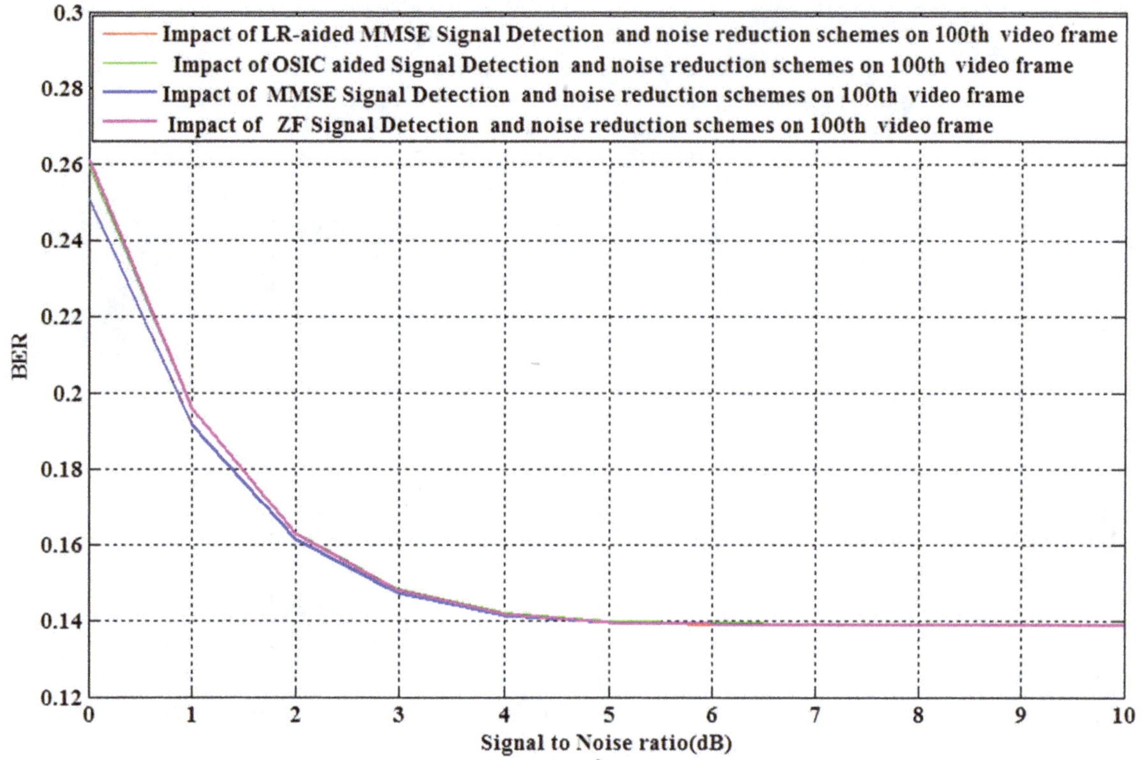

Figure 4. *BER performance of MIMO MCCDMA system with implementation of various signal detection and 2D median filtering schemes on 100th captured video frame.*

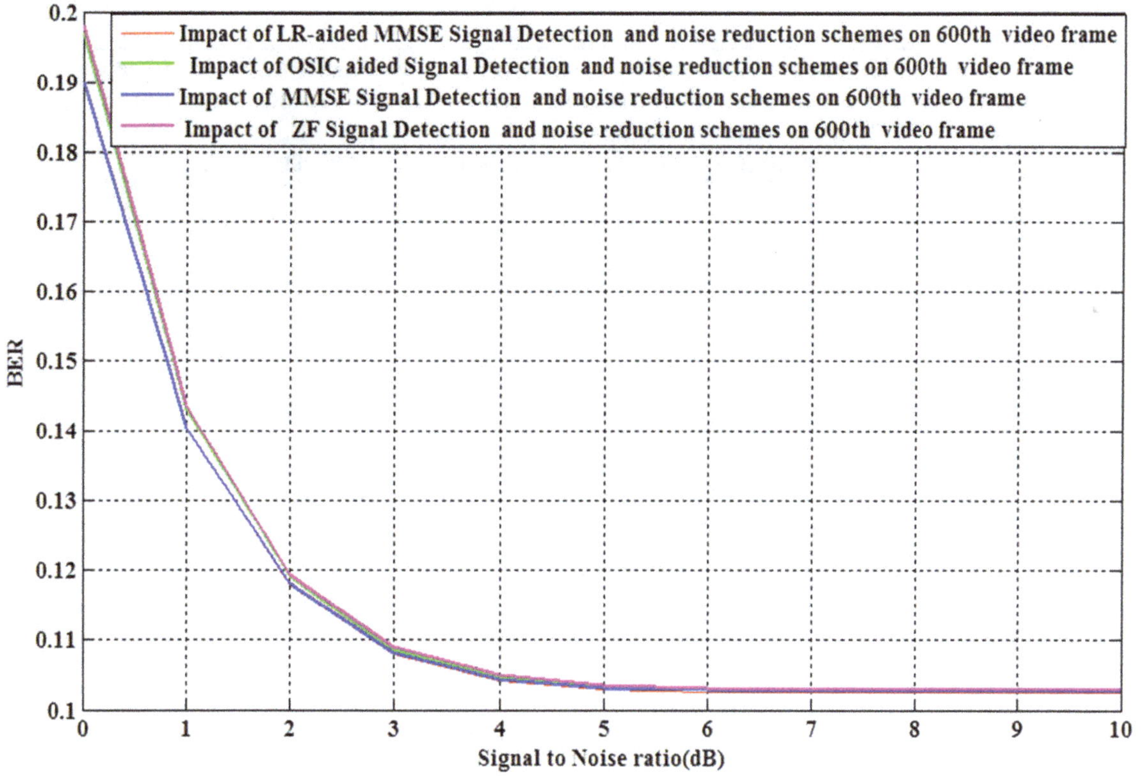

Figure 5. *BER performance of MIMO MCCDMA system with implementation of various signal detection and 2D median filtering schemes on 600th captured video frame.*

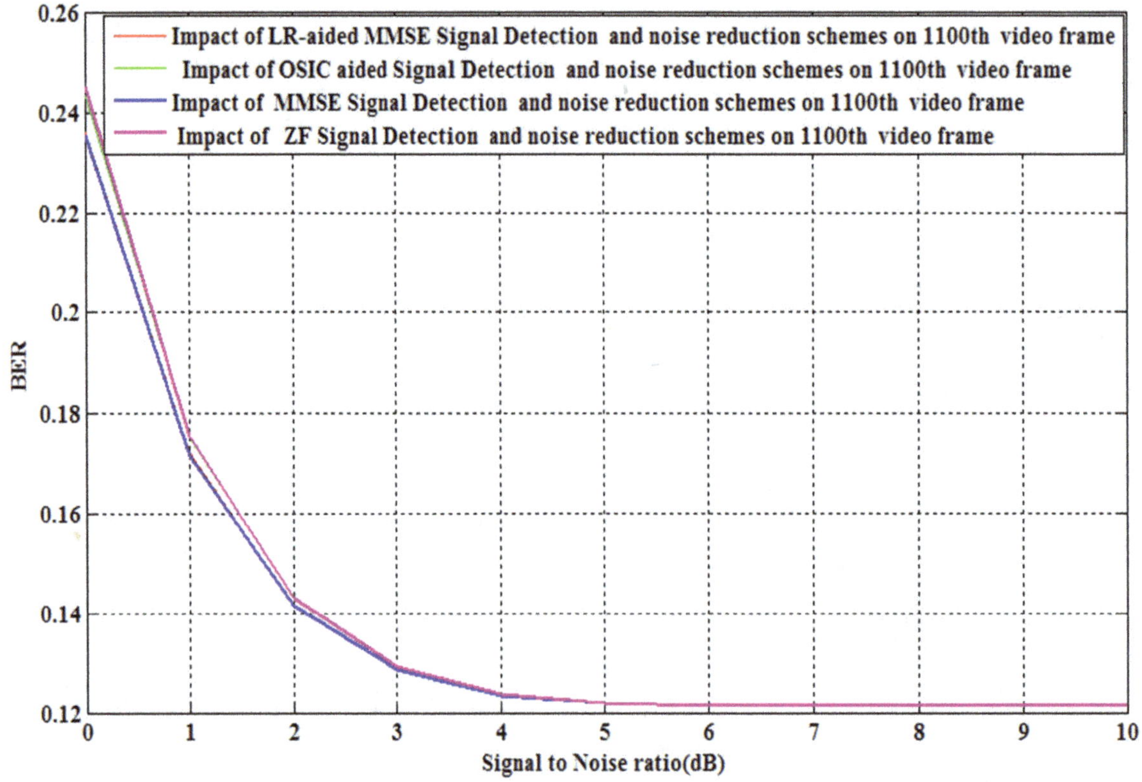

Figure 6. *BER performance of MIMO MCCDMA system with implementation of various signal detection and 2D median filtering schemes on 1100th captured video frame.*

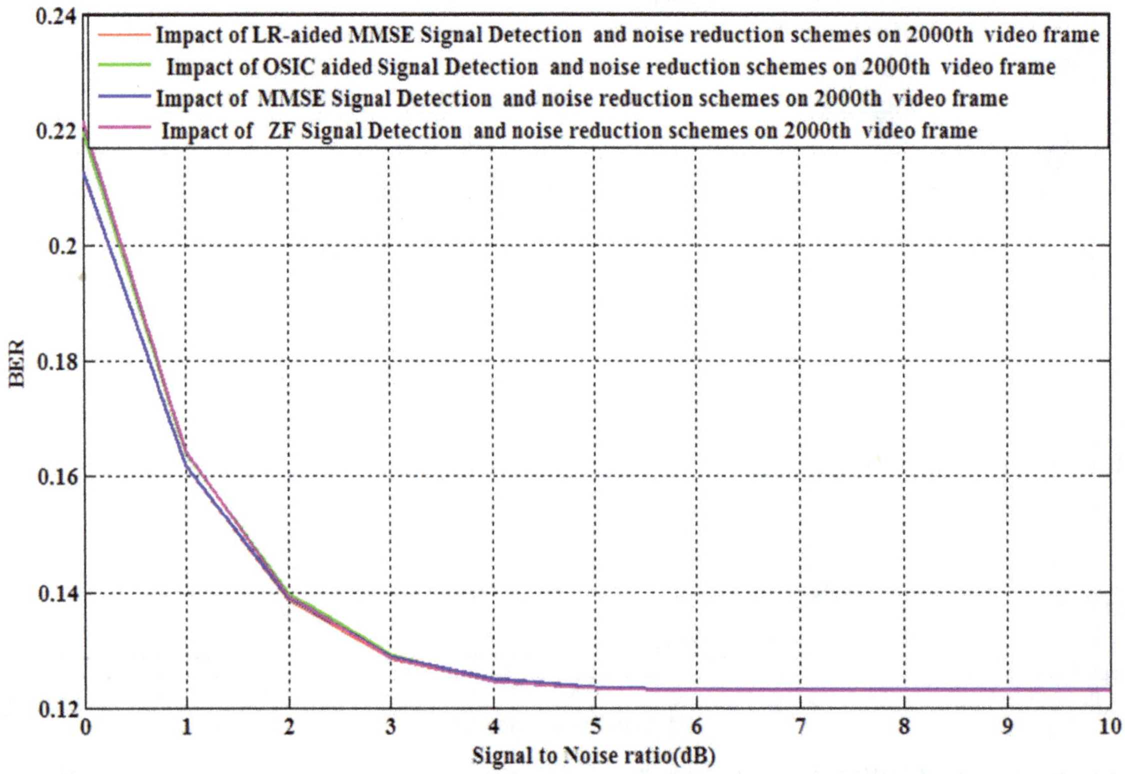

Figure 7. *BER performance of MIMO MCCDMA system with implementation of various signal detection and 2D median filtering schemes on 2000th captured video frame.*

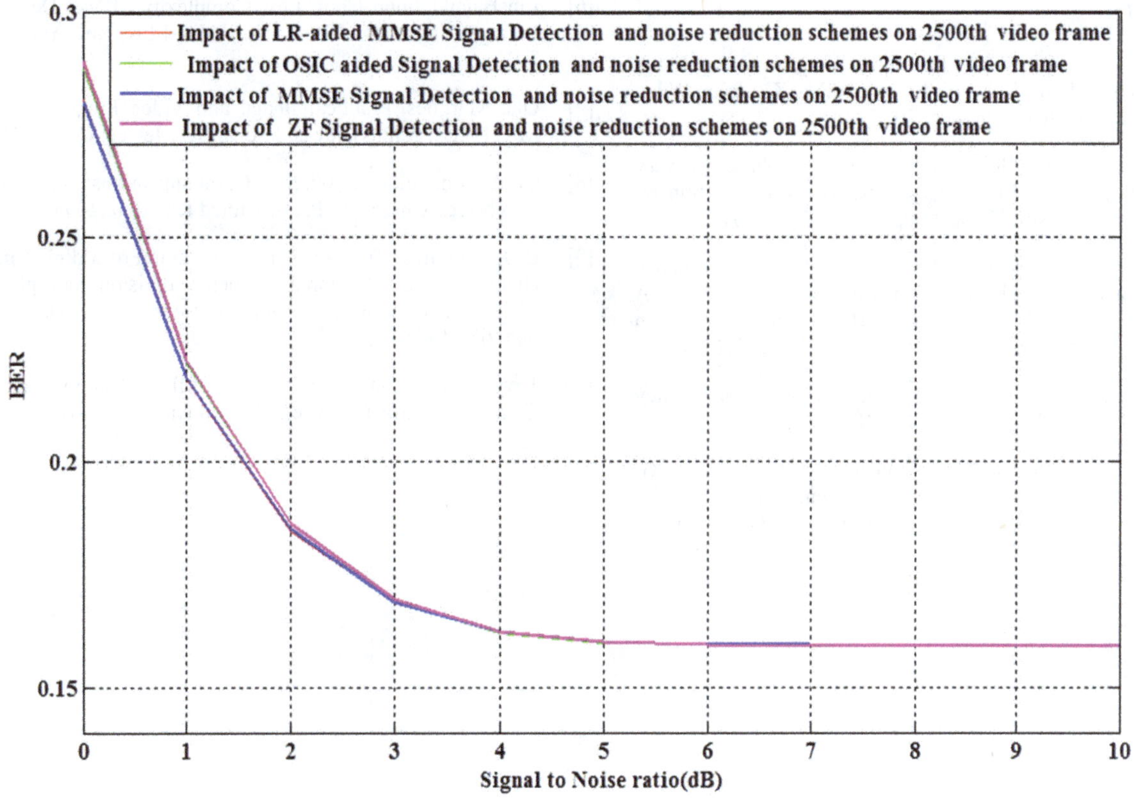

Figure 8. *BER performance of MIMO MCCDMA system with implementation of various signal detection and 2D median filtering schemes on 2500th captured video frame.*

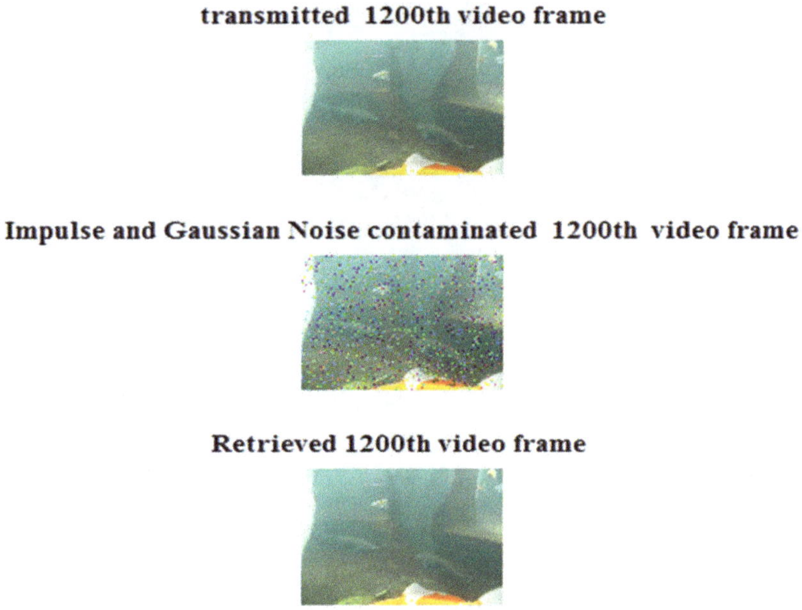

Figure 9. *Performance indicator of MMSE signal detection aided MIMO MCCDMA system for a typically assumed captured 1200thvideo frame at SNR value of 5 dB.*

5. Conclusions

In this paper, the performance of MIMO MCCDMA wireless communication system has been investigated using various signal detection techniques. The results show that MMSE signal detection scheme is better than other signal detection scheme in retrieving video signal transmitted over impulsive and Gaussian noise contaminated channel. As MC CDMA radio interface technology exploits advantages of both OFDM and CDMA and a great emphasis is being given on video communication, such MCCDMA radio interface technology can be used in 5G compatible dual polarized Massive MIMO antenna system.

References

[1] Lajos Hanzo, Yosef (Jos) Akhtman, Li Wang and Ming Jiang, "MIMO-OFDM for LTE, Wi-Fi and WiMAX", John Wiley and Sons Ltd, United Kingdom, 2011.

[2] RuiFa and Pei Xiao,"Jointdata detection and phase recovery for downlink MC-2D-CDMA systems",IEEE Transaction on Communications, vol. 57, no. 9, pp. 2782-2789, 2009

[3] Antonis Phasouliotis and Daniel K.C. So, "Performance Analysis and Comparison of Downlink MIMO MC-CDMA and MIMO OFDMA Systems", IEEE Transactions on Wireless Communications, vol. 8, no.1, pp. 214-225,2009

[4] Lie-Liang Yang,"Multicarrier Communications", John Wiley and Sons Limited, United Kingdom, 2009

[5] Yong Soo Cho, Jackson Kim, Won Young Yang, Chung G. Kang,"MIMO-OFDM Wireless Communications with MATLAB", John Wiley and Sons (Asia) PTE Limited, Singapore, 2010

[6] Lin Baiand Jinho Choi,"Low Complexity MIMO Detection", Springer Science and Business Media, LLC, New York, USA, 2012

[7] Oge Marques,"Practical Image and Video Processing Using MATLAB", John Wiley and Sons, New Jersey, USA, 2011

[8] Goldsmith, Andrea, "Wireless Communications", First Edition, Cambridge University Press, United Kingdom,2005

[9] L. J. Cimini, Jr.," Analysis and simulation of a digital mobile channel using orthogonal frequency division multiplexing", IEEE Transactions on Communications., vol. COM-33, pp. 665–675,1985

[10] Jerry R. Hampton, "Introduction to MIMO Communications", Cambridge University Press, United Kingdom, 2014.

8

Array Antennas Comprised of Tetragons for RFID Applications

Manato Fujimoto, Yukio Iida*

Department of Electrical and Electronic Engineering, Faculty of Engineering Science, Kansai University, Osaka, Japan

Email address:
manato@jnet.densi.kansai-u.ac.jp (M. Fujimoto), iida@kansai-u.ac.jp (Y. Iida)

Abstract: The location estimation capability of a RFID (Radio Frequency Identification) technology is useful for indoor navigation systems on mobile robots. In this paper, we study a simple and small structural antenna at 2.5GHz band that can reduce the beam when installed on a mobile robot. An antenna with small occupation volume by a mechanical rotation is assumed. We used electromagnetic field simulation and experimental results to design a simple structural antenna. Our design culminates in two antennas comprised of Tetragon elements formed with wires. One is called Tetra-4, and is an antenna which has a lateral arrangement of four tetragons. Because the input impedance is considerably high, Tetra-4 is suitable for use in the high impedance system. When driving in 100 Ω balanced system, the gain of Tetra-4 is 8dBi. The half-power band width (HPBW) in the H-plane is 38 degrees, and is 0.44 times that of a dipole with a reflector. Another one is Yagi-Uda antenna comprised of two tetragon elements. When driving in 50 Ω balanced system, the gain is 11dBi. The HPBW is 40 degrees in the H-plane and is 36 degrees in the E-plane.

Keywords: Array Antenna, Tetragon, Loop Antenna, RFID, Location Estimation, Mobile Robot

1. Introduction

Recently, RFID (Radio Frequency Identification) technology has attracted attention around the world as a new approach to achieving ubiquitous environments [1]. RFID technology is used in various fields due to features such as having a simple composition, being inexpensive, making it very easy to store information in the memory of a RFID tag, etc. In particular, the estimation of the location of RFID tags has attracted attention from many researchers [2]-[7]. If the location estimation of RFID tags is achieved, it can be used in various applications (for example, indoor navigation systems on mobile robots, etc.) [8]-[11].

A large antennas are used though the experiment with the robot and the wheelchair is conducted [12], [13]. Because the robot is not only work of self-positional recognition, a small antenna is preferable. We also have done a lot of experiments and the simulations [5]-[7], [11], [14]. As a result, we feel the necessity of the antenna that doesn't disturb the movement of the robot strong. The antenna for the robot will be needed more and more in the future. Then, a simple and small antenna that can reduce the beam width for use in mobile robots is studied in this paper. The antenna with small occupation volume by a mechanical rotation is assumed.

We consider a passive tag that is supplied with power from the electromagnetic waves of a reader/writer. A passive tag requires a strong electromagnetic field above a certain level in order to obtain sufficient power. Therefore, for the array antenna considered here, the radiation range of the peak level is very important, unlike antennas using for normal communications.

When the direction in a horizontal plane is detected by arranging tag for a linearly polarized wave, the radiation pattern in a horizontal plane is important. The antenna with narrow beam width is necessary in a horizontal plane. When it is necessary to detect the vertical position, the antenna with narrow beam width is necessary also for the vertical direction.

In recent years, the electromagnetic wave applications has increased. Various antennas have attracted attention due to their simplicity, cost effectiveness, and ease of manufacturing, etc. [15]-[17]. In addition, there are many studies in relation to loop antennas [18]-[23]. These reports are mostly based on the basic loop antenna and Yagi-Uda antenna. In this paper, we will study an array antenna based on a tetragonal structure.

This paper is organized as follows. Section 2 explains basic antenna structures of Tetra-1, Tetra-2, and Tetra-4 using this paper. Section 3 presents basic characteristics of each antenna through electromagnetic field simulations using FDTD method. Section 4 shows experimental results of Tetra-4 antenna. Section 5 presents the Yagi-Uda antenna comprised of two tetragon elements. Finally, we conclude this paper in Section 6.

2. Basic Structure of Antennas

In this section, we will investigate the round loop antenna and radiation principle. The necessary current for electromagnetic radiation flows when the circumference is equal to one wavelength. Since the direction of the current reverses for each half wavelength, an electromagnetic wave of a linear polarization is radiated. We designed a suitable structure using the radiation principle, and performed electromagnetic field simulations and corresponding experimental analysis. In this study, we are primarily interested in the development of an antenna that uses a wire to form a narrowly focused beam. We groped the resonating structures based on their suitability for the radiation of electromagnetic wave. As one of results, we arrived at the tetragonal structure shown in Figure 1.

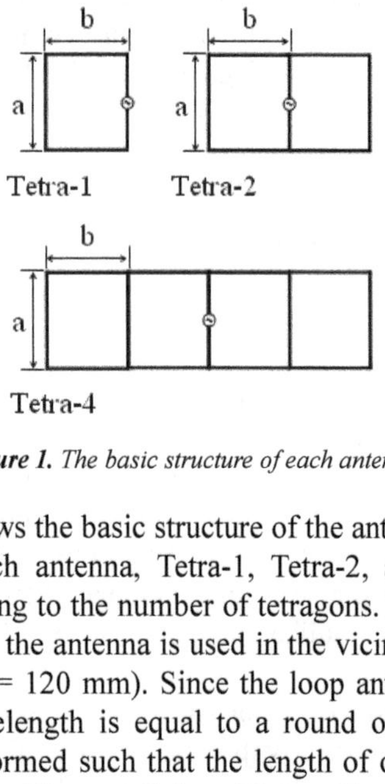

Figure 1. The basic structure of each antenna.

Figure 1 shows the basic structure of the antenna studied in this paper. Each antenna, Tetra-1, Tetra-2, and Tetra-4, is labeled according to the number of tetragons. In this paper, it is assumed that the antenna is used in the vicinity of 2.5 GHz (wavelength λ = 120 mm). Since the loop antenna resonates when the wavelength is equal to a round of the loop, the tetragon was formed such that the length of one loop is 120 mm. We set a =32mm and b = 28mm. The antennas were produced using copper wire of 2.0 mm in diameter.

3. Basic Characteristics through Electromagnetic Field Simulations

3.1. Simulation Method of the Electromagnetic Field

In this section, we investigate the basic characteristics of the antenna through electromagnetic field simulations using the FDTD (Finite-Difference Time-Domain) method. The structure of the antenna was discretized into 2 mm square cubes. Since the structure of each antenna is discretized into 2 mm blocks, the size of the antenna in these simulations is a = 32 mm and b = 28 mm for the tetragons.

Figure 2 shows the Tetra-4 antenna and the simulation region. We placed the electric wall at the YZ-plane at X = 0 and analyzed the space of the upper half. The FDTD cell size was $\Delta x = \Delta y = \Delta z$ = 2 mm. In order to simplify the representation, we used $X = x/\Delta x$, $Y = y/\Delta y$, and $Z = z/\Delta z$ as normalized coordinates.

In Section 2, we used the antennas shown in Figure 1. We assume that each antenna studied in this paper is attached to a mobile robot. From this assumption, the conductor plane should be placed at the back of the antenna. In Figure 2, the conductor board is placed at Z = 0 and the antenna surface is placed at Z = 30 (= $\lambda/4$).

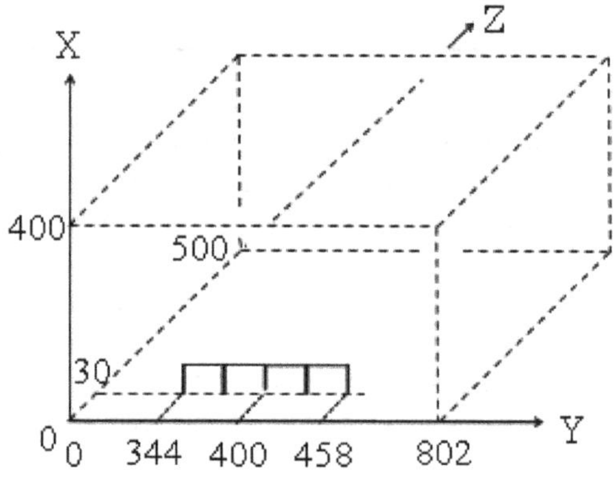

Figure 2. Tetra-4 antenna and analysis region.

3.2. Radiation Intensity and Beam Width

In the analysis area defined, we examined the electromagnetic field distribution on the line Y = 100 ~ 700 at Z = 360. The results are summarized in Figure 3. The vertical axis is given as P_r/P_{in}. Here, P_{in} is the active power that enters the antenna. P_r represents the electric field strength converted into the received power of the isotropic antenna and is the power that can be received when the isotropic antenna is placed at the observation line at Z = 360. We call P_r/P_{in} the radiation intensity. The gain of Hexa-4 is 3.4 dB higher than that of a dipole. The dipole here is an antenna with a reflector (conductor board).

The radiation intensity shows the performance of the antenna without any relation to the input impedance. The input impedance of the antenna is particularly important for achieving a suitable connection to the power feeding system. As for the power feeding system of the antenna, an unbalanced system at 50 Ω and 75 Ω is often used. The performance evaluation of an antenna that is independent of specific impedance is important if the antenna is required to operate with a variety of feed systems.

We considered the 3dB range of the Y coordinate which decreased by 3 dB from the maximum radiation intensity. The 3 dB range narrowed as the tetragon increased. The 3 dB range was 309 with Tetra-4, and was 66% of the 468 of a dipole.

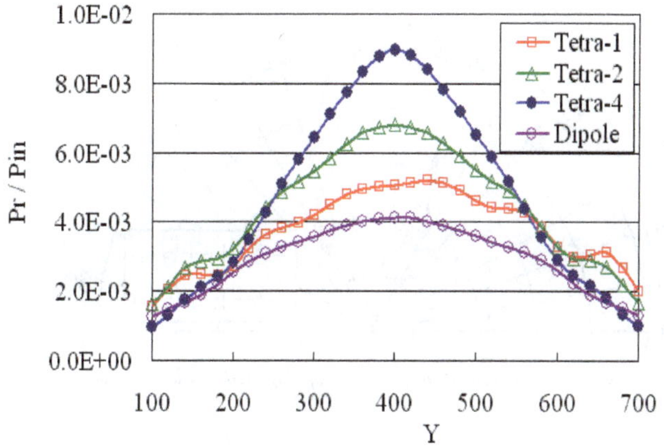

Figure 3. *Electromagnetic field distributions at Z=360. The vertical axis is the radiation intensity P_r/P_{in}, and the horizontal axis is the Y coordinate.*

Figure 3 is the result at frequency f_0 for the electromagnetic field, which is seen to strengthen at Y = 400, the center of the observation line. The frequency, f_0, of Tetra-1, 2, and 4 was 2.30, 2.25, and 2.25 GHz, respectively. A dipole antenna of 60 mm in length was also measured for comparison. The dipole was placed at the same position as the center of Tetra-4 as shown in Figure 2. The frequency f_0 of the dipole antenna was 2.20 GHz.

3.3. Various Characteristics

The simulated results for the relative frequency bandwidth (3dB), input impedance, radiation intensity P_r/P_{in}, antenna gain G, and the 3 dB range of beam width of Tetra-4 are given in Table 1 of Section 4 in comparison to those of a dipole. The real part of the input impedance of the dipole and Tetra-4 were 68 Ω and 492 Ω, respectively. Since the results were calculated using only the upper half, the actual input impedance is twice the value shown in the table. Moreover, the beam width is shown as a ratio to the distance, 329, between the antenna and the observation point. A real part of the input impedance of Tetra-1, Tetra-2, and Tetra-4 are 152 Ω, 318 Ω, and 592 Ω, respectively. The input impedance of Tetra-4 is very high with 592 Ω.

4. Experimental Results

In this section, we verify the operation of the Tetra-4 antenna through experimental measurements. Tetra-4 was installed in a balun with an impedance of 100 Ω. The power transmission was measured between broad band antennas placed 453 mm from our antenna. The gain of the antenna was 2.0 dBi. The measurement results are summarized in Table 1. The relative frequency bandwidth (3 dB) was 24%, and the real part of the input impedance was 360 Ω. The impedance is twice the value shown in the table since the balun converted an unbalanced 50 Ω line into a 100 Ω balanced line. The input impedance of Tetra-4 is considerably high with 360 Ω. Since we connected a 360 Ω load with a 100 Ω line, the return loss was 1.7 dB. The measured input impedance and the gain are values close to the calculation results shown in Table 1. The gain of the experimental result is 1 dB or more smaller. The difference corresponds to the return loss.

The dipole with a reflector was measured as well as the Tetra-4 antenna, and it can be seen from Table 1 that the gain was 5.3 dB. The input and gain of the measured antenna were a close match to those of the simulations given in Table 1.

Table 1. *A comparison of the calculated and experimental results. The upper part shows the calculated values while the lower part shows the experimental values.*

		dipole	Tetra-4
Computations	Relative bandwidth [%]	36	53
	Input impedance [Ω]	34+j7	246-j26
	Pr / Pin [dB]	-23.8	-20.5
	Gain G [dBi]	5.8	9.4
	3dB range (Y range / 329)	1.42	0.96
Experiments	Relative bandwidth [%]	37	24
	Input impedance [Ω]	37+j8	180+j87
	Gain G [dBi]	5.3	8.0
	HPBW [degrees]	87	38

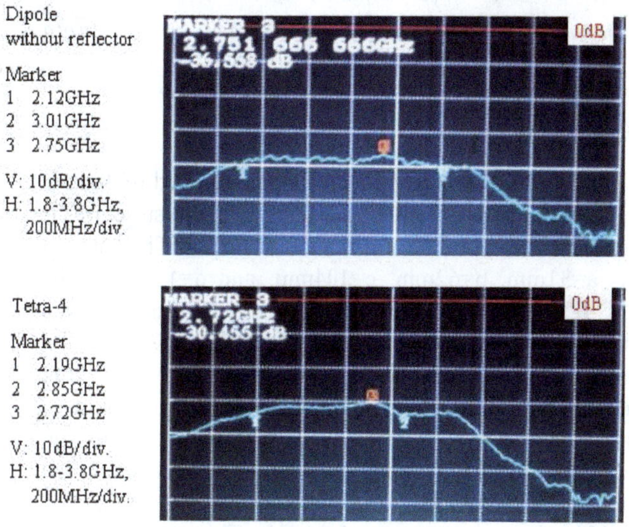

Figure 4. *Frequency characteristics of $|S_{21}|$. (a) Upper: A dipole without a reflector. (b) Lower: Tetra-4 antenna.*

Figure 4 shows a comparison of the power transmission characteristics of Tetra-4 and the half wavelength dipole antenna. Figure 4(b) shows the frequency characteristics of $|S_{21}|$ of Tetra-4. Here, S_{21} is a parameter of the scattering matrix. The frequency band (3 dB) was 2.19 - 2.85 GHz, and the flat level of $|S_{21}|$ was -30.5 dB. The half wavelength dipole antenna (without a reflector) was also measured, and had a gain of 2.0 dBi. The S_{21} characteristic is as shown in Figure 4(a). The frequency bandwidth (3 dB) was 2.12 - 3.01 GHz, and $|S_{21}|$ was -36.6 dB even when we read the largest point. Therefore, there was a 8 dB gain in Tetra-4, and this is confirmed by both the simulation and experimental data given in Table 1. As mentioned above, the experimental and simulated results are virtually identical.

Furthermore, Figure 5 shows the radiation patterns in the YZ plane (H-plane). Because a passive tag requires a strong electromagnetic field, only a strong range of the radiation is shown. The sidelobe is -25dB or less. The half-power beam width (HPBW) was 38 degrees for the Tetra-4 antenna, whereas it was 87 degrees for the dipole antenna. The HPBW of Tetra-4 was 44% that of the dipole. For XZ plane (E-plane), the HPBW of Tetra-4 was 60 degrees.

When the direction in a horizontal plane is detected by arranging tag for a linearly polarized wave, the radiation pattern in a horizontal plane is important. The antenna with narrow beam width is necessary in a horizontal plane. Tetra-4 can be used for this purpose.

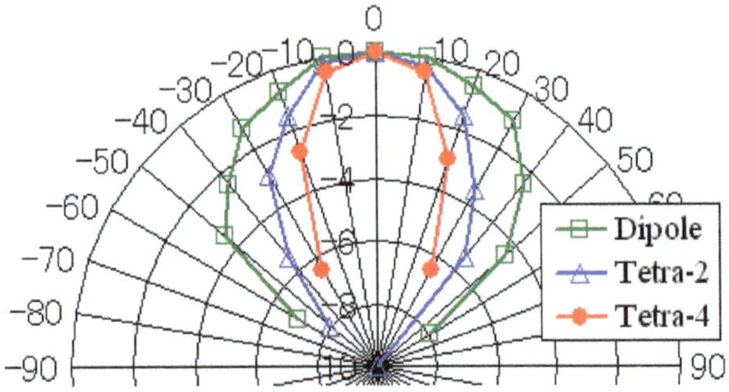

Figure 5. Experimental radiation patterns in the YZ plane. A comparison of a dipole with a reflector, Tetra-2, and Tetra-4.

5. Yagi-Uda Antenna

In this section, the Yagi-Uda antenna comprised of Tetra-1 elements is studied. Figure 6 shows the antenna structure. It consists of two Tetra-1 elements and a reflector. The antenna size is a=31mm, b=62mm, c=104mm, and d=128mm. The size of Tetra-1 far from the reflector is 10% smaller. This antenna is driven in balanced 50 Ω system. Figure 7 shows the experimental radiation patterns. Because a passive tag requires a strong electromagnetic field, only a strong range of the radiation is shown. The sidelobe is -30dB or less. The gain is 11dBi, with HPBW in the H-plane of 40 degrees and in the E-plane of 36 degrees. This antenna has the frequency bandwidth of 10% in frequency 2.54GHz. This antenna can be used to detect a vertical position in addition to the horizontal position.

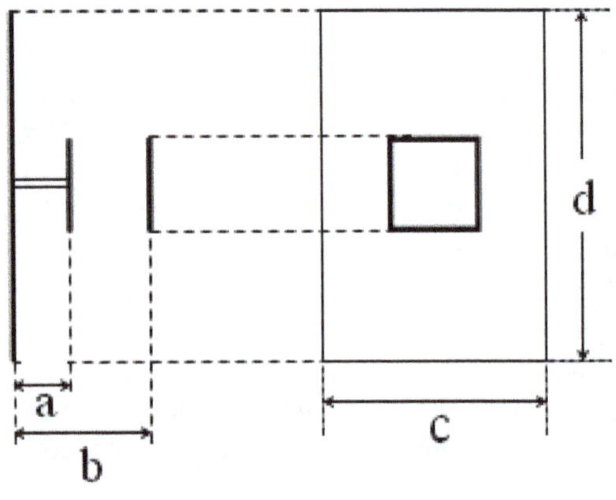

Figure 6. Configuration of Yagi-Uda antenna conformed with two Tetra-1 elements.

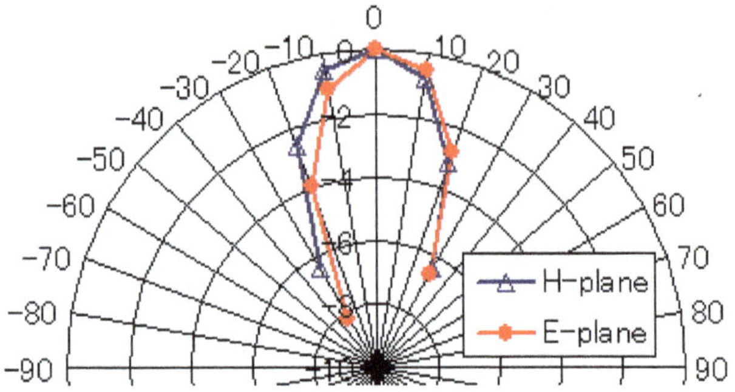

Figure 7. Experimental radiation patterns of Yagi-Uda antenna using two Tetra-1 elements.

6. Conclusions

In the field of location estimation using RFID technology, the antenna is a very important component. In this paper, a simple and small structural antenna was studied. Our design culminates in two antennas comprised of tetragon elements formed with wires.

One is an antenna, Tetra-4, which has a lateral arrangement of four tetragons made from conducting wire. Because the input impedance is considerably high, Tetra-4 is suitable for use in the high impedance system. Real part of the input impedance was 492 Ω in the simulation, and 360 Ω in the experiments. Antenna gain by the simulation is 9dBi. When driving in 100 Ω balanced system, the gain by the experiments was 8dBi. The experimental HPBW in the H-plane was 38 degrees, and was 0.44 times that of a dipole with a reflector.

Another one is Yagi-Uda antenna comprised of two tetragon elements. When driving in 50 Ω balanced system, the gain by experiments was 11dBi. The experimental HPBW was 40 degrees in the H-plane and was 36 degrees in the E-plane.

Acknowledgements

This research was partially supported by Grant-in-Aid for JSPS Fellows (No. 26·4643).

References

[1] Finkenzeller, G. K., RFID Handbook., Radio-frequency identification fundamentals and applications, John Wiley & Son, Chichester, 1999.

[2] Hähnel D., Burgard W., Fox D., Fishkin K., and Philipose M., "Mapping and localization with RFID technology," Proc. 2004 IEEE International Conference on Robotics and Automation, vol. 1, pp. 1015-1020, New Orleans, USA, Apr. 2004.

[3] Manzoor F., Huang Y., and Menzel K., "Passive RFID-based indoor positioning system, an algorithm approach," Proc. IEEE International Conference on RFID-Technology and Applications, Guangzhou, China, Jun. 2010.

[4] DiGiampaolo E., and Martinelli F., "A passive UHF-RFID system for the localization of an indoor autonomous vehicle," IEEE Transactions on Industrial Electronics, vol. 59, no. 10, pp. 3961-3970, Oct. 2012.

[5] Wada T., Hori T., Fujimoto M., Mutsuura K., and Okada, H. "An adaptive multi-range-sensing method for 3D localization of passive RFID tags," IEICE Transactions on Fundamentals of Electronics, Communications and Computer Sciences, vol. E95-A, no. 6, pp. 1074-1083, Jun. 2012.

[6] Fujimoto M., Wada T., Inada A., Nakamori E., Oda Y., Mutsuura K., and Okada H., "Localization of passive RFID tags by using broad-type multi-sensing-range (B-MSR) method," IEICE Transactions on Fundamentals of Electronics, Communications and Computer Sciences, vol. E95-A, No. 7, pp. 1164-1174, Jul. 2012.

[7] Fujimoto M., Wada T., Inada A., Mutsuura K., and Okada H., "Swift communication range recognition method for quick and accurate position estimation of passive RFID tags," IEICE Transactions on Fundamentals of Electronics, Communications and Computer Sciences, vol. E95-A, no. 9, pp. 1596-1605, Sep. 2012.

[8] Kulyukin V., Gharpure C., Nicholson J., and Pavithran S., "RFID in robot-assisted indoor navigation for the visually impaired," Proc. 2004 IEEE/RSJ International Conference on Intelligent Robots and System, vol. 2, pp. 1979-1984, Sendai, Japan, Oct. 2004.

[9] Gueaieb W., and Miah S., "An intelligent mobile robot navigation technique using RFID technology," IEEE Trans. Instrum. Meas., vol. 57, no. 9, pp. 1908-1917, Sep. 2008

[10] Zou J., and Wang L., "Research of navigation and positioning at local area based on RFID," Proc. 2010 International Conference on Computer Application and System Modeling, vol. 11, pp. 5-8, Taiyuan, China, Oct. 2010.

[11] Fujimoto M., Nakamori E., Tsukuda D., Wada T., Mutsuura K., Okada H., and Iida Y., "A new indoor robot navigation system using RFID technology," Proc. 4th International Conference on Indoor Positioning and Indoor Navigation (IPIN 2013), pp. 637-638, Montbéliard, France, Oct. 2013.

[12] Park S., and Lee H., "Self-recognition of vehicle position using UHF passive RFID tags," IEEE Transactions on Industrial Electronics, vol. 60, No. 1, pp. 226-234, Jan. 2013.

[13] Ahmad M. Y., and Mohan A. S., "Novel bridge-loop reader for positioning with HF RFID under sparse tag grid," IEEE Transactions on Industrial Electronics, vol. 61, No. 1, pp. 555-566, Jan. 2014.

[14] Nagao R., Fujimoto M., Tsukuda D., Nakanishi T., Wada T., Mutsuura K., and Okada H., "Routing and moving control method using passive RFID for robot navigation system," Proc. 8th Annual IEEE International Conference on RFID (IEEE RFID 2014), pp.68-69, Orlando, USA, Apr. 2014.

[15] Wang J., Sang L., Wang Z., Xu R., and Yan B., "A broadband quasi-yagi array of rectangular loops on LTCC," Int. J. RF Microw Comput—Aided Eng, vol.24, no.2, pp.196-203, Mar. 2014.

[16] Kim S. -G., Zepeda P., and Chang K., "Piezoelectric transducer controlled multiple beam phased array using microstrip rotman lens," IEEE Microwave Wireless Compon. Lett., vol.15, no.4, pp.247-249, Apr. 2005.

[17] Li M., and Luk K.-M., "A low-profile unidirectional printed antenna for millimeter-wave applications," IEEE Trans Antennas Propag., no.3, pp.1232-1237, Mar. 2014.

[18] Tsutsumi Y., Ito T., Hashimoto K., Obayashi S., Shoki H., and Kasami H., "Bonding wire loop antenna in standard ball grid array package for 60-GHz short-range wireless communication," IEEE Trans Antennas Propag, vol.61, no.4, Pt.1, pp.1557-1563, Apr. 2013.

[19] Haider N., Caratelli D., and Yarovoya G., "Circuital characteristics and radiation properties of an UWB electric-magnetic planar antenna for Ku-band applications," Radio Sci., vol.48, no.1, pp.13-2, Jan. 2013.

[20] Yuan X., Wong K. T., Xu Z., and Agrawal K., "Various compositions to form a triad of collocated dipoles/loops, for direction finding and polarization estimation," IEEE Sens J., vol.12, no.5—6, pp.1763-1771, May 2012.

[21] Yasin M. N. M., and Khamas S. K., "Measurements and analysis of a probe-fed circularly polarized loop antenna printed on a layered dielectric sphere," IEEE Trans Antennas Propag., vol.60, no.4, pp.2096-2100, Apr. 2012.

[22] Pal A., Mehta A., Mirshekar - Syahkal D., Deo P., and Nakano H., "Dual-band low-profile capacitively coupled beam-steerable square-loop antenna," IEEE Trans Antennas Propag., vol.62, no.3, pp.1204-1211, Mar. 2014.

[23] Gupta S., and Mumcu G., "Dual-band miniature coupled double loop GPS antenna loaded with lumped capacitors and inductive pins," IEEE Trans Antennas Propag., vol.61, no.6, pp.2904-2910, Jun. 2013.

Modeling and Simulation of Phase Noise Effect on 256-QAM

Bourdillon Odianonsen Omijeh[1], Ejioeto Evans Ibara[2]

[1]Department of Electronic & Computer Engineering, University of Port Harcourt, Port Harcourt, Nigeria
[2]Centre for Information and Telecommunications Engineering, University of Port Harcourt, Port Harcourt, Nigeria

Email address:
bourdillon.omijeh@uniport.edu.ng (B. O. Omijeh), omijehb@yahoo.com (B. O. Omijeh), ibaraevans@yahoo.com (E. E. Ibara)

Abstract: This paper models the effect of phase-noise on a 256-QAM Modulator using Computer Aided Design tool called Matlab/Simulink. By remodeling and varying the noise parameter in the AWGN channel of an expert–system based simulink model, and studying the impact of these variations on the BER of the system, values were recorded for every instance of simulation that was run before and after the addition of the noise. It was found that Phase Noise has an enormous effect on QAM modulators; an effect that grows strong with an increase in the density (version) of the QAM modulators. This implies that higher order QAM are prone to phase noise and hence error, but transmits more data.

Keywords: QAM, BER, Phase Noise, PNLD, Noise, AWGN

1. Introduction

Phase noise is the fundamental limitation in the performance of a system [1]. Technological advancement often had been driven by the demand for capacity, the limitations and challenges with existing systems, and the struggle to reduce noise effect. A good telecommunication system is measured in terms of its capacity, coverage capability, and the error rate determined at its output. There is more often a trade-off in trying to create a balance between these telecommunication necessities. For instance, an improvement in coverage may result in a noisy signal at the output. Also, to improve the quality of signal at the output of a transmission system, the use of telecommunication system with poorer bandwidth capabilities may just be ideal. The essence of digitizing telecommunication systems, in fact, is to reduce noise and thus the error rate in the received signal [2]. There are different kinds of noise-intrinsic and extrinsic. There are also different kinds of modulation schemes, each with its limitations in terms of noise effects.

2. Related Works

QAM is an acronym report demonstrates the effects of phase noise in a 256-QAM system using modeling and for Quadrature Amplitude Modulation. It is one of the many modulation schemes in telecommunication systems. A brief review of works done on the subject of this research is as follows.

2.1. Theoretical Background

A Modulation schemes convey message signals by varying a parameter of a carrier signal in accordance to the modulating or message signal. QAM as a modulation scheme does same, but conveys two messages each corresponding to a carrier signal whose amplitude is varied in response to the message being conveyed. According to [3] "QAM mixes two sine waves that are 90 degrees out of phase with one another". This means that, for a QAM scheme there are two carrier signals whose amplitudes are altered in course of transmitting the modulating or message signals. [4] describes QAM as a hybrid modulation scheme in which both "amplitude-shift keying (ASK) and phase-shift keying (PSK)"are varied.

QAMs are predecessors to the phase oriented modulation schemes ranging from BPSK (Binary Phase-Shift Keying), QPSK to 16-PSK. There are different orders of QAM each defined in terms of its prefix value. An increasing order of QAM indicates an increase in the number of information bits that the QAM scheme can convey per symbol. Below in figures 1 & 2 [5]" are constellation diagram for 64 QAM and 256 QAM.

Fig. 1. Constellation diagram for 64-QAM [5].

Fig. 2. Constellation diagram for 256-QAM [5].

A run through different QAMs from a lower to higher order indicates that as the points on the constellation gets denser, the distance between them decreases, and the points thus, become more susceptible to noise and other corruption [6]. The result is a higher bit error rate (BER). Using the mathematical expressions in eq 1, Freeman (2006) showed the relationships between the information bits per symbol, the bit error rate (BER), and the Symbol Error Rate (SER). Log M is an excerpt from this expression that represents the M-ary number of the QAMs or its predecessors. Equation (2) represents an expression for computing the precise M-ary or prefix number as described in the work of [7]. It is possible to therefore to determine the information bits per symbol for different QAM schemes as in 2 bits per symbol for 4QAM, 4 bits per symbol for 16QAM, 6 bits per symbol for 64QAM, and 8 bits for 256QAM respectively.

$$BER = (1/\log M) \times SER \tag{1}$$

Where M = M-ary value prefixing QAM and indicating the version

$$K = \log M, \tag{2}$$

Digital QAM "is constructed using two M-ary baseband signals (called i(t) and q(t)) modulating the two quadrature carriers. For example, in 16-QAM, both i(t) and q(t) are 4-ary digital baseband signals, which means each one of them can assume one of four possibilities" [11]. This results in 4 x 4 = 16 possible carrier symbols shown in the constellation diagrams in Fig. 3 and 4.

2.2. Effects of Phase Noise

Noise in transmission is defined as an unintended or unwanted signal introduced into a transmission system. Phase noise is expressed as:

- Loss of sensitivity in radar and communication
- Lack of definition in imagine
- Higher bit error rate (BER) in transmission

The effect of phase noise on an ideal 16-QAM constellation, for instance is as depicted in the figures below. Figure 3 & 4 indicates and shows the effect by small amount of phase jitter. Such shifts can be disastrous for more complex constellations than 16-QAM.

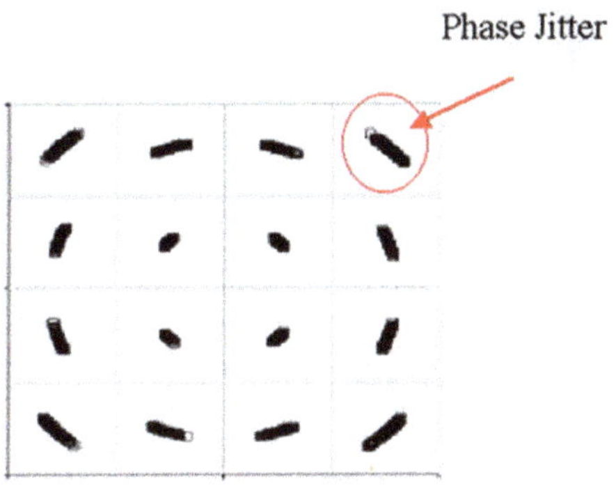

Fig. 3. Phase jilter [5].

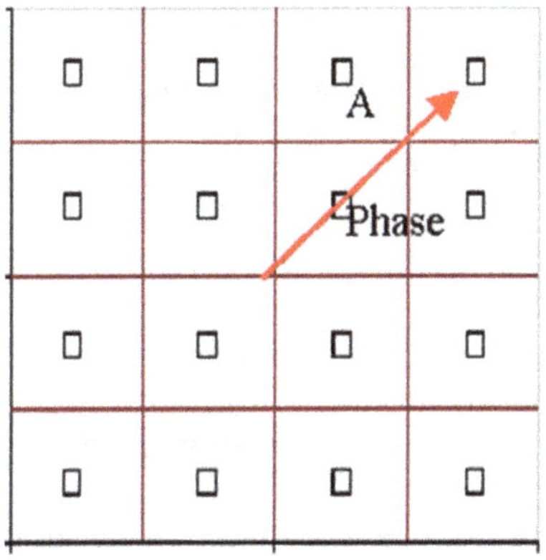

Fig. 4. Phase jilter [5].

The amplitude of a phase noise "can be expressed in RMS volts (Ao/√2). Phase noise is typically expressed in units of dBc/Hz at various offsets from the carrier frequency. This is usually referred to as Phase Noise Level Density (PNLD)" [8] and can be expressed mathematically as in eq 3.

$$PNLD = -10 \log I/Io \quad (3)$$

Where, I = noise intensity level in dB, Io = 10^-12 is a reference noise intensity level

"QAM schemes are enormously efficient in terms of spectrum but its demodulation is difficult particularly in the presence of noise" [9]. Some areas of application of QAM are cable TV, satellites, cellular telephone systems, Wi-Fi, and wireless local-area networks (LANs).

The aim of this work is to model the effect of phase noise on 256-QAM using computer aided design tool Matlab Simulink [10].

3. Design Methodology

The study of the phase noise effect is done using an existing model in Simulink. This model was designed to show the effect of phase noise in 256-QAM. First, as a control, it was ensured that the AWGN oriented noise does not have an effect on the modulated signal. This was done by attaching two scatter plot before and after the AWGN channel, and then increasing the value of signal to noise ratio (Es/No in dB) in this channel to 100dB. The Simulink model is as shown in figure 5 below.

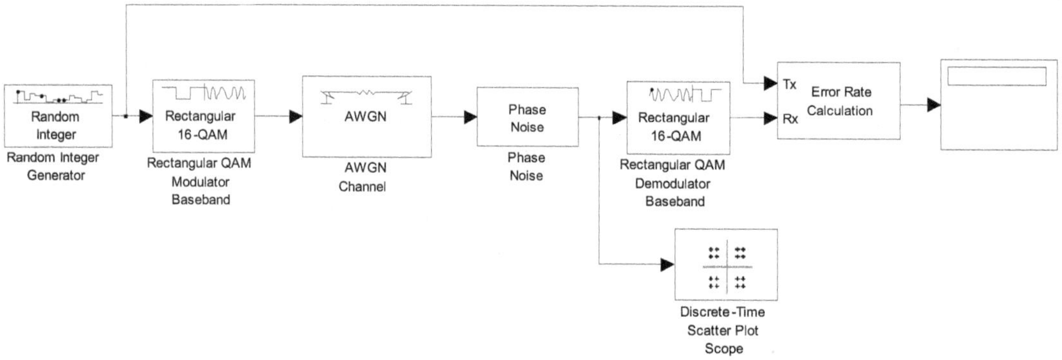

Fig. 5. *Phase Noise effect in 256-QAM (model in Simulink).*

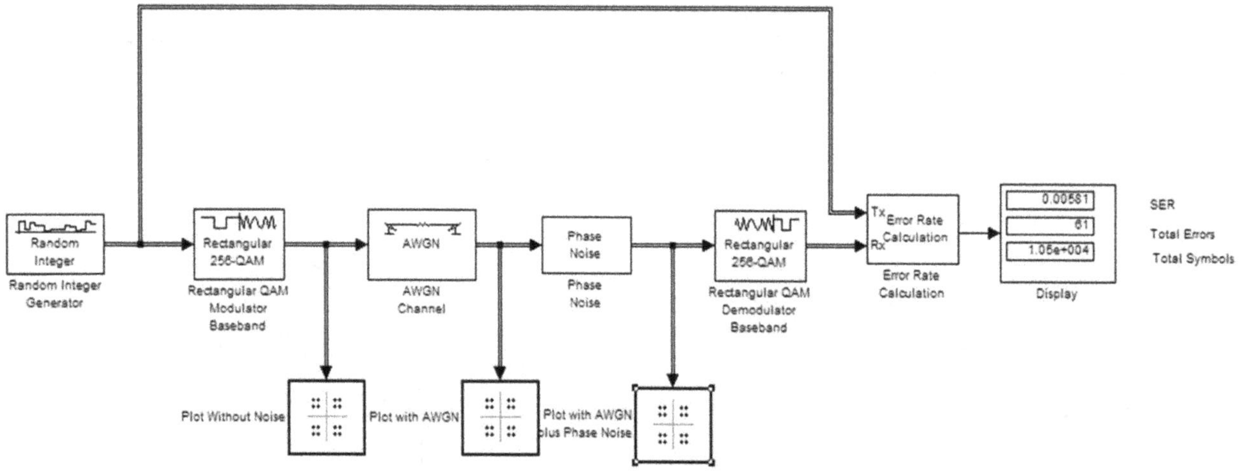

Fig. 6. *Phase Noise effect in 256-QAM showing scatter plot before and after AWGN.*

The following block sets were used;

- "A baseband 256-QAM modulator
- An additive white Gaussian noise (AWGN) channel
- A source of phase noise
- A baseband 256-QAM demodulator
- An error statistic calculator
- A display icon that shows the error statistics while the simulation runs
- A scatter plot that shows the received signal, including the phase noise"[

4. Simulation Results and Discussion

The following parameters were used in each block for the simulation:

4.1. "Random Integer Generator"

"Blocks:

a. M-ary number = 256

b. Initial seed = 12345

c. Sample time = 0.001s
d. Output status = Frame-based
e. Number of samples per frame = 500
f. Output data type used = unit 8"[12].

4.2. "Rectangular QAM Block"

a. "M-ary number = 256
b. Input type = Integer
c. Constellation ordering = Binary
d. Normalization method = Average power
e. Average power = 1 watts
f. Phase offset = 0 rad
g. Output data type = Single".

4.3. "Rectangular QAM Demodulation Baseband Block"

a. "M-ary number = 256
b. Output type = Integer
c. Constellation ordering = Binary
d. Normalization method = Average power
e. Average power = 1 watts
f. Phase offset = 0 rad
g. Output data type = Unit 8".

4.4. "AWGN Channel"

a. "Initial seed = 54321
b. Mode = signal-to-noise ratio (Es/No)
Es/No = 100 dB
c. Input signal power = 1 watts
d. Symbol period = 0.001s"

4.5. "Error Rate Calculation Block"

a. "Receive delay = 0
b. Computation delay = 0
c. Computation mode = entire frame
d. Output data to = port"

4.6. "Discrete-Time Scatter Plot Scope"

In min X-axis = 1.5
Max X-axis = 1.5
Min Y-axis valve = 1.5
Max Y-axis valve = 1.5
"In-phase x-axis label = In-phase Amplitude
Quadrature Y-axis label = Quadrature Amplitude
Scope position = [32 306 240 240]
Title = scatter plot".

"The same parameters were set for the three scatter plot scopes we have in the model. They were all set to "open scope at start of simulation". In this case, the scopes open on their own immediately simulation starts".

4.7. "Configuration Parameters"

"Simulation time: Start time: 0.0 stop time: 10
Solver = Ode 45
Relative tolerance = ie-3 = 0.001
Absolute tolerance = auto"

4.8. "Solver Options"

"Type = variable step
Max step size = auto
Min step size = auto
Initial step size = auto
Zero crossing control = use local settings".

4.9. Results of the Error Rate Calculation Block

The results of the Error Rate Calculation Block "helped to give numerical analysis of the effect of the phase noise in the QAM signal for each substituted phase noise density value". This result is represented in table 4.1.

Table 1. *Simulation results for a frequency offset of 200Hz.*

Time	Total Symbol	Total Errors	SER
1	1500	1414	0.9426667
2	2500	2339	0.9356
3	3500	3261	0.9317143
4	4500	4185	0.93
5	5500	5106	0.9283636
6	6500	6029	0.9275385
7	7500	6968	0.9290667
8	8500	7892	0.9284706
9	9500	8810	0.9273684
10	10500	9728	0.9264762

4.10. Research Findings

The Error Rate Calculation Block gave varying results as the values of the PNLD were varied from -20dBc/Hz to -120dBc/Hz.

Table 2. *Result of varied PNLD values with error rate.*

PNLD (dBc/Hz)	Number of Symbols Compared	Number of Errors that Occurred	Error Rate
-20	10500	10100	0.9616
-40	10500	8828	0.8408
-60	10500	1456	0.1387
-80	10500	0	0
-100	10500	0	0
-120	10500	0	0

The first value obtained for PNLD of -20dBc/Hz gave a high rate of error occurrence (of 0.9619). The number of errors that occurred at that noise level was also high (equal to 10100).

As the noise level was reduced to -40dBc/Hz there was a significant reduction (value obtained equals 0.8408) in the rate at which error was occurring in the calculation. The number of error calculation at that level was significantly low (value equals 8828) as well.

When the noise level was further reduced to -60dBc/Hz, the number of errors that occurred at that point was seen to be: 1456 (which is significantly low when compared to the previous two results).

The same thing applies to the calculated rate at which the error was occurring. The value of the error rate obtained for -60dBc/Hz noise level was: 0.1387.

When the noise level was reduce to -80dBc/Hz and down to -120dBc/Hz, zero error was calculated by the Error Rate Calculation Block.

5. Conclusion

From this research it is evident that Quadrature amplitude modulation is seriously affected by phase noise, especially in when the densities are high in the phase noise level. For critical data transmission, a lower order QAM can be used since higher order QAM are prone to higher error rates but transmits more data. Higher data rates are good for transmission but may eventually increase transmission time because of the increase in bit error rate with noise as this may mean that more packets with error will have to be resent.

References

[1] Dickstein, L. (2012) White Paper. *Introduction to Phase Noise in Signal Generators.* [Online] Available from: http://www.gigatronics.com/uploads/document/AN-GT140A-Introduction-to-Phase-Noise-in-Signal-Generators.pdf. [Accessed: 10th August 2015].

[2] Biebuma. J.J. Omijeh, B.O, Nathaniel. M.M (2014): Signal Coverage Estimation Model for Microcellular Network Propagation *IOSR Journal of Electronics and Communication Engineering (IOSR-JECE), Vol.9, Issue 6 PP 45-53, 2014.*

[3] Freeman, R.L. (2008). Technology and Engineering. *Radio System Design for Telecommunication.* [Online] Available from: http://books.google.com.ng/books?isbn=0470050438.

[4] Kumar, S. (2015) Technology and Engineering. *Wireless Communications Fundamental & Advanced Concepts.* [Online] Available from: https://books.google.com.ng/books?isbn=8793102801. [Accessed: 10th August 2015].

[5] Kang, H. & Kim, J. (2005). *Constellation Mapping Apparatus and Method US 20050220205 A1.* http://www.google.com/patents/US20050220205. [Accessed: 7th August 2015].

[6] Parihar, G. S. & Singh, P. (2014) *A project Report on Synchronization Techniques for OFDM, 2008-09 Session.* http://www.ni.com/white-paper/3896/en/. [Accessed: 7th August 2015].

[7] Ahmad, A. (2003) *Data Communication Principles for Fixed and Wireless Networks.* [Online] Available from: https://books.google.com.ng/books?isbn=1402073283. [Accessed: 10th August 2015].

[8] Efurumibe, E. L. & Asiegbu, A. D. (2012) Computer-Based Study of the Effect of Phase Noise in256-Quadrature Amplitude Modulation Using Error-Rate Calculation Block - A Comparative Study. *Journal of Information Engineering and Applications.* [Online] 2(11). P.35. Available from: www.iiste.org. [Accessed: 8th August 2015].

[9] Frenzel, L. (2012) Electronic Design. *Understanding Modern Digital Modulation Techniques.* [Online]. Available from: http://electronicdesign.com/communications/understanding-modern-digital-modulation-techiques. [Accessed: 16th August 2015].

[10] Wetzheng Wang (1996): Communication toolbox user guide, www.mathworks.com.

[11] www.fetweb.ju.edu.jo

[12] www.mathworks.com

10

Spectrum Sharing in Cognitive Radio Work Using Goodput Mathematical Model for Perfect Sensing, Zero Interference and Imperfect Sensing, Non Zero Interference

Ojo Festus Kehinde[1], Fagbola Felix Adetunji[2]

[1]Department of Electronic and Electrical Engineering, Ladoke Akintola University of Technology, Ogbomoso, Nigeria
[2]Department of Works & Physical Planning, University of Lagos, Yaba, Lagos, Nigeria

Email address:
fkojo@lautech.edu.ng (F. K. Ojo), ffagbola@unilag.edu.ng (F. A. Fagbola)

Abstract: In recent years, there has been increase in the demand for spectrum allocation as wireless communication witness rapid growth on a daily basis. Literatures have established that this spectrum is scarce and more than 70% of the available spectrum is not utilized optimally. This paper proposes a model called Goodput model by which the under-utilized spectrum can be shared effectively between primary users and the secondary users without causing harmful interference between the users and also find solution to the problem of increasing demand for spectrum allocation on an already scarce spectrum. Goodput is a mathematical modeling in which the total amount of primary and secondary data that is successfully delivered per unit time can be used as performance index. Compared to other models, the Goodput model with zero interference, perfect sensing and imperfect Sensing, non-zero interference is used to determine the secondary users that will be able to use unoccupied portion of radio frequency channel of primary users with different values of data probability of arrival. Result shows that, the point of intersection between np and $N(1-p)$ is the optimum point of interference, where $n, N, and\ p$ are number of secondary users, number of primary users, and Probability of data arrival rate respectively. Below the optimum point of interference (left side of the point), all the secondary users will transmit opportunistically without interference. However, above the optimum point of interference (right side of the point), there will be interference between any secondary users that attempt to transmit.

Keywords: Cognitive Radio, Spectrum Sharing, Primary User, Secondary User, Goodput

1. Introduction

Radio frequency spectrum can be defined as the entire spectrum of electromagnetic frequencies used for communication and broadcasting of radio, radar and television. Radio frequency (RF) is the rate of oscillation around 3kHz to 3,000GHz which corresponds to the frequency of radio waves and the alternating current which carry radio signals (Wikipedia, 2014). Cognitive radio is thereby a better technique to fulfill the utilization of radio frequency spectrum as most communication systems go wireless. Both licensed and unlicensed users can use the frequency spectrum using cognitive radio technique. License users are the primary users while the unlicensed users are the secondary users. Primary users allocate the spectrum to the secondary users on demand without degrading its own performance using spectrum sharing techniques, (Varaka, et al. 2013). Distribution of the spectrum among the secondary users according to the usage cost is called Spectrum sharing, (Varaka, et al. 2013). It is the simultaneous usage of a specific radio frequency band in a specific geographical area by a number of independent users. Also, spectrum sharing in cognitive radio network allows cognitive users to share the spectrum bands of the licensed–band users. However, the cognitive radio users have to restrict their transmitting power so that the interference caused to the licensed-band users is kept below a certain threshold, (Haykin, 2005). International body such as International Telecommunication Union (ITU) harmonize usage of radio frequency spectrum allocation, while national bodies such as Nigerian Communication Commission (NCC) and Nigerian Broadcasting Commission (NBC) assign the bands and license to service providers such as GSM service providers, Radio &

Television stations , internet service providers (ISP) etc. Interestingly, a portion of radio frequency spectrum is dedicated to industrial scientific & medical (ISM) bands which are unlicensed, protective and non-interference. Therefore, the issue of Spectrum sharing cannot be over emphasized in today wireless communication; in fact it brings improvement in spectrum congestion, provision of additional bandwidth capacity, improvement in quality of service, and provision of emergency network when the existing one is not operational due to destruction of infrastructures; hence a need for proper consideration of sharing techniques.

2. Related Works on Spectrum Sharing

According to Wigglinski, (2009), there are three physical dimensions to share the spectrum; these are time (S), frequency (Hz) and space (m^2). Also, there are several ways of classifying spectrum sharing. Licensed spectrum sharing which is sub divided into Horizontal and Vertical. Unlicensed spectrum sharing which is sub divided into horizontal and single system. The classifications are shown in Figure 1.

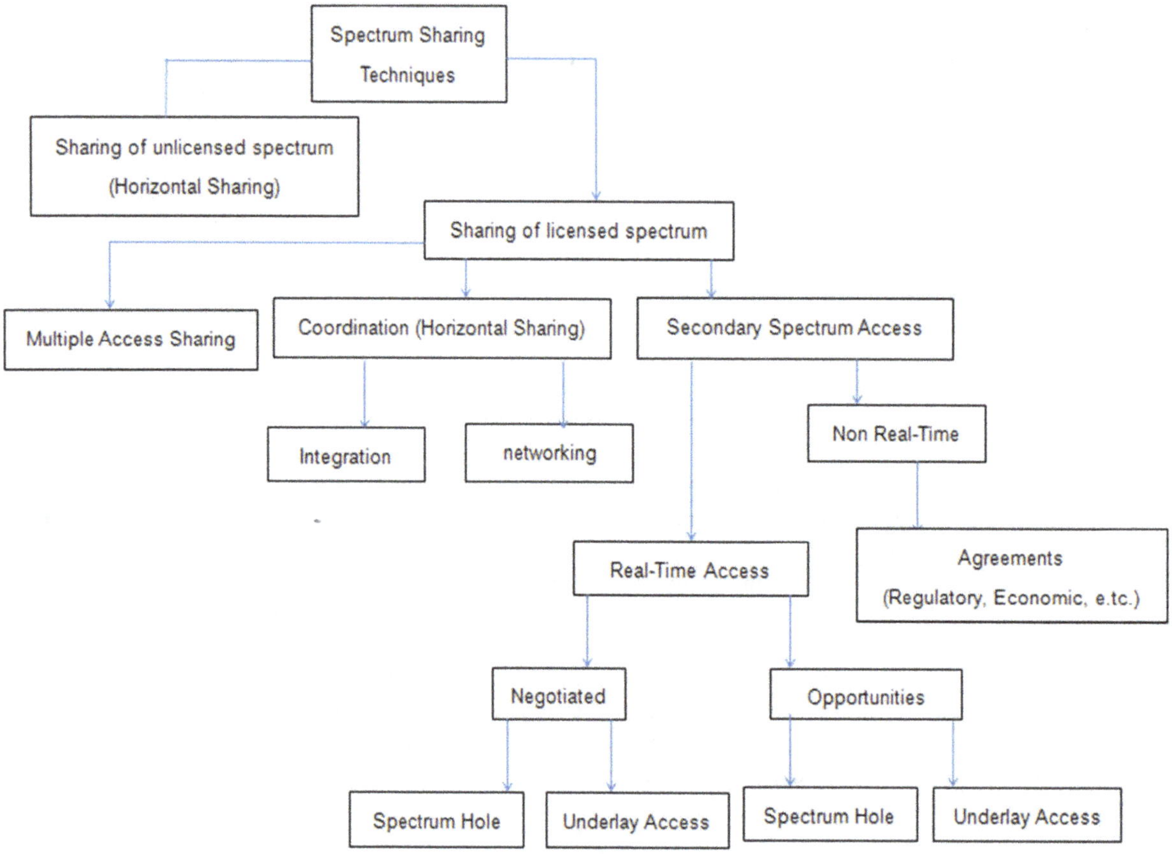

Source: (Wikipedia, 2014)

Figure 1. *Classification of Unlicensed Spectrum Sharing.*

A game theoretical model to analyze dynamic spectrum sharing was presented by Jiand*et al*, (2007). Game theoretical dynamic spectrum sharing has been extensively studied for more flexible, efficient and fair spectrum usage through analyzing the intelligent behaviors of network user equipped with cognitive radio devices. The authors described the game models for dynamic spectrum sharing for various networking scenarios such as belief assisted pricing and auction based spectrum sharing game. Distributed Spectrum Allocation via local bargaining proposed a local bargaining approach to achieve distributed conflict free spectrum assignment adapted to network topology charges, (Cao *et al*, 2005).

Also a study of repeated Spectrum Sharing Game Model was carried out by (Etkin, 2005) where the spectrum sharing problem among multiple secondary users for interference constrained wireless systems in a non – cooperative game frame work. Their study is focused on investigating self enforcing spectrum sharing game rules and the corresponding game efficiently measured in total throughput obtained from available spectrum resources. (IEEE DySPAN, 2005).

3. The System Model

The system model adopted in this work is presented in Figure 2. *Pu* are the primary users, *Su* are the secondary users and the circles around the secondary nodes are sensing regions where primary users are perfectly detected while the colored circles are different sub channels. The dotted lines are corresponding primary/secondary users that do not have data to transmit (OFF).

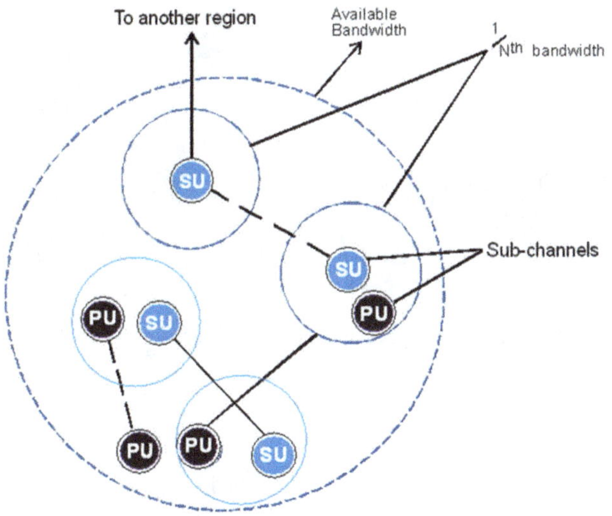

Figure 2. *The system model.*

For perfect sensing, certain resources are considered which is equally shared among N primary users (primary transmitter- receiver pairs). Each primary radio is licensed to transmit on a channel that spans $(^1/_N)th$ of the available bandwidth. Assuming the data traffic arrives at each primary user with a probability of arrival P. The channel is also to be used opportunistically by n secondary users in order to allow higher spectral efficiencies. Assuming that delay intolerant data arrives at the secondary users with the same arrival probability P, and the transmission at the primary and secondary users take place at the rate R, the primary users can reliably transmit their data at rate R. Then the secondary users may be considered lost if there is no free channel for secondary transmission; or more secondary users select the same unoccupied licensed channel.

4. Modeling of Perfect Sensing and Zero Interference

The sum Goodput of the primary users is (assuming no collision between primary users)

$$G_p^{sum} = \mathrm{R \times N \times P} \quad (1)$$

While the sum Goodput of the secondary users depends on the number of unoccupied sub channels and could be written as:

$$G_S^{sum} = \sum_{i=1}^{N} \binom{N}{i} p^{N-i}(1-P)^{\mathrm{i}}\, G_S^{sum}(\mathrm{i}) \quad (2)$$

Where $G_S^{sum}(\mathrm{i})$ is the secondary goodput given that i of the N primary users do not have data to transmit (OFF).

If we condition on the number of secondary users who have data to transmit (ON) we could express $G_S^{sum}(\mathrm{i})$ as

$$G_S^{sum}(\mathrm{i}) = \sum_{j=1}^{n} \binom{n}{j} p^j (1-P)^{\mathrm{n-j}}\, G_S^{sum}(\mathrm{i,j}) \quad (3)$$

Where G_S^{sum} is the secondary goodput, given that i subchannels are unoccupied and j secondary users are ON

$$G_S^{sum}(\mathrm{i, j}) = \mathrm{R}\,(1 - \tfrac{1}{i})^{j-i} \quad (4)$$

If we substitute equation (4) into equation (3), we have

$$G_S^{sum}(\mathrm{i}) = \sum_{j=1}^{n} \binom{n}{j} p^j (1-P)^{\mathrm{n-j}}\, \mathrm{R}\,(1 - \tfrac{1}{i})^{j-1} \quad (5)$$

The sum goodput is the addition of equations (1) and (5)

$$G_S^{sum} = \mathrm{RNP}\,[1 + \tfrac{n}{N}\sum_{i=1}^{N} \binom{N}{i} p^{N-i}(1-P)^{\mathrm{i}}(1 - \tfrac{\mathrm{P}}{\mathrm{i}})^{\mathrm{n-i}}] \quad (6)$$

At the optimum, $\frac{dG}{dn} = 0$, then,

$$nP = N(1-P) \quad (7)$$

Therefore when $nP > N(1-P)$ there will be no interference. This is a case of perfect sensing, zero interference scenario.

5. Modeling of Imperfect Sensing and Non Zero Interference

In order to model and determine the effect of sensing and interference tolerance on the goodput maximum number of secondary users, it is required to determine the steady state probabilities of the detected state (i.e. including sensing errors).

The steady state probability observed without the true knowledge of primary user activity and therefore, sensible to sensing errors are:

$$P^S_{(ij)} = \sum_{n=0}^{N_C} b(i,n)\; P_{(nj)} \quad (8)$$

Average number of primary users and secondary users (i.e. average served traffic) is computed as:

$$N_P = \sum_{S_{ij}} i \cdot P_{ij} \quad (9)$$

$$N_S = \sum_{S_{ij}} j \cdot P_{ij} \quad (10)$$

Average number of sensed primary users

$$N_p^i = \sum_{S_{ij}} i \cdot P_{ij}^S = \sum_{S_{ij}} i \left[\sum_{n=0}^{N_i} bin \cdot P_{(nj)} \right] \quad (11)$$

The blocking probability for secondary users can be computed as:

$$P_B^{SU} = \sum_{i=0}^{N_C} \sum_{j=N_C-i}^{N} P^S_{(ij)} = \sum_{i=0}^{N_C} \sum_{j=N_C-i}^{N_C} \left[\sum_{n=0}^{N} (b_{in})\, P_{nj} \right] \quad (12)$$

Interruption probability can be computed as:

$$P_{INT.} = 1 - \frac{T_S^{served}}{T_S\left(1 - P_B^{SU}\right)} = 1 - \frac{N_S}{\frac{\lambda_S}{\mu_S}\left(1 - P_B^S\right)} \quad (13)$$

$$\text{Where } P_S^{served} = T_S\left(1 - P_B^{SU}\right)\left(1 - P_{INT.}\right) \quad (14)$$

P_S^{served} is the average number of secondary users.

Also, given a maximum possible number of collisions to be C, the collision probability ratio (interference probability) in the state S_{ij} as

$$P_{C(ij)} = \frac{N_C^{upper\ band}(ij)}{C} \quad (15)$$

Error sensing interruption probability can be computed as:

$$P_{INT.} = 1 - \frac{T_S^{served}}{T_s\left(1 - P_B^{sa}\right)} = 1 - \frac{N_S}{\frac{\lambda_s}{\mu_s}\left(1 - P_B^S\right)} \quad (16)$$

Where

$$T_S^{served} = T_S\left(1 - P_B^{SU}\right)\left(1 - P_{INT.}\right) \quad (17)$$

T_S^{served} is the average number of secondary users.

Also, given a maximum possible number of collision to be C, the collision probability ratio (interference probability) in state S_{ij} is given as:

$$P_C(ij) = \frac{N_{C(ij)}^{upper\ based}}{C} \quad (18)$$

Average interference probability,

$$P_C = \sum P_{c(ij)} P_{(ij)} \quad (19)$$

Average interference probability and sensible to sensing error =

$$P_C \times P_{ij} = P_{imp} \quad (20)$$

Average interference probability for sensing error =

$$P_C \times P_{ij} = P_{imp} \quad (21)$$

The goodput under the imperfect sensing, non-zero inference tolerance is given as:

$$G_{IMP}^{SUM} = P_N C'N\left[1 + \frac{n}{N}\sum_{i=1}^{N}\binom{N}{i} P_{IM}^{N-i}\left(1 - P_{IM}\right)^i\left(\frac{P_{iM}}{i}\right)^{n-1}\right] \quad (22)$$

In order to optimize this imperfect sensing, non-zero interference tolerance, we use Newton's or Gradient method of optimization.

$$\frac{dG_{IM}^{SUM}}{dn} = P_{IM}C'\left[\frac{1}{N}\sum_{i=1}^{N}\binom{N}{i}P_{IM}^{N-i}\left(1 - P_{IM}\right)^i\left(1 - \frac{P_{IM}}{i}\right)^{n-1} + \frac{n}{N}\mathrm{In}\left(1 - \frac{P_{IM}}{i}\right)\sum_{i=1}^{N}\binom{N}{i}P_{IM}^{N-i}\left(1 - P_{IM}\right)^i\left(1 - \frac{P_{IM}}{i}\right)^{n-1}\right]$$

$$\frac{dG_{IM}^{SUM}}{dc} = \frac{P_{iM}C'}{N}\sum_{i=1}^{N}\binom{N}{i}P_{IM}^{N-i}\left(1 - P_{iM}\right)^i\left(1 - \frac{P_{iM}}{i}\right)^{n-1}\left[1 + n\,\mathrm{In}\left(1 - \frac{P_{iM}}{i}\right)\right] = 0 \quad (23\ \&\ 24)$$

Since

$$\frac{P_{iM}C'}{N}\sum_{i=1}^{N}\binom{N}{i}P_{iM}^{N-i}\left(1 - P_{iM}\right)^i\left(1 - \frac{P_{iM}}{i}\right)^{n-1} \neq 0 \quad (25)$$

$$\Rightarrow 1 + n\sum_{i=1}^{N}\mathrm{In}\left(1 - \frac{P_{iM}}{i}\right) = 0 \quad (26)$$

$$n = \frac{-1}{\sum_{i=1}^{N}\left[\mathrm{In}\left(1 - \frac{P_{iM}}{i}\right)\right]} \quad (27)$$

Also, at the optimum, $\frac{dG}{dn} = 0$, then,

$$nP = N(1\text{-}P) \quad (28)$$

Therefore when nP < N(1-P), there will be interference (29)

The optimum number of secondary user on N primary users in order to minimize interference, interruption or collision is given as:

$$n_{optim} = \frac{1}{\sum_{i=1}^{N}\mathrm{In}\left(\frac{i}{i - P_{iM}}\right)} \quad (30)$$

Therefore, the optimum goodput is

$$G_{opt}^{SUM} = PC'N\left[1 + \frac{n_{opt}}{N}\sum_{i=1}^{N}\binom{N}{i}P^{N-i}\left(1 - P\right)^i\left(1 - \frac{P}{i}\right)^{n_{opt}-1}\right] \quad (31)$$

6. Simulation Parameters

The simulation parameters used for the modeling in this paper are stated below:

Number of Primary Users, N = 9

Probability of data arrival rate P =0.1, 0.15, 0.20, and 0.25

Rate of data transmission, R = 9

R = log {1/ (1-P)}

Goodput (performance index) G = (N * (1-P)) /P

7. Results & Discussion

7.1. For Perfect Sensing & Zero Interference Simulations

In this case we made some assumptions; It is assumed that there is no collision between primary users; there are ten (10) primary users with different probability of arrival *P* then the goodput of the primary users is simulated and presented as shown in Figure 3.

The plot of Normalized Goodput against the fraction numbers of primary licensed users is shown in Figure 3; when the fraction of licensed primary users is 0.1 and the probability of transmission data arrival $P = 0.10$, then the normalized goodput is 0.53. Also, when the fraction of primary licensed users is 0.2; and the probability of transmission data arrival $P = 0.20$, the normalized goodput is 0.15. From the result obtained, it can be established that the lower the number of primary licensed users with lower probability of arrival the higher the normalized goodput G. This satisfies equation (6).

Simulation of the numbers of secondary users that can transmit when the numbers of secondary users that has data to transmit varies.

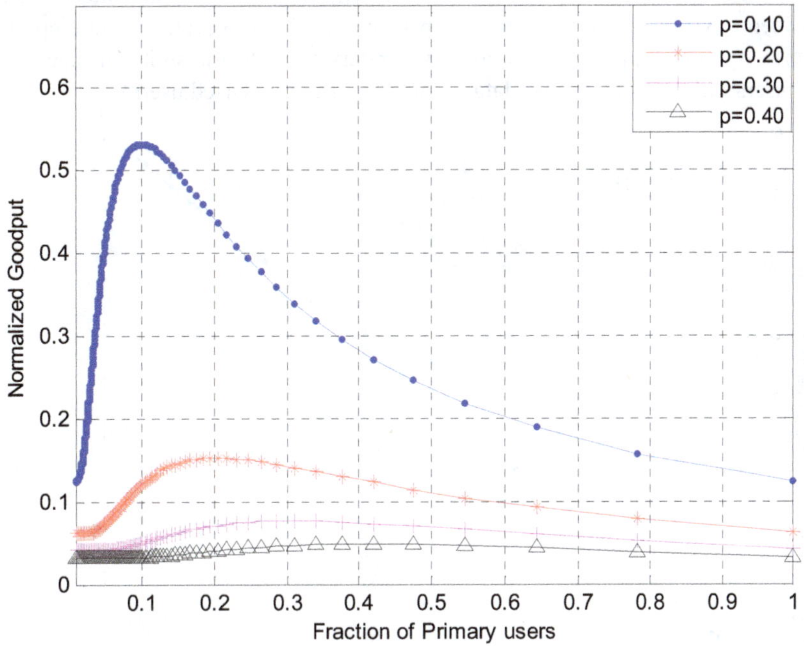

Figure 3. *Normalized Goodput against Fraction of Primary users.*

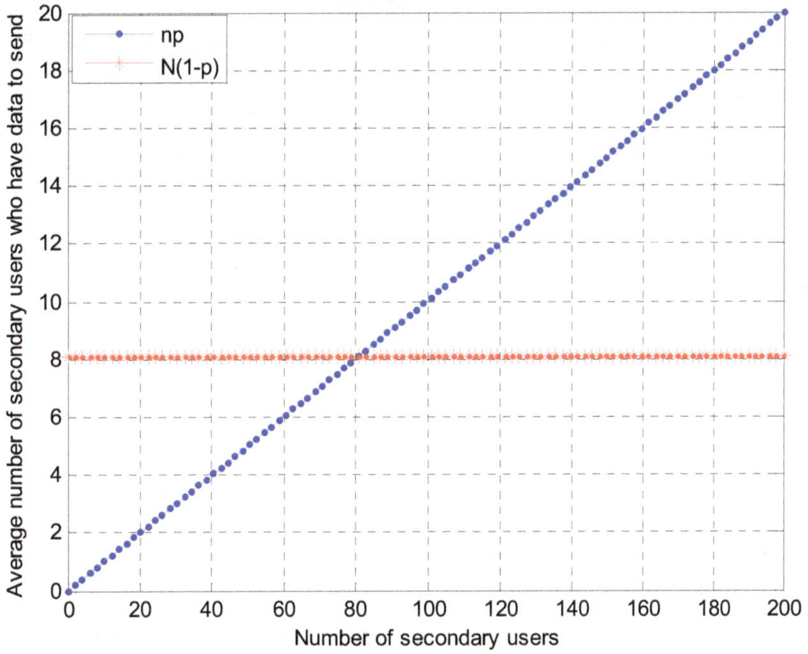

Figure 4. *Average number of secondary users who have data to transmit against number of secondary users* ($P = 0.1$).

The number of secondary users that have data to transmit is set at 20 and the total number of secondary users is 200 as shown in Figure 4; then the number of secondary users that can transmit without interference can be determined considering probability of arrival of transmission $P, (P = 0.1)$. It is noted that the point of intersection between np *and* $N(1-p)$ is the optimum point of interference between the licensed primary users and the opportunistically secondary users. It implies here that, below the optimum point of interference, all the secondary users will transmit opportunistically without interference with the licensed primary users and above the optimum point of interference, there will be interference between any secondary users that transmit and the licensed primary users. It shows that for $P = 0.1$, only 8 secondary users out of 20 that have data to transmit will opportunistically transmit without interference since they lie below the optimum point of interference. At $P = 0.2$, the simulation result presented in Figure 5 shows that only 7 secondary users out of 40 average numbers of secondary users who have data to send will transmit without interference with the licensed primary users. When the probability of arrival of transmission data increases to 0.3, ($P = 0.3$), and the average number of secondary users who have data to transmit increases to 60 as shown in Figure 6. It is clearly shown that only 6 secondary users out of 60 average numbers of secondary users who have data to send will transmit without interference with the licensed primary users. However, any attempt for the remaining 52 secondary users to transmit will cause harmful effect on the primary users (licensed users).

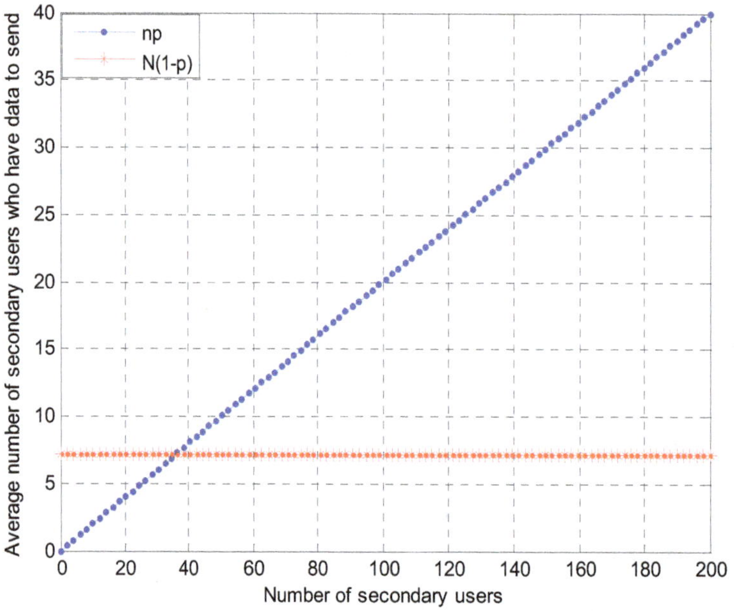

Figure 5. *Average number of secondary users who have data to transmit against number of secondary users* ($P = 0.2$).

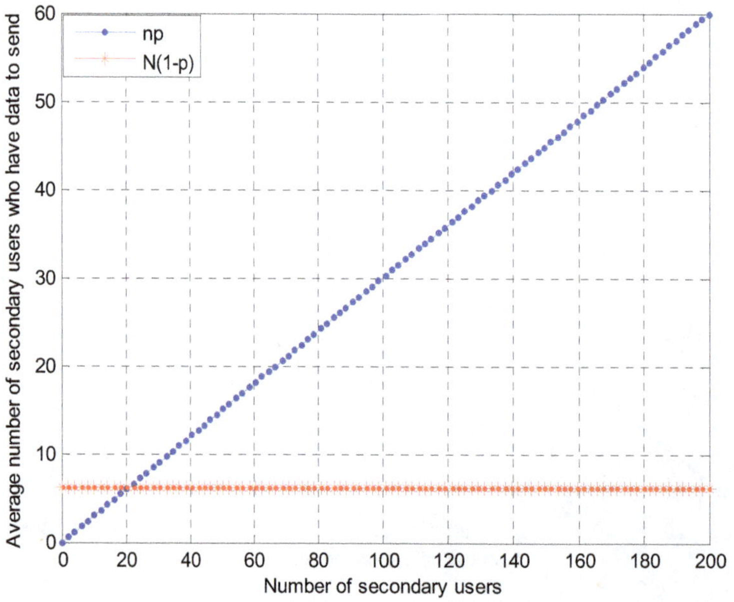

Figure 6. *Average number of secondary users who have data to transmit against number of secondary users* ($P = 0.3$).

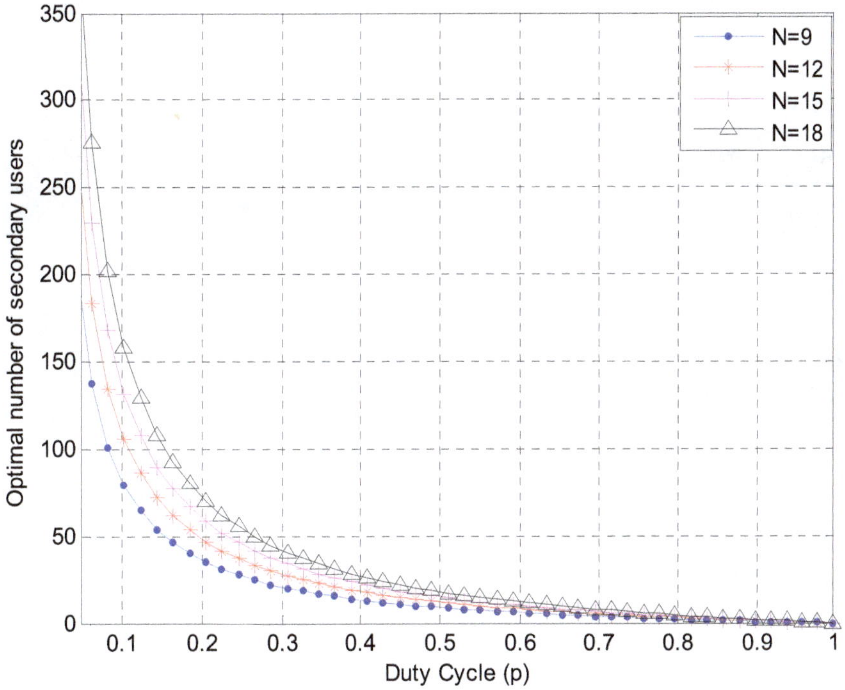

Figure 7. *Graph of Optimum Number of Secondary Users against Duty Cycle (p).*

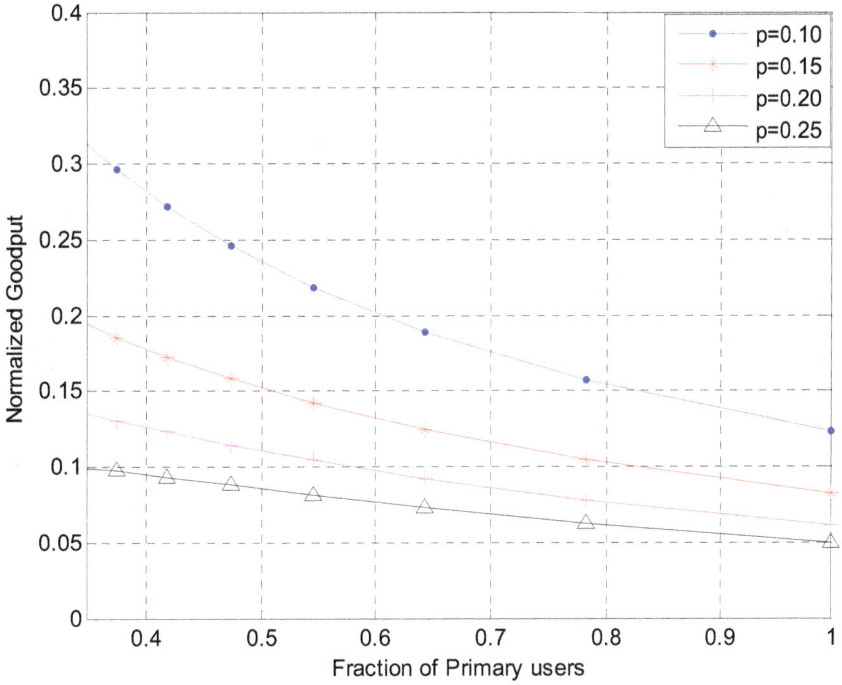

Figure 8. *Graph of Normalized Goodput against Fraction of Primary Users.*

7.2. For Imperfect Sensing & Non Zero Interference Simulations

Case 1:

With different numbers of N (number of licensed primary users) and duty cycles, the optimum number of secondary users show that the number of opportunistically secondary users increases when N increases with a very low value of duty circle as shown in figure 7.

This satisfies equation (27) which shows the number of secondary users that can opportunistically use a specific number of licensed primary users.

Case 2:

In figure 8, we have different values of Probability of arrival of transmission data P and fraction of primary users, the goodput of licensed primary users largely depend on the probability of arrival of transmission data P.

Normalized goodput has the highest value when P = 0.1 and the lowest when P= 0.25 with the fraction of licensed primary users of 0.2 and 0.78 respectively.

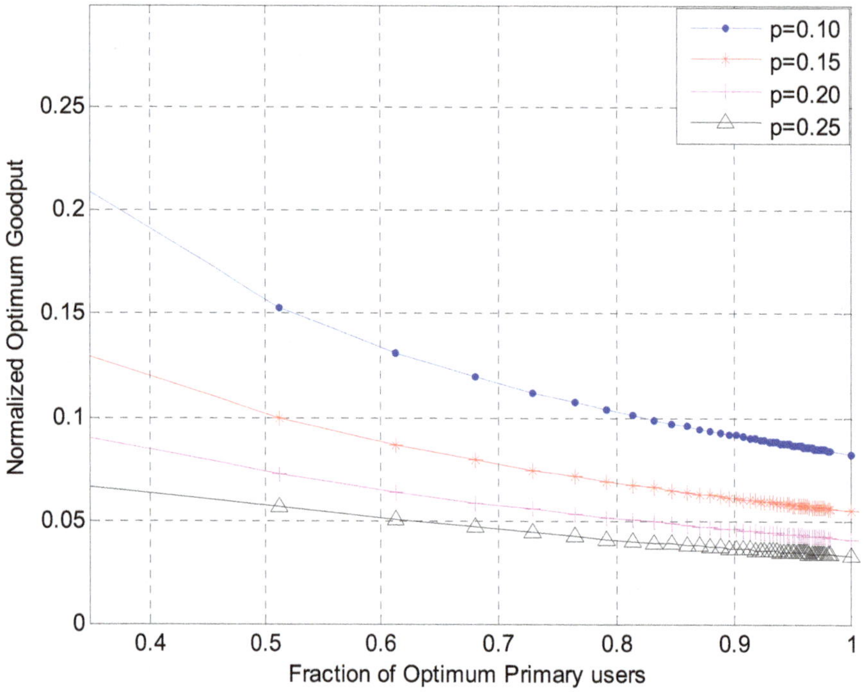

Figure 9. *Graph of Normalized Optimum Goodput against Fraction of Optimum Primary Users.*

Case 3:

Figure 9 is the normalized goodput plotted against fraction of primary users.

With different values of P, it is observed that the lower the value of P the higher the goodput. It also follows that the normalized goodput has the highest value when P = 0.10, with a few numbers of licensed primary users.

Figure 9 satisfies equation (31).

8. Conclusion

The results presented in this paper shows the application of Goodput mathematical model for perfect sensing, zero tolerance and imperfect sensing, non zero tolerance for spectrum sharing in cognitive radio network. It is observed that for high Goodput, perfect sensing and zero interference tolerance, the probability of arrival of signal transmission among the primary and secondary users must be very low in order to have appreciable number of opportunistic secondary users. Also, secondary users are not allowed to transmit above the optimum point of interference between the licensed primary users and the opportunistically secondary users as proposed by Goodput mathematical model.

References

[1] http://en.wikipedia.org/wiki/ Cognitive_ Radio (Cognitive Radio Communication and Networks: Principles and Practice) retrieved on 24/09/2014 8:18 PM.

[2] http://en.wikipedia.org/wiki/Electromagnetic_Spectrum, retrieved on 8/1/2014 12:02 PM.

[3] A. M. Wigglinski, M. Nekovee, and Y. N T. Hou, (2009), "Cognitive Radio "*IEEE VTS*, chap.6.

[4] Zhu Jiand and K. J. Ray Liu, (2007), "Cognitive Radio for Dynamic Spectrum Access" IEEE *Communications Magazine*, pp. 84-94.

[5] L. Cao and H. Zheng, (2005) "Distributed Rule-Regulated Spectrum Sharing" *IEEE Journal on* Selected *Areas of Communications*, vol. 20, no.3, pp.150.

[6] R. Etkin, A. Parekh, and D. Tse, (2005) "Spectrum Sharing for Unlicensed bands"*Proc. IEEE Dy SPAN, pp.159.*

[7] Varaka Uday Kanth, Kolli Ravi Chandra, Rayala Ravi Kumar, (2013) "Spectrum Sharing In Cognitive Radio Networks", IJETT Journal, vol. 4, issue 4,pp.1172.

[8] Haykin, S. (2005) "Cognitive Radio: Brain Empowered Wireless Communications" *IEEE Journal on Selected Areas of Communications*, vol.23, no.2, pp. 201-220.

[9] Xavier Gelabert, Oriol Sallent, Jordi Pérez-Romero, Ramon Agustí, (2010) "Spectrum sharing in cognitive radio networks with imperfect sensing: A discrete-time Markov model" pp. 2522-2532.

[10] Ana Luz Mendiguchia Gonzalez, Massimiliano Picole and Silvia Colabrese, (2005) "*Cognitive Radio: Meaning and Application", IEEE journal on Cognitive Radio,* vol.20, no.4, pp.120.

[11] *Sudhir Srinivasa and Syed Ali Jafar, (2008) "Cognitive Radio Networks: How much Spectrum Sharing is Optimal"pp. 1-4.*

Dual-band dipole antenna for 2.45 GHz and 5.8 GHz RFID tag application

Yanzhong Yu[1, 2, *], Jizhen Ni[1], Zhixiang Xu[1]

[1]College of Physics & Information Engineering, Quanzhou Normal University, Fujian, China
[2]Key Laboratory of Information Functional Materials for Fujian Higher Education, Fujian, China

Email address:
yuyanzhong059368@gmail.com (Yanzhong Yu)

Abstract: In this paper, a dual-band dipole antenna for passive radio frequency identification (RFID) tag application at 2.45 GHz and 5.8 GHz is designed and optimized using HFSS 13. The proposed antenna is composed of a bent microstrip patch and a coupled rectangular microstrip patch. The optimal results of this antenna are obtained by sweeping antenna parameters. Its return losses reach to −18.7732 dB and −18.2514 dB at 2.45 GHz and 5.8 GHz, respectively. The bandwidths (Return loss <=−10 dB) are 2.42~2.50 GHz and 5.77~5.82 GHz. And the relative bandwidths are 3.3% and 0.9%. It shows good impedance, gain, and radiation characteristics for both bands of interest. Besides, the input impedance of the proposed antenna may be tuned flexibly to conjugate-match to that of the IC chip.

Keywords: RFID, Dipole Antenna, Dual Frequency, Tag

1. Introduction

In the last few years, the wireless identification and communications technology has been developing rapidly [1]. Radio frequency identification (RFID) is obtaining a growing interest to wirelessly track and identify objects due to its cheapness and reliability [2-4]. Now RFID finds many applications in lots of areas like pallet tracking, electronic toll collection, parking lot access, information industry, medical and defense [5-7]. Generally speaking, in RFID system different frequency spectrums are allocated to different countries or regions. For instance, 840.5~844.5 MHz and 920.5~924.5 MHz in China, 865~867 MHz in India, 902~928 MHz in Argentina and America, 866~869 MHz and 920~925 MHz in Singapore, and 952~955 MHz in Japan, and so on [6, 7]. With the fast development of global economy, the exchanges of products made from different countries have become more and more frequent. In order to make the tag attached to the products effective, the RFID system should operate in dual-band or multi-band regime [8-12]. An RFID system consists of a transponder (tag), which stores an identification code, and of a detector (reader) that is capable to retrieve the identity of the tags through a wireless wave. The passive RFID tag usually is composed of a tag antenna and an IC chip [1, 2]. Tag antenna is a key part of RFID systems. Therefore, a design of dual-frequency or multi-frequency tag antenna is of great importance for RFID application in different countries. In the current paper, a novel dual-frequency tag antenna, i.e., 2.45 GHz and 5.8 GHz, is presented. It is composed of a bent microstrip patch and a rectangular microstrip patch. The rest of the present paper is organized as follows. Antenna design is described in Section 2. Section 3 presents the analysis and optimization results of the designed antenna. We give a brief conclusion in the last Section 4.

2. Antenna Configuration

The configuration of the proposed dual-band dipole antenna is illustrated in Fig. 1. This antenna works at 2.45 GHz and 5.8 GHz for RFID tag application. The basic structure of the designed antenna is composed of a bent and a rectangular microstrip patch. The common material FR4_epoxy, with relative dielectric constant $\varepsilon_r = 4.4$, loss tangent $\tan\delta = 0.017$ and thickness h =0.5 mm, is considered as substrate. The IC chip is connected to the bent microstrip patch, and the power exchanges between the bent and rectangular patch by coupling through the gap with length of a and width of e. The initial sizes of the proposed antenna are given as follows: the length and width of the rectangular microstrip is a =33 mm and g =2 mm; the length and width of the bent microstrip patch is b =8 mm and c =1 mm, respectively; the space of d =4 mm, and the gap width of e =1 mm.

The proposed antenna is analyzed and optimized by using Ansoft simulation software High Frequency Structure Simulator (HFSS) [13]. This work is given in Section 3.

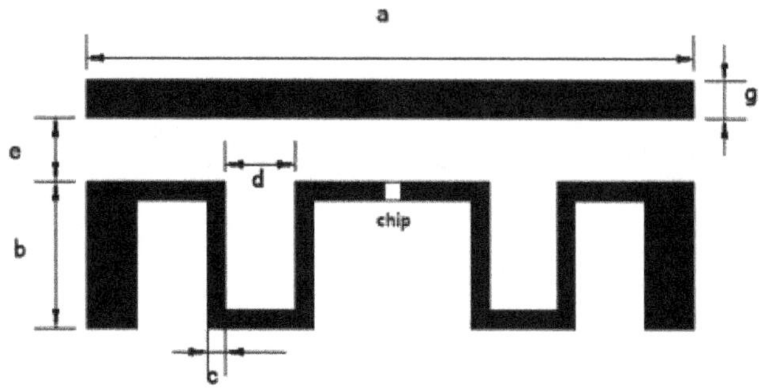

Fig. 1. *Configuration of dual-band tag antenna.*

3. Analysis and Optimization

HFSS is an interactive software package for calculating the electromagnetic behavior of a structure. It is a famous electromagnetic simulation tool used for antenna design, and the design of complex RF electronic circuit elements including filters, transmission lines, and packaging. In the present paper, the HFSS is thus employed to analyze and optimize the proposed antenna. Fig. 2 shows the return loss S11 of the initial antenna as a function of frequency. It can be seen clearly from Fig. 2 that the initial antenna has two resonating frequency points within 1~7 GHz, the first one at 2.39 GHz (S11=-10.6107 dB) and the second at 5.86 GHz (S11=-33.6484 dB). Neither of them is located at the expectation frequency points of 2.45 GHz and 5.8 GHz. The relative bandwidths (S11<=-10dB) of the first and second resonating frequencies are (2.40-2.39)/2.39=0.42% and (5.88-5.84)/5.86=0.68%, respectively. Obviously, they do not satisfy the requirements of RFID applications.

Proper impedance matching must be considered in antenna design. Common antenna is usually connected to coaxial cable with impedance of 50 Ω or 75 Ω. Consequently the value of input impedance of common antenna is adjusted to 50 Ω or 75 Ω. However in RFID system the tag antenna is linked to the IC chip whose input impedance may be an arbitrary value [14] and is thus no longer 50 Ω or 75 Ω. In order to achieve the purpose of maximum power transmission and improvement tag performance, proper impedance matching between the tag antenna and the IC chip is of considerable importance in RFID applications. Because the design and manufacturing of new IC chip is a big and costly venture, the input impedance of tag antenna is tuned to conjugate match to a specific IC chip available in the market [6]. The IC chip is usually a capacitive element and its input impedance is a complex value. We suppose that the input impedance of IC chip is Z_{chip}=(25-j136) Ω in our work. The value of tag antenna impedance is a function of frequency, as depicted in Fig. 3. One can observe that the values of input impedances are Z_1=(15.4+j160.7) Ω at 2.45 GHz and Z_2=(20.8+j103.3) Ω at 5.8 GHz. The purpose of conjugate matching between tag antenna and the IC chip do not achieve obviously. Accordingly, it can be found from Figs. 2 and 3 that both resonating frequency points and input impedances do not meet the design requirements yet. Therefore, the initial sizes of tag antenna must be optimized. Next, the effects of antenna sizes on their performances are analyzed.

In order to optimize the performances of initial antenna, the relevant parameters, like rectangular microstrip patch length a and width g , bent microstrip patch length b and space d , and the gap width e , are examined. Firstly, to investigate the influence of rectangular microstrip patch length a on antenna performances, one can assume that others parameters of initial antenna remain unchanged and parameter a varies from 30 mm to 36 mm only. Fig. 4 displays the return loss S11 as a function of frequency. It can be observed that within 1~7 GHz the larger the length a is, the lower are two harmonic frequencies. The effect of length a on antenna input impedance is depicted in Fig. 5. It can be found that at 2.45 GHz, the value of resistance increases from 13 Ω to 26 Ω and the reactance adds from 64 Ω to 237 Ω; and at 5.8 GHz, the resistance varies from 16 Ω to 20 Ω and the reactance changes from 16 Ω to 154 Ω, when length a varies from 30 mm to 60 mm. Therefore, a conclusion can be drawn that the values of resistance and reactance rise with the increase of length a and grow faster and faster.

Secondly, the effect of width g on antenna performances is investigated, as illustrated in Figs. 6 and 7. One may find that the return loss S11 within 1~7 GHz and input impedance remain unchanged nearly when the value of width g varies from 1 mm to 3 mm.

Now, let us examine the effect of bent microstrip patch length b on resonant frequency and input impedance, as shown in Figs. 8 and 9, respectively. One can observe from Fig. 8 that the harmonic frequency decreases with length b increasing. We can also find from Fig. 9 that not only at 2.45 GHz but also at 5.8 GHz the values of resistance and reactance increase as the length b increases.

Figs. 10 and 11 show the influence of bent microstrip patch space d on return loss and input impedance. Within 1~7 GHz the resonant frequency decreases with the increase of space d . Moreover, the change quantity of the second resonant frequency is larger than that of the first one. At 2.45 GHz, the real and imaginary parts of input impedance increase slowly as space d raise from 2 mm to 3 mm; but they also decrease slowly when the value of space d increases from 3mm to 4mm. Similarly, at 5.8 GHz the real component of input impedance is almost unvaried, and the imaginary component increases from -31 Ω to 86 Ω when space d varies from 2 mm to 4 mm.

Lastly, the influence of gap width e is analyzed as depicted in Figs. 12 and 13. Within 1~7 GHz the first harmonic frequency rises but the second does not change almost while the gap width e increases from 0.5 mm to 1.5 mm. At 2.45 GHz, the imaginary part of input impedance reduces from 250 Ω to 122 Ω but the real part remains unchanged when increasing the space d from 0.5 mm to 1.5 mm. And the change of input impedance at 5.8 GHz is very small.

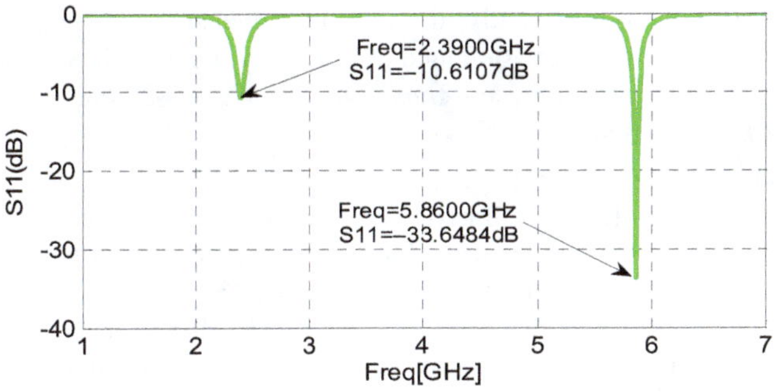

Fig. 2. *Return loss S11 of initial antenna.*

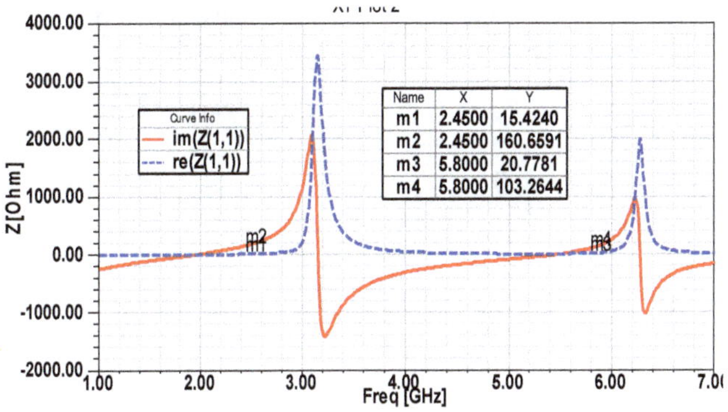

Fig. 3. *Input impedance of initial antenna.*

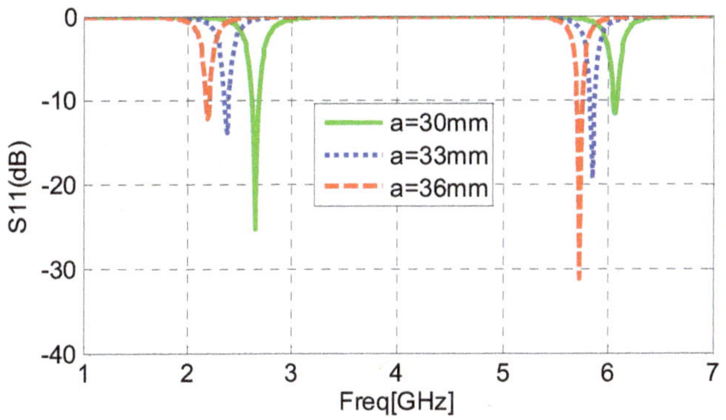

Fig. 4. *Influence of length a on resonating frequency.*

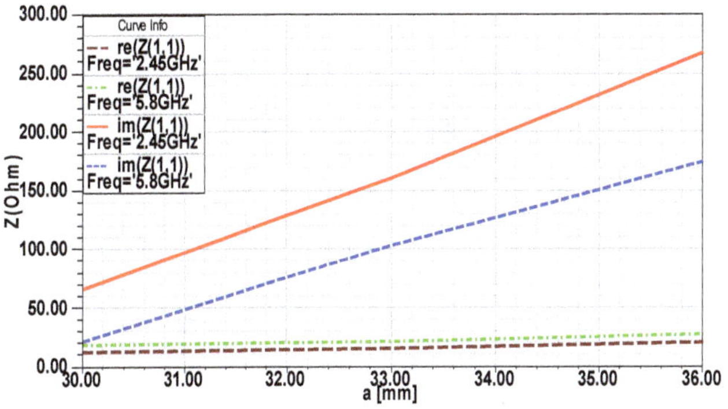

Fig. 5. *Effect of length a on input impedance.*

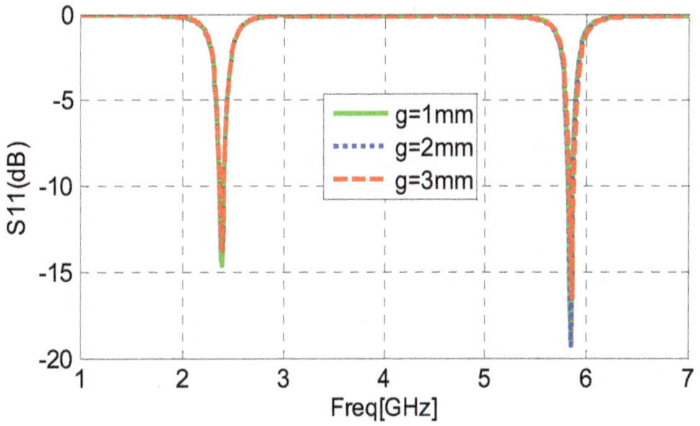

Fig. 6. *Return loss S11 with different width g.*

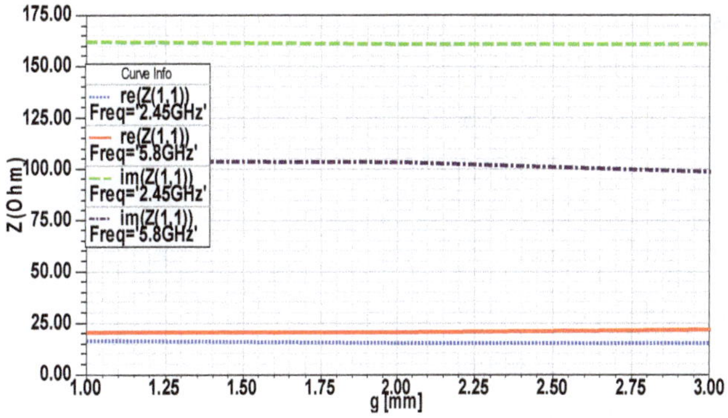

Fig. 7. *Input impedance with different width g.*

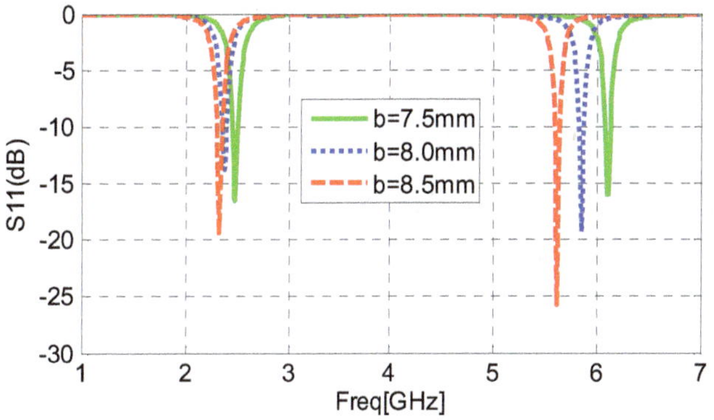

Fig. 8. *Influence of length b on resonating frequency.*

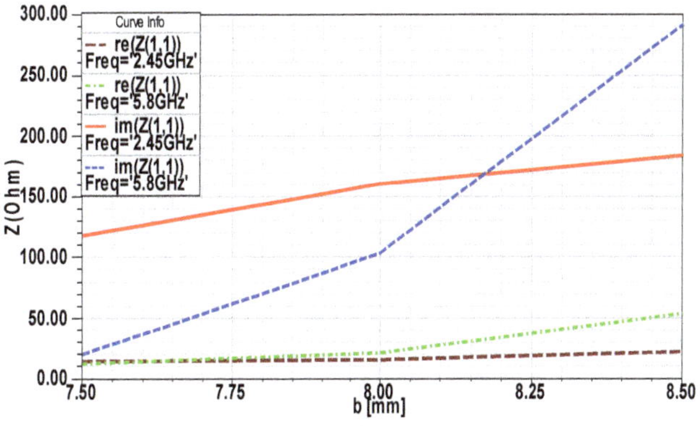

Fig. 9. *Effect of length b on input impedance.*

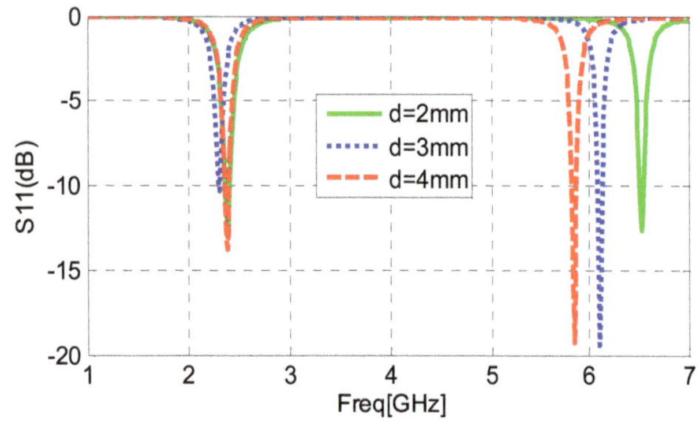

Fig. 10. Influence of space d *on resonating frequency.*

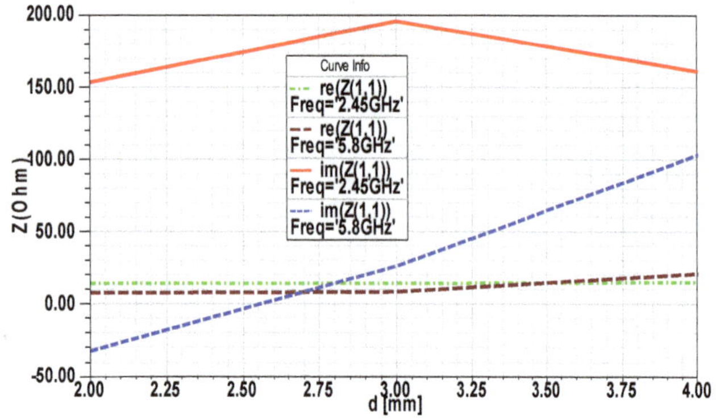

Fig. 11. Effect of space d *on input impedance.*

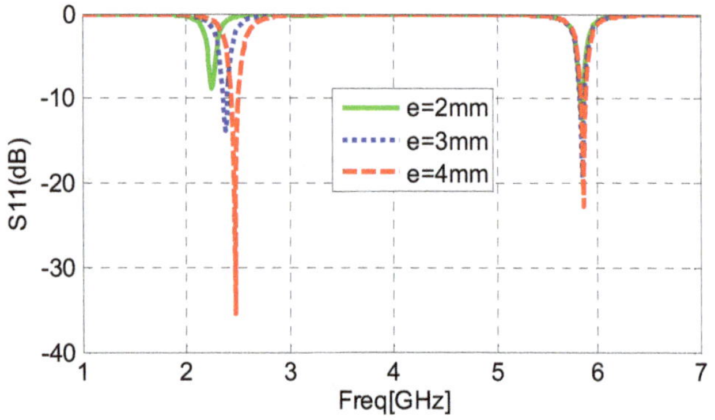

Fig. 12. Influence of gap width e *on resonating frequency.*

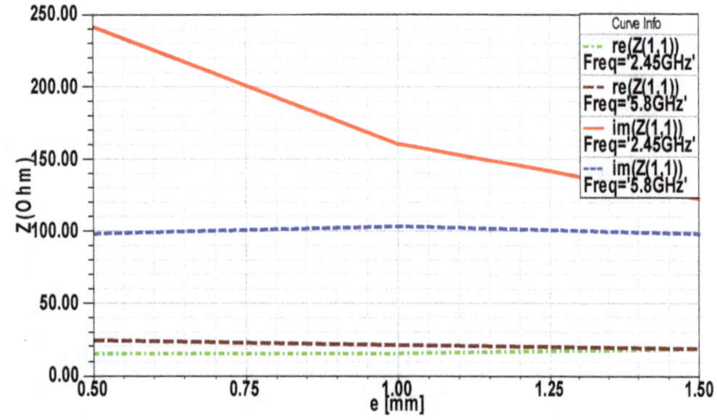

Fig. 13. Effect of gap width e *on input impedance.*

On the basis of analysis above, we find that the gap width e only affects on the first resonating frequency. We thus can drop the second resonating frequency point to 5.8 GHz by increasing the rectangular microstrip length a. Then the first resonating frequency can be tuned to 2.45 GHz by changing the value of gap width e. In a word, by optimizing the sizes of initial antenna repeatedly, the optimal dimensions of the proposed tag antenna are obtained at last. They are as follows: $a = 34.7mm$, $b = 8mm$, $c = 1mm$, $d = 4mm$, $e = 3.5mm$, $g = 2mm$, $h = 0.5mm$.

The return loss S11 as a function of frequency for the optimized tag antenna is illustrated in Fig. 14. It can be seen clearly that at the range of 1~7 GHz the antenna has two resonating frequency points, one at 2.45 GHz (S11=-18.7732 dB) and the other at 5.8 GHz (S11=-18.2514 dB). For the first resonating frequency of 2.45 GHz, the absolute bandwidth (S11<=-10dB) reaches to 80 MHz (2.42~2.50 GHz) and the fractional bandwidth is (2.50-2.42)/2.45=3.3%. For 5.8 GHz, the absolute and fractional bandwidths are 50 MHz and 0.9% respectively.

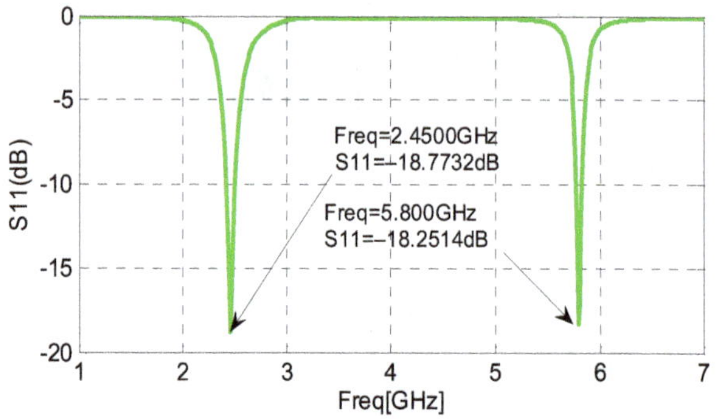

Fig. 14. Return loss of optimized tag antenna.

Fig. 15 presents the input impedance of the optimized tag antenna. It can be observed from Fig. 15 that the values of input impedance are Z1=(30.8+j132.1) Ω at 2.45 GHz and Z2=(20.2+j140.2) Ω at 5.8 GHz. Both are close to the conjugate value of Zchip=(25-j136) Ω, which is an assumed value of input impedance of the IC chip. This mean it has a good conjugate match between the IC chip and tag antenna. Generally, the resistance of most of RFID IC chips varies from 10 Ω to 30 Ω, and the reactance limits in the range of (100~300) Ω. Therefore, the purpose of conjugate matching can be achieved only by tuning the parameter a or b. This may meet the impedance matching requirement of most of RFID IC chips.

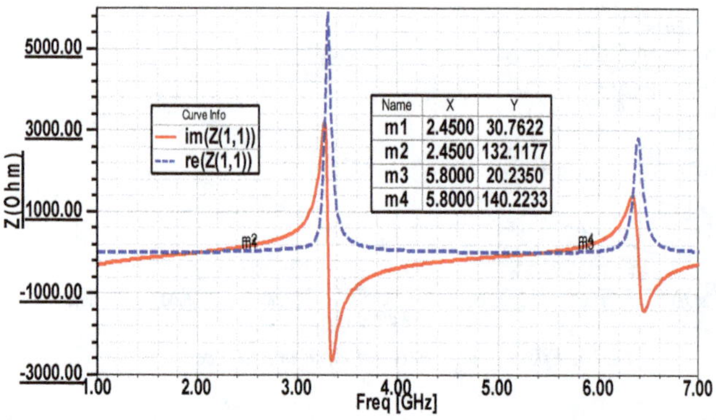

Fig. 15. Input impedance of optimized tag antenna.

Antenna gain is an important index to measure antenna electric performance. The larger the total gain is, the farther the read range of RFID system become. The 3D gain patterns of the optimized tag antenna at 2.45 GHz and 5.8 GHz are displayed in Figs. 16 and 17, respectively. The values of maximum gain are 2.68 dBi at 2.45 GHz and 3.27 dBi at 5.8 GHz. These performance indexes meet the requirements of RFID applications. In order to observe the pattern more clearly, Fig. 18 illustrates the 2D radiation patterns in xz-plane ($\text{phi} = 0^{\circ}$) and yz-plane ($\text{phi} = 90^{\circ}$) at 2.45GHz. The radiation in xz-plane is maximum and omnidirectional, and yet in yz-plane is a poor directionality. The similar chart for 5.8 GHz is shown in Fig. 19. It can be seen from Fig. 19 that in xz-plane the radiation exhibits an approximation to omnidirectional pattern, but in yz-plane the sidelobe appears at $\theta = 0^{\circ}$ or $\theta = 180^{\circ}$.

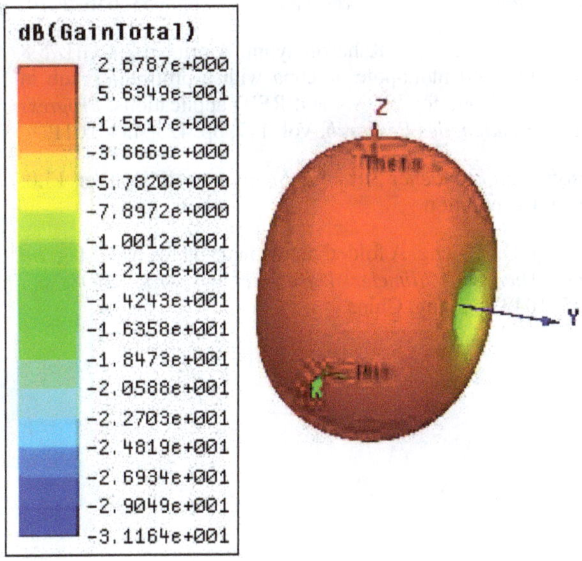

Fig. 16. 3D gain pattern at 2.45 GHz.

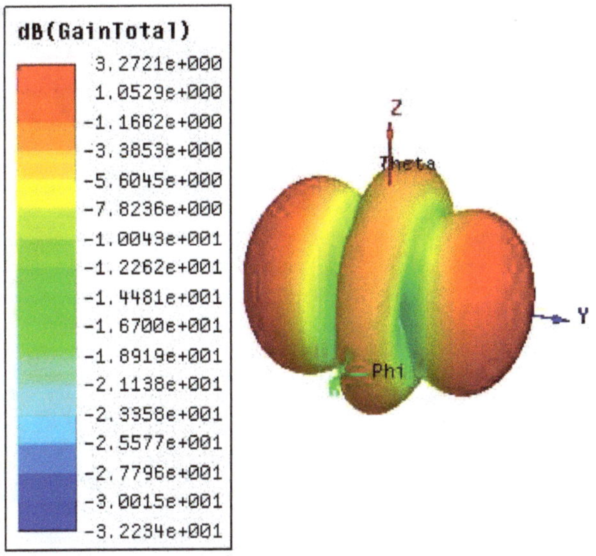

Fig. 17. 3D gain pattern at 5.8 GHz.

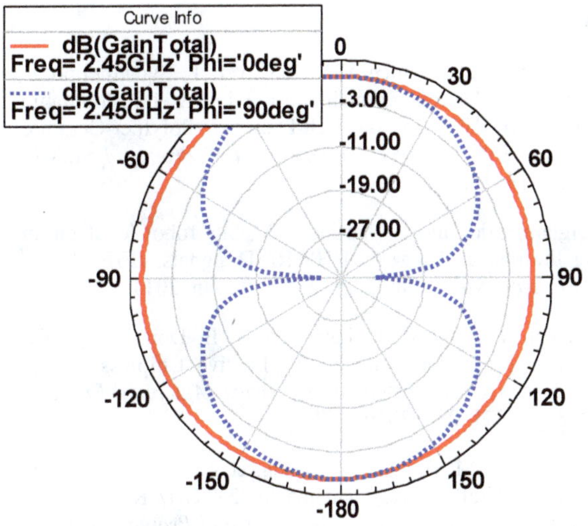

Fig. 18. 2D radiation patterns at 2.45 GHz.

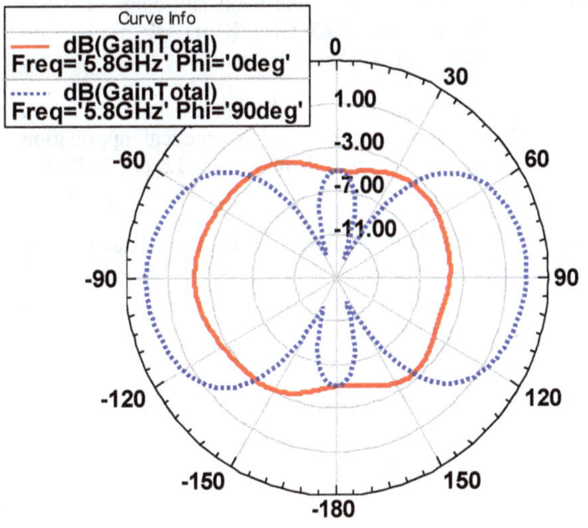

Fig. 19. 2D radiation patterns at 5.8 GHz.

4. Conclusions

With the rapid development of RFID technology and applications, the operating frequency of tag antenna has been developing to microwave wavebands (2.45GHz, 5.8GHz). The design of tag antenna at these wavebands has become more and more important. In the present paper, the tag antenna with dual-band work frequency, i.e. 2.45GHz and 5.8 GHz, is proposed. Its performances are analyzed and optimized by using the HFSS. The simulation results demonstrate the designed tag antenna satisfy the application requirements of RFID system. It can find applications in traffic or logistics management.

Acknowledgements

This work was supported by the Key Project of Science and Technology Department of Fujian Province (No. 2012H0035), the Key Project of Quanzhou City Science and Technology Program (No. 2011G14), and the Key Discipline of Electronic Science and Technology.

References

[1] A. Ali Babar, T. Björninen, V.A. Bhagavati, L. Sydänheimo, P. Kallio, and L. Ukkonen. Small and Flexible Metal Mountable Passive UHF RFID Tag on High-Dielectric Polymer-Ceramic Composite Substrate. *IEEE Antennas Wirel. Propag. Lett.* Vol. 11, pp. 1319–1322, 2012.

[2] E. Digiampaolo, and F. Martinelli. Mobile robot localization using the phase of passive UHF RFID signals. *IEEE Trans. Ind. Electron.* Vol. 6, No. 1, pp. 365–376, Jan. 2014.

[3] S. M. Hu, Y. Zhou, C. L. Law, and W. B. Dou. Study of a Uniplanar Monopole Antenna for Passive Chipless UWB-RFID Localization System. *IEEE Trans. Antennas Propag.*, Vol. 58, pp. 271–278, 2010.

[4] H. T. Chou, T. M. Hung, N. N. Wang, et.al. Design of a near-field focused reflect array antenna for 2.4 GHz RFID reader applications. *IEEE Trans. on Antennas and Propag.*, Vol. 59, pp. 1013–1018, 2011.

[5] S. Jeon, Y. Yu, and J. Choi. Dual-band slot-coupled dipole antenna for 900MHz and 2.45 GHz RFID tag application. *Electronics Letters*, Vol. 42, No. 22, Oct. 2006.

[6] K.V. S. Rao, P. V. Nikitin, and S. F. Sander. Antenna design for UHF RFID tags: a review and a practical application. *IEEE Trans. Antennas Propag.*, Vol. 53, No. 12, pp. 3870–876, Dec. 2005.

[7] Raviteja, Chinnambeti; Varadhan, Chitra; Kanagasabai, Malathi; Sarma, Aswathy K.; Velan, Sangeetha. A fractal-based circularly polarized UHF RFID reader antenna. *IEEE Antennas Wirel. Propag. Lett.*, Vol. 13, pp. 499–502, 2014.

[8] Dongho Kim, and Junho Yeo. Dual-Band Long-Range Passive RFID Tag Antenna Using an AMC Ground Plane Digiampaolo, F. Martinelli. *IEEE Trans. Antennas Propag.*, Vol. 60, No. 6, pp. 2620–2626, June 2012.

[9] C. Varadhan, J. K. Pakkathillam, M. Kanagasabai, R. Sivasamy, R. Natarajan, and S. K. Palaniswamy. Triband Antenna Structures for RFID Systems Deploying Fractal Geometry. *IEEE Antennas Wirel. Propag. Lett.*, Vol. 12, pp. 437–440, 2013.

[10] Young-Ho Suh, and Kai Chang. A high-efficiency dual-frequency rectenna for 2.45- and 5.8-GHz wireless power transmission. *IEEE Trans. Microw. Theory & Tech.*, Vol. 50, No. 7, pp. 1784–1789, 2002.

[11] A. T. Mobashsher, M. T. Islam, and N. A. Misran. A Novel High-Gain Dual-Band Antenna for RFID Reader Applications. *IEEE Antennas Wirel. Propag. Lett.* Vol. 9, pp. 653–656, 2010.

[12] J. R. Panda, and R. S. Kshetrimayum. A printed 2.4 GHz/5.8 GHz dual-band monopole antenna with a protruding stub in the ground plane for WLAN and RFID applications. *Progress In Electromagnetics Research*, Vol. 117, pp. 425–434, 2011.

[13] *Ansoft High Frequency Structure Simulator* (HFSS), Ver. 13.0, Ansoft Corporation.

[14] B. Yang, Q. Y. Feng. A folded dipole antenna for RFID tag. *Int. Conf. Microw. Millimeter Wave Technol. Proc.* 2008, pp. 1047–1049, Nanjing, China.

12

Research on Remote Monitoring System of Temperature and Humidity Based on GSM and ZigBee

Zhu Hong-xiu, Sun Zhi-yuan, Hu Yuan-zhou

School of Mechanical Electronic & Information Engineering, China University of Mining & Technology (Beijing), Beijing, China

Email address:

353642795@qq.com (Sun Zhi-yuan), zhx@cumtb.edu (Zhu Hong-xiu), 1028197153@qq.com (Hu Yuan-zhou)

Abstract: This paper is about the temperature and humidity control system based on Short Messaging Service (GSM) and wireless networks (ZigBee technology). The system can be used in home network communication by software IAR Embedded Workbench as ZigBee development platform. After the user send short message to Short Messaging Service TC35 module, master microcontroller completes the communication between GSM module and ZigBee coordinator, then ZigBee coordinator control Internal family ZigBee terminal nodes according to the content sending by the single-chip computer. On the one hand, the ZigBee coordinator receives data collected by temperature and humidity sensor DHT11 and Short Messaging Service module sent to the user's phone in the form of short message via coding by the master microcontroller, on the other hand, the single-chip computer controls household appliances switch by the triode switch. Through debugging and experimental research of the hardware and software system, we concluded that the system can realize the remote monitoring of indoor temperature and humidity and remote control household appliances reliably, which laid a foundation of further implement on the intelligent of the household environment.

Keywords: Short Messaging Service Module, Remote Monitoring, ZigBee, Temperature and Humidity

1. Introduction

With the improvement of people's living standards, the demand of perceiving and regulating the indoor environment is growing. The main communication technology of remote monitoring system includes a network telephone, network, etc, who have a lot of shortcomings in anti-interference. This system uses Short Messaging Service (GSM). Users send control instructions and receive feedback information by mobile phone. Compared with other communication technology, Short Messaging Service (GSM) technology has strong ability of anti-interference and high reliability. This system uses the ZigBee wireless network technology to construct home communication network. ZigBee terminal module connects to external sensors and home appliance controller. The ZigBee coordinator helps implement home network communication. This paper uses ZigBee wireless network technology and GSM short message technology to complete the remote monitoring of indoor temperature and humidity and remote control of household appliances, which laid a foundation of further implement on the intelligent of the household environment.

2. The System Hardware Design

The remote monitoring system of temperature and humidity based on GSM and ZigBee consists of three major areas: short message module, main control center module and ZigBee wireless communication module. Master control center module is responsible for receiving and analyzing short message sent by the user's mobile phone, then the contents parsed out can be transmitted to ZigBee coordinator. ZigBee coordinator controls ZigBee terminal nodes according to the content sending by the single chip. Indoor terminal node transmit the data collected by ZigBee coordinator to the master microcontroller. After encoded, the data should be sent to the user's phonere by Short Messaging Service module in the form of short message. The system composition block diagram is showed in figure 1.

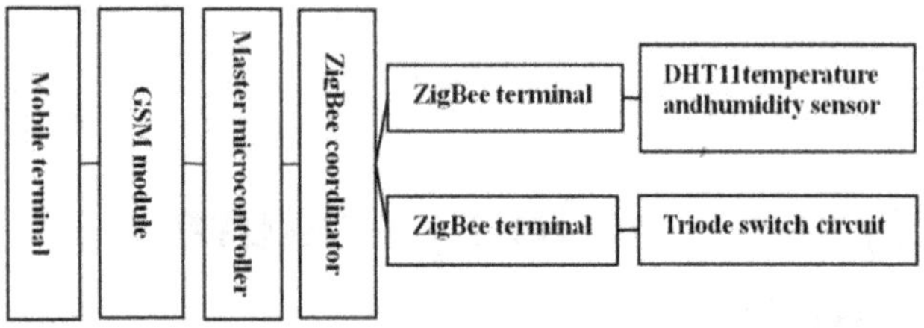

Figure 1. The over all plan of monitoring system.

Siemens TC35 module is be used as the GSM module, which is a GSM modem using AT commands format and to communicating with PC or master singlechip via RS232 serial port.

ZigBee is a short-range wireless communication technology, which collects from each terminal node on the network and communicates with GSM module through the master singlechip. Coordinator node be consisted of three parts: ① CC2430 minimum system; ② antenna circuit; ③ serial port communication circuit. Terminal node is used to communicate with coordinator, collecting data and meanwhile controlling household appliances switches.

DHT11 is a compound sensor to measure temperature and humidity, whose output signal is calibrated digital signal. DHT11 only has 4 pins, including the power, ground, data line, and empty leg. Temperature and humidity sensor interface circuit is shown in figure 2.

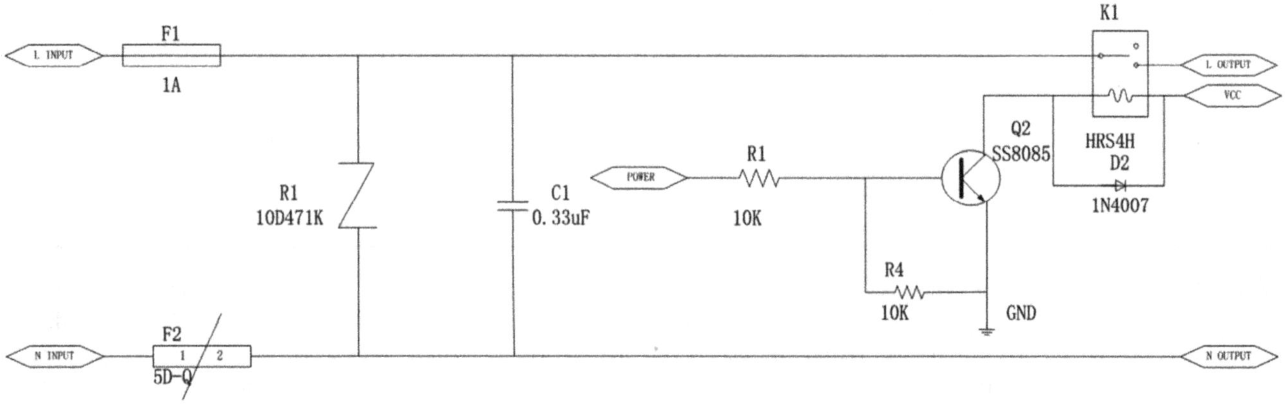

Figure 2. Temperature and humidity sensor.

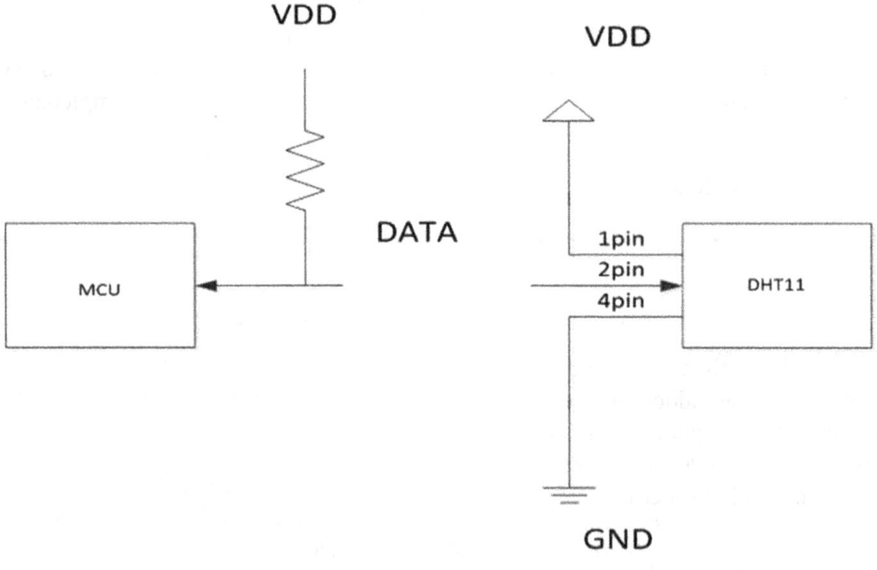

Figure 3. The switch control circuit sensor interface circuit.

Indoor electrical switch control circuit principle is shown in figure 3. In the figure, the L, N on the left side is 220 v ac input, the L, N on the right side is 220 v ac output. We control the level of the power switch port instead of controling electric equipment using the triode switch function. triode disconnects, the relay cuts off and electric equipment shuts down. When the power output high electricity, triode conducts, tentacles of relay conducts and electrical equipment opens.

3. The System Software Design

3.1. The Software Overall Design Scheme

In the System users send control instructions to GSM short message module through mobile phone. When GSM module receives the control command, the corresponding commands should be sent to the master microcontroller STC89C51, the master microcontroller parses the corresponding command, then ZigBee coordinator receives the parse instructions. ZigBee coordinator sends the corresponding instruction to each ZigBee terminal nodes. Terminal nodes control electric switch or measure parameters that need to be monitored (temperature and humidity), finally the results can be sent to mobile phone. The system total process is shown in figure 4.

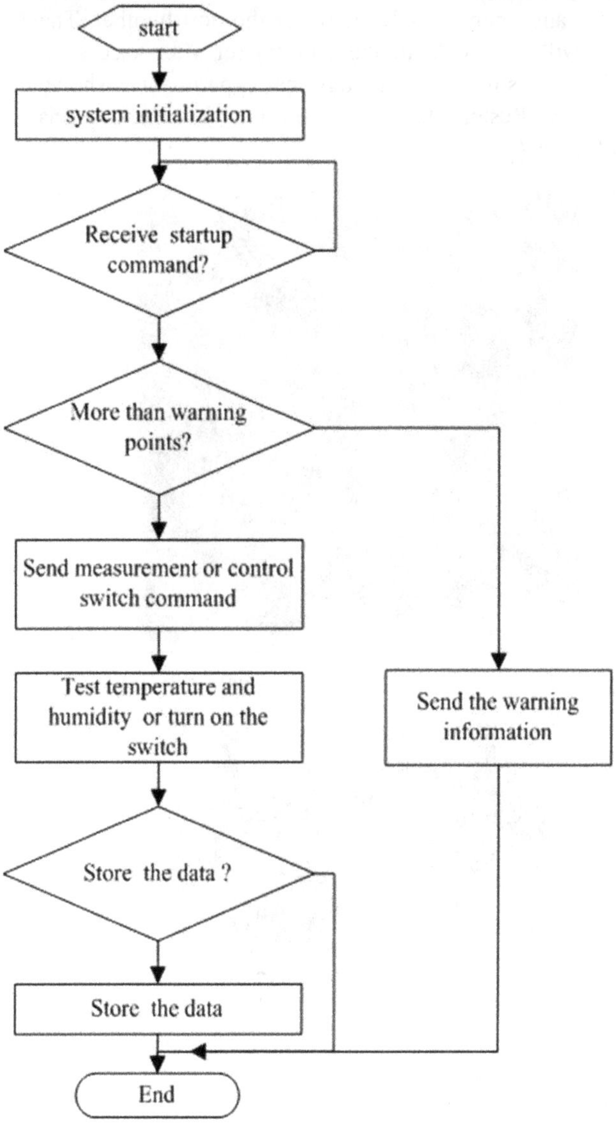

Figure 4. The flow chart of the system.

3.2. The GSM Module Software Design

The function of module is realize sending the temperature and humidity data to the users remotely in the form of short message. The design of GSM module software mainly includes the following two parts: 1, the GSM module initialization: (1) set up center address number (2) set text mode (3) text location Settings 2,the GSM module sends and receives messages. After powered on, the whole system initializates, serial of communication opens and codes inside the module, a text string generates, the information is send by corresponding AT command. Serial communication closes after sending information.

3.3. The ZigBee Module Software Design

The system uses the CC2430 chip of TI company, TI company provides a complete agreement in accordance with ZigBee protocol stack, that helps form the ZigBee network, the services and applications of the ZigBee agreement. Protocol stack is divided into four layers: the physical layer and MAC layer, network layer and application layer. Software design is mainly aimed at the application layer of the application of writing, calling the corresponding service to the operation of the underclass.

Coordinator for the wireless network also wants to ensure the normal operation of the network, data updating and the network maintenance, etc.

The terminal nodes as same as the coordinator node, need a corresponding initialization before use. After initialization, terminal nodes searches area network and sent the requests of join network to the coordinator. The Coordinator allows it to join the network. The terminal nodes acheive a temperature and humidity data immediately, send data to the coordinator and recycle into the next temperature and humidity collection. If the system sets low power, the terminal nodes can be done in a temperature and humidity data acquisition and send later into dormancy until the next again awakens when temperature and humidity data acquisition, and then enters the second data collection and send.

3.4. The Master Single Chip Microcomputer Software Design

Master microcontroller in the system completes communication between the GSM module and the ZigBee coordinator. Software design consider how to realize the communication with the GSM module and how to exchange data with ZigBee coordinator. According to the above, master Single Chip Microcomputer software design seperates the following two parts: 1, the software design of communication between the master Single Chip Microcomputer and TC35 module; 2, the software design of communication between the master Single Chip Microcomputer and ZigBee coordinator.

3.4.1. The Software Design of Communication Between the Master Single Chip Microcomputer and TC35 Module

In this system, the Single Chip Microcomputer communicates with ZigBee coordinator by simulate URAT with the interrupt method, because there is only one URAT on the singlechip used in communication with Short Messaging Service module. P3.5 of the master microcontroller is used as the simulation sender pin; P3.3 is used as the receiving pin; the format bits of the transmission data a are ten; eight bits in the middle is used to deposit information data.

The communication baud rate is set to 9600 BPS and the time required to transmit one bit data is about 0.1 ms. Crystal oscillator of Master microcontroller is 11.0592 MHz and the time required to transmit one bit data happens to be an integer instruction cycle. Calculation process is as follows

$$S = \frac{1000000/9600}{12/11.0592} = 6$$

When there is data communication between the master microcontroller and the coordinator, the terminal subprogram starts working to complete the communication simulate URAT.

Simulate URAT communication program includes a URAT initialization program, the URAT dispatch interrupt routine, the URAT reception interrupt routine and receiving program.

3.4.2. The Software Design of Communication Between the Master Single Chip Microcomputer and ZigBee Coordinator

The master microcontroller connects with GSM module through the serial port and parses command message of GSM module sent to ZigBee coordinator. Meanwhile, it also needs to send the temperature and humidity transmitted by ZigBee coordinator to GSM module, the flow as shown in Figure 5.

4. Test and Result Analysis

4.1. A serial Port Test Program and Single Chip Microcomputer Programs Load Debugging Experiment

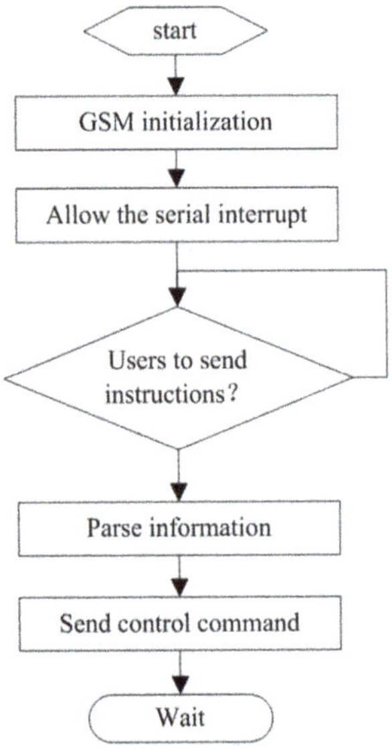

Figure 5. *Single Chip Microcomputer and GSM communication flow chart.*

First debugging TC35 module, the prepared SIM is inserted into card slot of TC35 and the TC35 is connected to the computer through the URAT. Then debugging software serially and dialling a number, if it is possible to get through, the test is successful. Loading the compiled program into the singlechip, running status of the program can be observed depending on the indicator light on development board.

4.2. ZigBee Module Serial Debugging Experiment

Experimental procedure: 1. Coordinator EB-Pro downloaded to the development board A is chosen as a coordinator to connect with the computer by USB cable; 2. End Device EB-Pro downloaded to the development board B is chosen as a terminal device to transmit data wirelessly to the coordinator and connect with the computer also by USB cable; 3. Electrify two development board, open the URAT debugging assistant, set communication parameters at 9600,8N1 and open the URAT, select the port number. The terminal will send data to the coordinator after successful networking. Result of ZigBee terminal experiment is shown in Figure 6. Result of ZigBee coordinator experiment is shown Figure 7.

Figure 6. *ZigBee terminal experimental results.*

Figure 7. *ZigBee coordinator experimental results.*

4.3. The Mobile Testing System

Program debugged successfully should be downloaded to the coordinator, terminal nodes and the master microcontroller. Insert SIM card and power on. When the first indicator on the TC35 module, the serial begins to connect. When the second light flashes, TC35 module begins to register. The third light flash means registration done. Mobile phone demands ZigBee coordinator "adding nodes", then diode circuit and DHT11 sensor circuit come into the net, the corresponding light of terminal nodes flash, remote monitoring can be achieved. First of all, control the diode. After electric starting system, another user mobile phone send "k1" to SIM card as shown in figure 8, then open relay, led shines, as shown in figure 9. In the same way, after send control commands "k0" to the SIM card, relay and diode shut down.

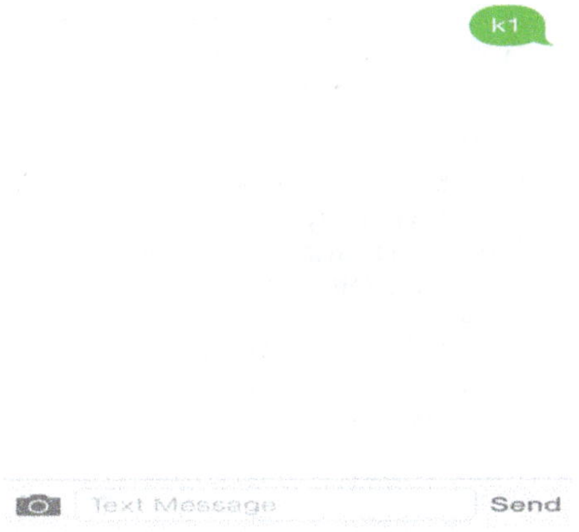

Figure 8. *Send command kl.*

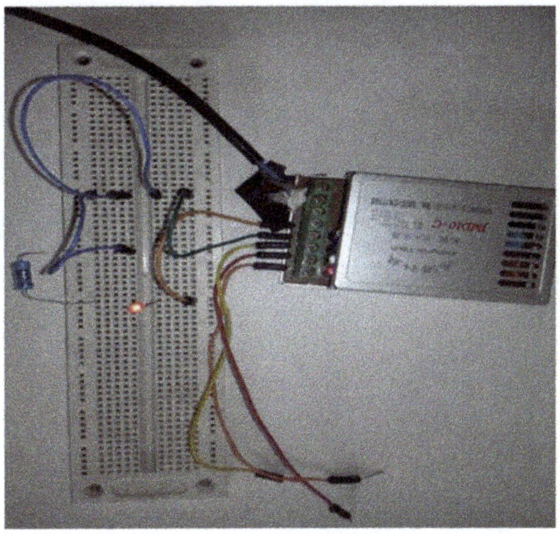

Figure 9. *Diode shine.*

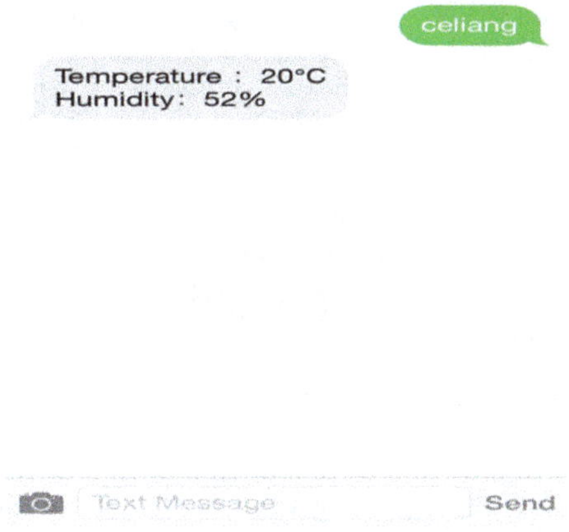

Figure 10. *Collect temperature and humidity.*

The indoor temperature and humidity collected by phone sending a short message should be sent back, as shown in Figure 10. After the test, mobile phone successfully collected indoor temperature and humidity and controlled the light of diode, meaning the system run successfully.

4.4. The Experiment Data Recording and Analysis

The indoor temperature and humidity collected by phone sending a short message and the indoor temperature and humidity measured by temperature and humidity meter should be recorded. The system measures per hour, totally six times. The indoor temperature and humidity collected by the system and the indoor temperature and humidity measured by temperature and humidity meter are shown in table 1.

Table 1. *Temperature and humidity collected by the system and by temperature and humidity meter.*

Measuring temperature and humidity of the time	Temperature of mobile phone reception (°C)	Relative humidity of mobile phone reception (RH %)	Temperature of temperature and humidity meter (°C)	Relative humidity of temperature and humidity meter (% RH)
8:00	24.5	63	24.7	64
9:00	25.3	62	25.6	63
10:00	26.0	61	26.4	60
11:00	27.1	56	27.6	57
12:00	28.6	55	29.0	54
13:00	29.9	50	30.4	49
Mean value	26.9	57.8	27.3	57.8

In conclusion, the temperature and humidity collected by this remote monitoring system of temperature and humidity based on GSM and ZigBee is consistent with the actual temperature and humidity. In terms of temperature monitoring, measurement results of this system compared with the actual temperature has absolute error of 1%. This result has great reference value.

5. Conclusion

Remote Monitoring System of Temperature and Humidity Based on GSM and ZigBee successfully implements communication between the indoor wireless sensor networks and mobile network., the indoor temperature and humidity remote monitoring and electric control. Compared with traditional cable data acquisition, the system is easy to form, saves space, needs low power consumption and convenient function extension, so it has a good application prospect.

References

[1] Liu C L. Design and Implementation of System Based on the Intelligent Household [D]. Jilin: jilin university, 2015.

[2] Shen C. Five of the bottleneck of the development of smart home [J]. China's public security, 2014, (8): 52-54.

[3] Ye J P. Research on Intelligent Monitoring System of Temperature and Humidity Based on GSM [D]. Xi'an: Xi'an University of Technology, 2009.

[4] Gao S W, Wu C Y,Yang C, et al. ZigBee Technology Practice Tutorial—— Scheme of Wireless Sensor Network Based on CC2430/31 [M]. Beijing: Beijing University of Aeronautics and Astronautics Press, 2009: 27-30.

[5] Liu Q L,Jiang J C,Yang. Application of Zigbee Wireless Sensor Network Technology on Natural Gas Multi-purpose Station [J]. Instrument Technique and Sensor, 2007 (01): 20-21.

[6] TI company database. TI. CC2430 Data Sheet (rev. 2.1) [DB/OL]. (2007-5-30) [2011-1-5]. http://www.ti.com/.

[7] Wang M Y. Temperature sensor DS18B20 applications [J]. Cotton processing technology, 2007 (6): 2-24.

[8] Yu Z H. The smart home control system based on ZigBee research [D]. Shanghai: Shanghai university, 2013.

[9] Shi J F, Zhong X X, Chen S, et al. Wireless sensor network structure and characteristics of the analysis [J]. Journal of chongqing university (natural science edition), 2005, 28 (2): 16-19.

[10] Xue Y H, Song B Y, Zhou F Y, et al. Building an Intelligent Home Space for Service Robot Based on Multi-Pattern Information Model and Wireless Sensor Networks [J]. Intelligent Control and Automation, 2012, 03: 90-97.

[11] Liang S, Hu Y, Wang K Z, et al. The landslide early warning system based on wireless sensor network (WSN) design [J]. Journal of sensors, 2010, 23 (8): 1184-1188.

An Entropy-Based MIMO Array Optimization for Short-Range UWB Imaging

Li Zhi[1], Zhang Jianwen[2], Shen Yu[1], Hu Jun[1]

[1]Department of Communication and Command, Chongqing Communication College, Chongqing, China
[2]Department of Clinical Medicine, Chongqing Medical and Pharmaceutical College, Chongqing, China

Email address:
lizhi_cs@126.com (Li Zhi), macsea@126.com (Zhang Jianwen)

Abstract: A novel approach to design the position of linear Multiple-Input Multiple-Output (MIMO) array elements for short-range UWB imaging is proposed. The proposed method uses Particle Swarm Optimization (PSO) algorithm to determine the optimal MIMO antenna array topologies that can provide minimum entropy of the reconstructed image. According to the approach, a MIMO array is optimized with minimum entropy indicating high focusing quality.

Keywords: Array Design, UWB, MIMO, Entropy, PSO

1. Introduction

As an emerging technology, short-range UWB imaging has the ability to penetrate through obstacles with high precision and low electromagnetic radiation. It has many applications, such as through wall imaging (TWI) [1], security inspection [2] and ground penetrating radar (GPR) [3]. By taking the advantages of MIMO technique [4, 5], short-range UWB imaging can reduce lots of antennas, which releases the total cost and weight of the sensor system.

In short-range UWB imaging application, array design is one of the key techniques, and it can influence target detection, parameters estimation and imaging performance such as the resolution and sidelobe level of the forming image. Usually, for imaging cases, resolution in the range direction can be achieved by transmitting wide band signals, and resolution in the cross-range direction can only be achieved by varying the illumination over the field of view of the sensor system. But for short-range application, the traditional pattern design method that based on plane wave is limited for the space-variance effect, so the imaging is determined by both wide band signals and short-range effect, which means that the resolution in range direction and cross-range direction is crossed. Thus, how to design a proper array transmitting specified signals (e.g. the frequency band for TWI is from 250 MHz to 3 GHz [6]) is on focus. In some experiments [7, 8], a uniform linear array is used to form the image. It is simple but not the best configuration. In some other papers, a random distributed array is discussed that it could reduce the effect of ghost images [9, 10]. But not all random patterns produce good results [9].

As the short-range effect and wide band property, it is difficult to design an optimized MIMO array. But on the other hand, the antenna array is used for high quality images, so it is direct and feasible to optimize array by the imaging performance. In this paper, the optimization of a linear MIMO array for short-range UWB imaging application is investigated. To find the best positions of antenna elements within the array, which provide desired resolution and sidelobe level, the PSO algorithm is used. By using PSO with proper optimization objective, which is the entropy of image that indicates the focusing quality and constraints on the fixed number of antenna elements and the probable maximal baseline, we can determine the proper antenna element positions.

This paper is organized as follows. The antenna array, imaging method and Point Spread Function (PSF), which is defined as the image distribution of a point scatter over the area of interest, are described in section 2. In section 3, the proposed approach is illustrated. In section 4, the optimized MIMO array and its corresponding results are shown. Finally, conclusions can be found in section 5.

2. Imaging

2.1. Antenna Array

In a conventional narrow-band radar system with a linear array, the cross-range resolution is solely determined by the size of the array aperture, and the sidelobe level is associated with the space between elements. When the number of elements is given, the conflict between the aperture size and the elements space corresponds to the conflict of the mainlobe width and the sidelobe level, which should be a trade-off. However, in UWB radar systems, the cross-range resolution is related not only to the baseline size, but also the bandwidth and the pulse shape of the system [11, 12, 13]. Hence, the cross-range resolution is a function of not only the array geometry, but also the spectrum of the radiated signal. Furthermore, the sidelobe level at range direction and cross range direction is also influenced by the pulse shape. As a result, the desired resolution may be achieved by modifying the transmitted pulse spectrum to achieve a certain PSF, as in [14], or by modifying the antenna topology, or both. In this paper, we only concentrate on the influence of antenna array topology to the PSF, so we fix the bandwidth and the pulse shape, and optimize the resolution by adjusting the antenna elements positions.

To obtain the image with real aperture requires large number of elements and multiple channels in transceiver. To avoid too much cost, electronics fabrication problems, and large data flow for real-time imaging, the antenna array must be thinned. However, aperture thinning generally causes the raise of sidelobe level. Thinned arrays have been designed and used as a means of obtaining maximum resolution with a minimal number of elements, using the coarray as a means of quantifying performance [11]. A good alternative array thinning can be MIMO approach. It was demonstrated in [12] and [15] that for a MIMO antenna array, the equivalent aperture is the spatial convolution of the transmitting antenna array aperture and the receiving antenna array aperture. What's more, each frequency responds to one equivalent aperture, which forms a disperse coarray [15, 16]. As a result, the antenna aperture size and element spacing of the effective array can be much larger and denser than those of the monostatic antenna array, providing better mainlobe and sidelobe control. The advantages of using MIMO antenna arrays are also demonstrated in [5]. So in our program, a MIMO array is applied.

2.2. Imaging Algorithm

In this paper, the image is formed by Back Projection (BP) algorithm, which can be calculated with high resolution in time domain and makes no geometrical approximations [17, 18]. BP algorithm is very suitable for short-range UWB imaging, such as TWI and GPR [4, 17].

If t denotes fast time, $\mathbf{x}_{tm}$ denotes the location of transmitter m, $m=1,...,M$, $\mathbf{x}_{rn}$ denotes the location of receiver n, $n=1,...,N$, and $s_{mn}(t)$ represents the signal transmitted by transmitter m and received by receiver n, then according to the BP algorithm the amplitude at the image pixel $\mathbf{x}_q$ is given by

$$\mathrm{I}(\mathbf{x}_q)=\sum_{m=1}^{M}\sum_{n=1}^{N} w_{rn}\cdot s_{mn}\left(\frac{\left\|\mathbf{x}_{tm}-\mathbf{x}_q\right\|+\left\|\mathbf{x}_q-\mathbf{x}_{rn}\right\|}{\upsilon}\right) \tag{1}$$

Where w_{rn} is the weight of receiver n, υ is the velocity of the electromagnetic wave in the ambient medium. The coherent sum is performed via time-shifting the signal obtained by each antenna pair and then adding across all antenna pairs to integrate the value at the image space pixel $\mathbf{x}_q$. This time shift and sum sequence is repeated for all the image space pixels. For an ideal point target, the calculated image is its PSF [19].

3. Methodology

Particle Swarm Optimization (PSO) is a biologically inspired computational search and optimization method developed in 1995 by Eberhart and Kennedy based on the social behaviors of birds flocking or fish schooling [20]. PSO has been used by many applications of several problems and has a great success. PSO is a sort of evolutionary algorithm utilizing swarm intelligence to achieve the goal of optimizing a specific fitness function, and it operates on a model of social interaction between independent particles [20]. PSO consists of a swarm of particles, where one particle represents a potential solution. Exploration is the ability of a search algorithm to explore different region of the search space in order to locate a good optimum. Exploitation, on the other hand, is the ability to concentrate the search around a promising area in order to refine a candidate solution [21]. With their exploration and exploitation, the particles of the swarm fly through hyperspace and have two essential reasoning capabilities: memory of their own best position -- local best, and knowledge of their neighborhood's best -- global best. With a velocity, each particle could change its position to explore in the solution space.

According to the PSO algorithm, the position $x_k(i)$ and velocity $v_k(i)$ of the particles are updated at each iteration i according to the velocity and position update equations given by

$$\begin{aligned} v_k(i+1) = v_k(i) + c_1\cdot\xi_1\cdot(localbest_k(i) - x_k(i)) \\ + c_2\cdot\xi_2\cdot(globalbest(i) - x_k(i)) \end{aligned} \tag{2}$$

$$x_k(i+1) = x_k(i) + v_k(i) \tag{3}$$

Where $k=1,2,...,K$, k is the total particle number in the swarm, $x_k(i)$ and velocity $v_k(i)$ are the position and velocity vectors, and $localbest_k(i)$ and $globalbest(i)$ represent the local and global best position in the swarm. ξ_1, ξ_2 is the random variable between 0 and 1, and c_1, c_2 is the optimization parameter which is dependent on the particle's movement and usually set to be 2, that is $c_1 = c_2 = 2$. In each dimension, the particle has a limited maximum velocity $V_{\max}$, and if the velocity exceeds $V_{\max}$, the velocity is set to be $V_{\max}$.

The entropy of a radar image is an indicator of its focusing quality. As the image is blurred, the uncertainty in the location and dimension of a target increases. In this paper, the Shannon entropy was calculated to measure the focusing quality. The best focusing quality is achieved when the entropy of the reconstructed image is minimized [22, 23], and the corresponding position of the array element is regarded as optimal.

The focusing quality of the image depends on the array element positions during the reconstruction process, and it can be used as a metric. This metric measures the level of uncertainty in a random variable. Let r be a discrete random variable with a probability density function $p(r)$. According to Shannon's definition [24], the entropy of r is given by

$$H = -\sum p(r)\log(p(r)) \tag{4}$$

The entropy of a digital image with W intensity levels is defined as

$$H = -\sum_{w=1}^{W} \frac{\psi_w}{\psi} \log(\frac{\psi_w}{\psi}) \tag{5}$$

Where ψ_w are the pixel number corresponding to the wth intensity level on the image and ψ is the total number of pixels in the image. The image that has the minimum entropy has the optimal focusing quality. The entropy of the image is used as the fitness function of PSO algorithm to find the corresponding array element positions with some specific constraints.

The proposed design method can be described as follows.

a) Set the parameters of the imaging scenery and array constraints, such as the element number of the MIMO array and the probable maximum length of the array.
b) Generate the initial population with random array element position $x_1, x_2, \ldots x_k, \ldots x_K$, where K is the size of the particle swarm. The particle k and the velocity at iteration i are denoted as $x_k(i)$ and $v_k(i)$. Here $i = 0$ at the beginning. Initially all of the particle velocity is assumed to be zero. Thus the particles has the initial status $x_k(0)$ and $v_k(0)$.
c) Evaluate the fitness function value for each particle as $f[x_1(i)], f[x_2(i)], \ldots f[x_k(i)], \ldots f[x_N(i)]$. The fitness function is based on the entropy of the image and $f[x_k(i)] = -H$, H is calculated by equation (5).
d) For each particle k, compare its fitness function value $f[x_k(i)]$ with $P_{best}(k)$, which is the best fitness function value of particle k at all previous iteration (Initialize as $P_{best}(k) = -\infty$). If $f[x_k(i)] > P_{best}(k)$, that means $f[x_k(i)]$ is better than $P_{best}(k)$, then let $P_{best}(k) = f[x_k(i)]$ and $localbest_k(i) = x_k(i)$.
e) For each particle k, compare its fitness value $f[x_k(i)]$ with G_{best}, which is the best fitness function value of all particles at all previous iteration(Initialize as $G_{best} = -\infty$). If $f[x_k(i)] > G_{best}$, that means $f[x_k(i)]$ is better than G_{best}, then let $G_{best} = f[x_k(i)]$ and $globalbest_k(i) = x_k(i)$.
f) Update $x_k(i),\ k = 1,2,\ldots,K$ and $v_k(i),\ k = 1,2,\ldots,K$ according to equation (2) and (3).
g) Check whether the current solution is convergent. This iteration process continues until all particles convergence the same solution or the iteration i reaches a specific number. The convergence is determined by the termination criteria that the difference between the current solution and the previous one is very small. If not convergent then step (c) is repeated by updating iteration $i = i + 1$.

4. Results

In the following short-range UWB imaging simulation, the configuration of a linear MIMO array is optimized by PSO algorithm with the corresponding image entropy as its cost function. The MIMO array has 3 transmitting elements and 10 receiving elements. The aperture of the linear MIMO array is limited to shorter than 2 meters. The transmitting signal is orthogonal stepped frequency continuous wave (SFCW) [25] with a center frequency of 1.5 GHz and a bandwidth of 2 GHz, and the signal duration is 1 μs. The imaging area is 8 meters along azimuth and 4 meters along range. An ideal point target at the center of the imaging area is used as the scatter, and the PSF formed by BP algorithm is used to calculate the entropy.

The optimization is executed on MATLAB platform. Fig. 1 shows the optimized MIMO array, in which red * represents transmitting antenna element, blue dot represents receiving antenna element. The topology indicates that the MIMO array has a transmitting array aperture expanding to the largest length as well as the receiving array aperture. This result is consistent with the coarray theory that the outer elements provide better resolution [13, 15]. Fig. 2 shows the corresponding PSF. Rather than a uniform array, the distributed elements provide sidelobe with low level. The entropy changes following the iteration as shown in Fig. 3. From the figure it can be seen that the entropy declines to lower and lower level, then close to steady. The focusing quality gets better and better following with the iteration.

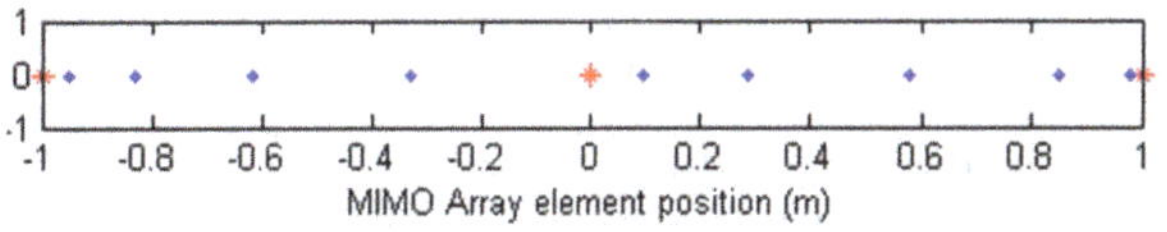

***Fig. 1.** the optimized MIMO array topology.*

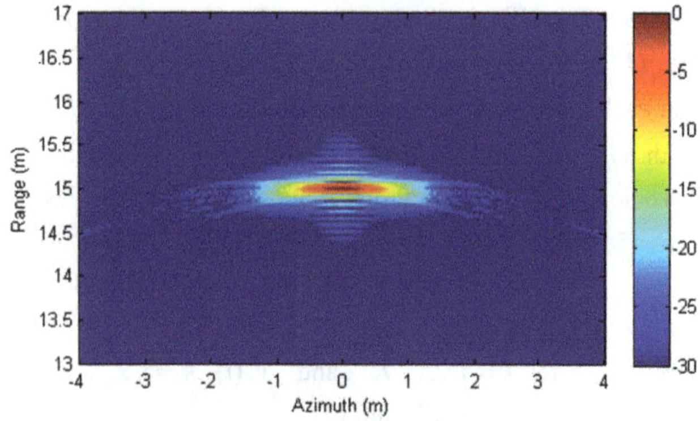

Fig. 2. PSF formed by the optimized MIMO array.

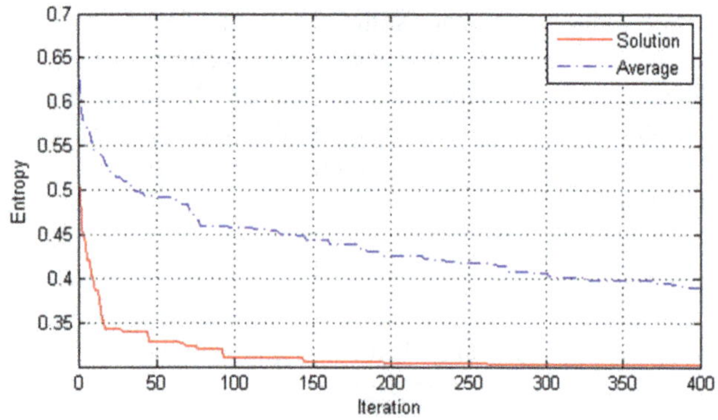

Fig. 3. Entropy changes following iteration.

5. Conclusion

A novel approach to design MIMO array element positions for short-range UWB imaging application is presented in this paper. The results show that the proposed approach has the potential to find better positions than random array. The proposed approach is available for short range application regardless of the space variance phenomenon. Moreover, rather than the Shannon's entropy, the approach could be also applied by changing the fitness function of the PSO due to the purpose of the imaging system. In 3D imaging cases, this approach could be used for plane MIMO array optimization.

References

[1] Edward J. Baranoski. Through-wall imaging Historical perspective and future directions [J]. Journal of the Franklin Institute. 2008, 345: 556-569.

[2] Stanley E. Borek. An Overview of Through the Wall Surveillance for Homeland Security [J]. Applied Imagery and Pattern Recognition Workshop, 2005.

[3] Laila Sakkila, Charles Tatkeu, Yassin El Hillali. UWB short range radar for road applications [J]. Physical and Chemical News, 2012, 64, p20-29.

[4] Zhuge Xiaodong. Short-range ultra-wideband imaging with multi-input multi-output arrays [D], Delft University of technology, 2010.

[5] A. Martinez-Vazquez, UWB MIMO Radar Arrays for Small Area Surveillance Applications [C], Proc. of 2nd EuCAP, 2007.

[6] L. Frazier, MDR for Law Enforcement [J], IEEE Potentials, Vol. 16, No. 5, pp. 23-26, 1998.

[7] Genyuan Wang, Imaging Through Unknown Walls Using Different Standoff Distances [J], IEEE Transactions on Signal Prossing, Vol 54, No. 10, Oct 2006, pp. 4015-4025.

[8] Francesco Soldovieri, A Multiarray Tomographic Approach for Through-Wall Imaging [J], IEEE Transactions on Geoscience and Remote Sensing, Vol, 46, No. 4, April 2008, pp. 1192-1199.

[9] Allan R. Hunt. Image Formation Through Walls Using a Distributed Radar Sensor Array [C]. Applied Imagery and Pattern Recognition Workshop, 2003.

[10] Calvin Le, Traian Dogaru, Lam Nguyen, Marc A. Ressler. Ultra-wideband (UWB) Radar Imaging of Building interior: Measurements and Predictions [C]. IEEE Transactions on Geoscience and Remote Sensing, 2009, vol. 47. 1409-1420.

[11] B. Yang, UWB MIMO Antenna Array Topology Design Using PSO for Through Dress Near-field Imaging [C], EuMA, Oct 2008, pp. 1620-1623.

[12] A. G. Yarovoy. Comparison of UWB Technologies for Human Being Detection with Radar [C], Proc. of 4th EuRAD, 2007, pp. 295-298.

[13] J. L. Schwartz, Ultrasparse, Ultrawideband Arrays, IEEE Trans. on Ultrasonics, Ferroelectrics, and Frequency Control [J], Vol. 45, No. 2, March 1998, pp. 376-393.

[14] M. Ciattaglia, Time Domain Synthesis of UWB Arrays [C], Proc. of 2nd EuCAP, 2007.

[15] Zhi Li, Tian Jin, Bo Chen, Zhimin Zhou. A coarray based MIMO array design method for UWB imaging [C]. IET International Radar Conference 2012, Glasgow, UK, 2012.

[16] Fauzia Ahmad, Saleem A. Kassam. Coarray analysis of the wide-band point spread function for active array imaging [J]. Signal Processing, Vol. 81, pp. 99-115, 2001.

[17] Cui G, Kong L, Yang J. A Back-Projection Algorithm to Stepped-Frequency Synthetic Aperture Through-the-Wall Radar Imaging [C]. Synthetic Aperture Radar, 2007. Apsar 2007. Asian and Pacific Conference on. 2007: 123 - 126.

[18] Chen A L, Wang D W, Ma X Y. An improved BP algorithm for high-resolution MIMO imaging radar [C].Audio Language and Image Processing (ICALIP), 2010 International Conference on. 2010:1663 - 1667.

[19] Ahmad F, Kassam S A. Coarray analysis of the wide-band point spread function for active array imaging [J]. Signal Processing, 2001, 81(1): 99-115.

[20] Dian Palupi Rini, Particle Swarm Optimization: Technique, System and Challenges [J]. International Journal of Computer Applications, vol. 14, No.1, January 2011.

[21] M. B. Ghalia, Particle Swarm Optimization with an Improved Exploration-Exploitation Balance [J], IEEE, vol. 3, 2008.

[22] J. Sok-Son,G. Thomas, Range-Doppler Radar Imaging and Motion Compensation [J], Artech House, Norwood, Mass, USA, 2001.

[23] Daniel Flores-Tapia, An Entropy-Based Propagation Speed Estimation Method for Near-Field Subsurface Radar Imaging [J], EURASIP Journal, 2010.

[24] Pun T. A New Method for Grey-Level Picture Thresholding Using the Entropy of the Histogram [J]. Signal Processing, 1980, 2(3): 223-237.

[25] Qu L, Yang T. Investigation of Air/Ground Reflection and Antenna Beamwidth for Compressive Sensing SFCW GPR Migration Imaging [J]. IEEE Transactions on Geoscience & Remote Sensing, 2012, 50(8): 3143-3149.

14

Implement of Face Recognition in Android Platform by Using Opencv and LBT Algorithm

Liela Khobanizad[1], Mahmood Khobanizad[2], Ahmad Houssien Bieg[3], Behrouz Vaseghi[2], Hamid Chegini[3]

[1]Telecommunication of Non-profit Institution of Higher Education, ABA, Abyek, Qazvin, Iran

[2]Electrical Engineering, Abhar Branch, Islamic Azad University, Abhar, Iran

[3]Non-profit Institution of Higher Education, ABA, Abyek, Qazvin, Iran

Email address:

L.khoubani@yahoo.com (L. Khobanizad), mahmoodkhob@yahoo.com (M. Khobanizad), hosseinbeig@yahoo.com (A. H. Bieg), Behrouz.vaseghi@yahoo.com (B. Vaseghi), hamidchegini26@gmail.com (H. Chegini)

Abstract: One way of consideration for identifying the human IS recognition of face by portable tools like mobile and tablet. One challenge is low power in portable android tools for face recognition (identification), so GPU must be used in software connection central Graphic processor which has a good function, compared to present processors in today portable android tools. Binary pattern (local) is one of the methods that are used for characteristic production and the image stratification. In this study, it is suggested to use connection and local binary pattern histogram algorithm to use optimum software open CV and using hardware platform android to identify the face.

Keywords: Face Recognition, Opencv, Android, LBT Algorithm

1. Introduction

Todays, recognition from face has a lot of utilities in commerce and security. The most information is gained from the face identification. To recognize the face, the entrance (input) image is identified and theexist information is also considered. Theexist information in The IB (In formation Bank) contains the known persons image characteristics. Face recognition has a lot of advantages in misdemeanant identification, security systems and other Issues, and so it is taken into account during the recent years. One of the technics for examining to identify the human being is to recognition of faces by portable equipment's.

Recognition is the usual action of human beings' every day job. Increasing of portable tools like computer and mobile causes more consideration to automatic processing on the images including biometric identification, recognition, human interaction and computer and multimedia management. For this reason, searches and developments have been conducting to recognize the faces. Face recognition is prior to other technics like, finger printing and iris. Besides being natural and obscure, the most important advantage of this recognition of face is that the image can be taken or covered from every distance.

Recognition has an important role in photographing, conserving a lot of volume of images in memory or web and increasing the security. One of the examining technics for human identification is face recognition with portable devices. Identification recognition from face image has developed during the recent years and is accompanied by the other devices like phone, mobile and tablet which have intelligent operator system. One the supported smart operation system is android in our age. Android is an open source system based on Linux kernel and Java programing language. Android 3rd part application use Java language and for communication with underlay they can use Java for program.

2. Face Recognition

One of the challenges of face recognition in android potables tools is low of processing ability in GPU, so the software conjecture should be used, because it has a good function. Regarding the exist processors in today android portable tools and the used connector and our kernel process in this study for processing the image in opencv is written in Java language and android compiler can know and use it.

The main applications of face recognition are in table 1. For each group, the sample applications are also listed. Our application of recognition system of individual identification recognition background is by using the portable android tools which are similar to mobile.

Table 1. *Main applications of face recognition.*

Driving ID card, immigration, national code passport, the voter's registration, people recognition and finding their files.	Face ID
Car accessibility, ATM and intelligent booth, computer and online accessibility and accessibility of online exams.	Control access
Flying security system, stadium fan scanning, computer security, files codding	Surveillance
Conserved value security, users	Smart card
Suspects warning and preventing crime, prosecution and suspect files survey, recognition and exploitation of crime face	Low Enforcement
Faces marking and Images gaining, faces classification automatic labeling	Face data base
Researching based on face, video segregating based on face identification	Multimedia management
Interaction games, Interaction computation	Human and computer interaction

Much software has been developed during the recent years to facilitate the processing applications, and the most famous ones are image processing tools boxes in Matlab. However, people who have experienced this tool believe that fulfillment speed of Matlabare annoyer and needs to a powerful processor. It is not an open source either.

2.1. Definition of Open CV

Machine vision is one of artificial intelligence that is searching for problem solution which human cannot do it by his/her vision and information analysis. Vision data are prepared in a machine by camera. Different information exploitation of the data and its analysis is the duty of machine vision science. The machine vision can be used everywhere that is necessary and can see instead of human. Using the camera and decision of camera can improve many of the situations that are difficult for human, including individual image (face) identification. Use the machines and camera influences in decreasing of cost fouls and development of system function. The machine vision can be considered one of the middle tools of different sciences. This branch of science has a close relationship with conceptions like image processing and/ or video processing. In many cases it is difficult to distinguish the two cases, however we can call the image processing technic the machine vision science or MVS.

Open source of a vision machine it is an open source library for computer vision that has the following address:

OpenCVhttp://sourceforge.net/projects/opencvlibrary/files/opencv-android

The program is written in Java and C++C and under Linux, windows, Macintosh is applicable. It also is active for means such as Matlab, Ruby and Python.

The aim of opener designing is processing particularly for prompt affairs. It can be used with multichannel processors. One aim of open CV is preparing a base of computer vision with simple applications, so that the individuals can make relatively complex vision programs. The open CV library includes more than 500 functions about different vision subjects.

Android: simply it is an operating system for mobiles and intelligent receivers which is supported by more than 30 companies. Android is an open source and the developers can write different programs by using SDK Android.

Android programming set or SDK Android including; Android libraries, stimulator, Android documents, and sample files and trading files which help the users to create programs. Now, this SDR is implemented on system with 32 bit or 64 bit which has Linux, windows or OSX mac. The written programs for android are reserved by apk suffix. During the recent years, LBPs display increasingly interest in image processing and computer vision. As a non-parametric technic, LBP, will pluralize the local structures so that compared to Pixel, they are better. The main charter is tics of LBP are:

Tolerance of lighting groovy changes and its simple calculations. Originally, LBP is used for analyzing of suggested contexture and a simple but strong technic for local structures definition.

It is used in applied programs such as face picture analysis, video recovery, environment modeling, vision and optical recovery, moving analysis, air image analysis, biomedical and distance assessment.

Face image analysis based on LBP is one of the most famous and successful application in recent years. It is a research topic and active in computer vision and has a lot of important application such as, interaction of human-computer, biometric recognition, security and surveillance and computer animation. LBP for image display is used in, image identification, image recognition, face situation (position) analysis, demographic classification (sex, race, age ...). The main LBP operator codes the image pixels with decimal and labels them which is called LBPs and then codes the local structure around each pixel. As it is shown in figure (1), the process is as follows:

Patch pixel is neighbored with eight, through fraction of central pixel, which can be compared with 3; the values that are strictly negative are coded zero or one for each pixel, B is used. To consider the tissue of different sizes, the next operators will be extended and use the neighborhood with different sizes.

6	5	2
7	6	1
9	8	7

1	0	0
1		0
1	1	1

1	2	4
128		8
64	32	16

Pattern = 11110001 **LBP** = 1 + 16 +32 + 64 + 128 = 241

Figure 1. *An example of basic operators.*

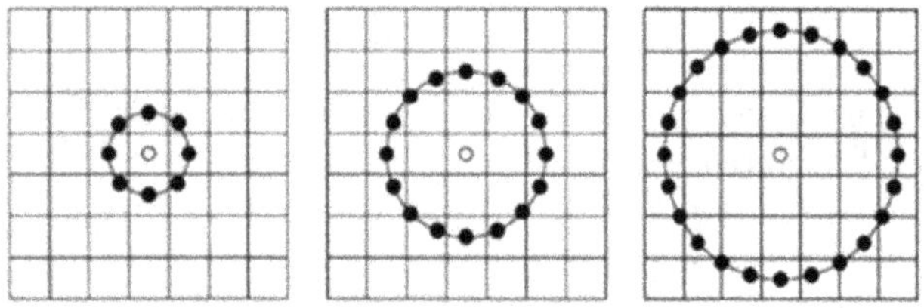

Figure 2. *Examples of LBP operator, circular neighbors (3,24), (8,1), (16,2).*

The local neighborhood, which $T = t(n_c, n_0, n_1, n_2, \dots n_{P-1})$ are a set of sample points with equal distance on the circle, and its center is labeled in pixel and the points that are not in these pixels, by using two ways in patting, are inputted and by this method it is possible to prepare each radius and sampling points near the neighborhood. Figure (2) shows some examples of LBP developed operator. In this figure P, R shows the P neighborhood on the circle with R radius. The neighbor pixels in LBP are defined on one $LBP_{P,R}(x_c, y_c) = \sum_{P=0}^{P-1} s(i_p - i_c)2^P$ circle LBP (X_c‘Y_c)

The main reason for this neighborhood definition is unchangeable evolution and description of tissue and sample. Generally, the result LBP (X_c, Y_c) can be explained as follows.

Where i_c, and i_p are the central pixel grizzly and p is around pixel with R ridus and s(x) is defined as follows:

$$xp = xc + Rcos(2\pi p/P), \quad s(x) = \{1 : x \geq 0$$

$$yp = yc + Rsin(2\pi p/P). \quad \{0 ; x < 0.$$

If the central pixel coordinates are (x_c, y_c) and the neighbor points the circle are (x_p, y_p), so by using these formulas, we can conclude them.

$$LBP^{n}_{P,R} = \min\{ROP(LBR_{P,R}, i)|, \quad i = 0,1, \dots, P-1\}$$

If the grizzly pixel value of center is n_c and the neighbor pixel grizzly is n_p and i=0…P-1, so the T tissue in pixel local neighborhood (x_c, y_c) can be defined as equation shows.

$T{=}t(n_c, n_0, n_1, n_2, \dots, n_{P-1})$;

$T{=}t(n_c, n_0-n_c, n_1-n_c, n_2-n_c, \dots, n_{P-1}-n_c)$; step 1

$T \approx t(n_c, n_0-n_c, n_1-n_c, n_2-n_c, \dots, n_{P-1}-n_c)$; step 2

$T \approx t((n_0-n_c), (-n_1-n_c), (n_2,-n_c) \dots (n_{P-1}-n_c))$; step 3

T

By using step 0. we can access the following stages.

As $t(n_c)$ defines the total brightness of the image and (n_i-n_c) is not defended on n_c, and $t(n_c)$ is not connected to the local image and does not have helpful into for tissue analysis, so we can omit it and the changes between enteral pixel and neighbor pixel is enough.

According to previous definition, LBP base operator is fixed in front of equal transmission of fixed grizzly sizes this process shows that the pixel intensity next to local pixel can be fixed. LBP labels histogram can use as tissue and template descriptor.

The $LBP_{(P,R)}$, s(x)={1;x≥0 {1;x<0 2^P produces different output values Which is like different 2^P composed by pixel P, and is in the format of neighborhood. In case of image rotation, the round pixels in each neighborhood will move along with the circle and result in different value of LBP. Of course, there is an exception and that is the templates with zero and 1. To neutralize the rotation, it is suggested to use unchangeable LBP against the rotation.

Where ROR is shift or location change bit by bit $LBP^{ri}_{(P,R)}$ is unchangeable templates occurrence statistic against the individual rotation which is similar special micro characters. However, it is shown that this operator cannot provide the segregation info necessarily. The reason the number of individual templates occurrence in $LBP^{ri}_{(P,R)}$ is changeable.

It is proved that the special templates have more info. Compared to the others it is possible to use the subset 2P for tissue and image template. There template are called invariant ones and are shown as $LBP^{ri}_{(P,R)}$.

A LBP is called uneventful in case of having two buses and bit to bit transmission from zero to 1 or vice versa. For example, 00000001 (bus zero) and 201110000 (bus) are uneventful, while 411001001 (bus) and 601110000 (bus) are not uneventful. It is observed that uneventful templates next to (8,1) are 90% of total templates and next to (16,2) in tissue images are 70% recently Shan and Gritti have surveyed uneventful templates validity and confirmed them. Especially, they used the face situations and Adaboost and their tests showed that by using LBP (8,2) operator 91/1% of the selected templates are uneventful. The integration of uneventful templates in a small store of LBP operator is less than 2P label. For example, the number of labels with pixel 8 neighborhood for LBP standard is 256 and for LBP^{U2} is 59.

The aim of f.o., is determine the position and size of face in digital images. Hadid et al, for the first time, used LBP to identify the face. To describe the faces with low resolution and clarification, the LBP operator (4,1) was used for overlapping small area. The SVM was used to discriminate the´faces and non-faces. Next, they suggested a hybrid technic to identify unlimited areas. Their technic was studied for main area of skin in an input image to prevent total image scanning.

Then the big strategy to small strategy was used to identify to be face or not to be face in scanned areas. In big step, LBP was used, but in the second one Zhao, Zhang was used.

LBP is also used as preprocessing technic on images. For example, Heusch and others used the LBP as preprocessing step to omit brightness effects. (fig. 3)

One advantage of LBP is that it can make the prompt image face analysis systems structure very attractive. Besides, it can enforce the hardware depended on design to calculate the LBP with high speed.

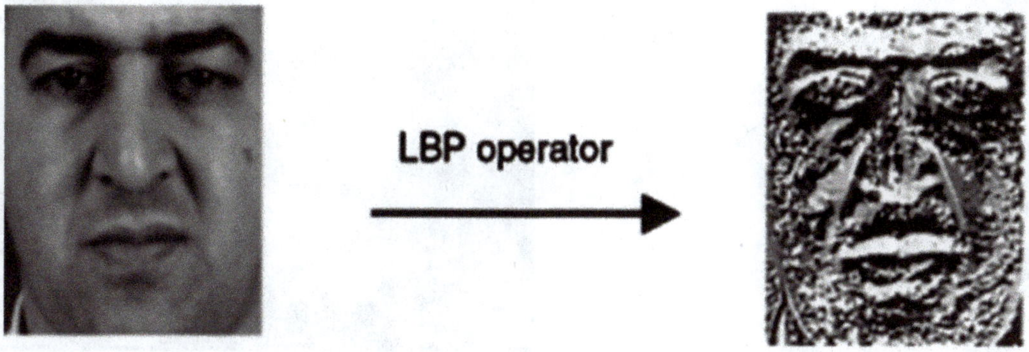

Figure 3. *The main image (Left) is processed by LBP (Right).*

Hadid et al, by using the F.O. based on LBP, could make a control system. In their system, a camera is on the door to take photos of video frames. In recent years, LBP could attracted many people and it could shows its effect on applied programs, especially face ID analysis, face ID position and demographic classification.

2.2. Finding

Today, the word "application" (App) is more usual then "software" for portable devices. The word App has the same meaning as "applied software". The android compilers output can create application with APK suffix, and android operator system can read this suffix. To use this application it should be installed in android tablet or mobile. It is used to face ID. The name of this necessary is manager open CV. If it is not installed, the application (we created) is not practicable. If it is not installed, the software will tell you that you should download this file from Google market and install it. The name of this file is Google play. Google play is a digital distribution service which contains multimedia and has an online market for music, film, book and android games and applications.

Update version of open CV manager is in the following link:

https://play.google.com/store/apps/details?id=org.opencv.engine&hl=en

There is another way of installing the prerequisite application, which is defined in the following like:

http://sourceforge.net/projects/opencvlibrary/files/opencv-android

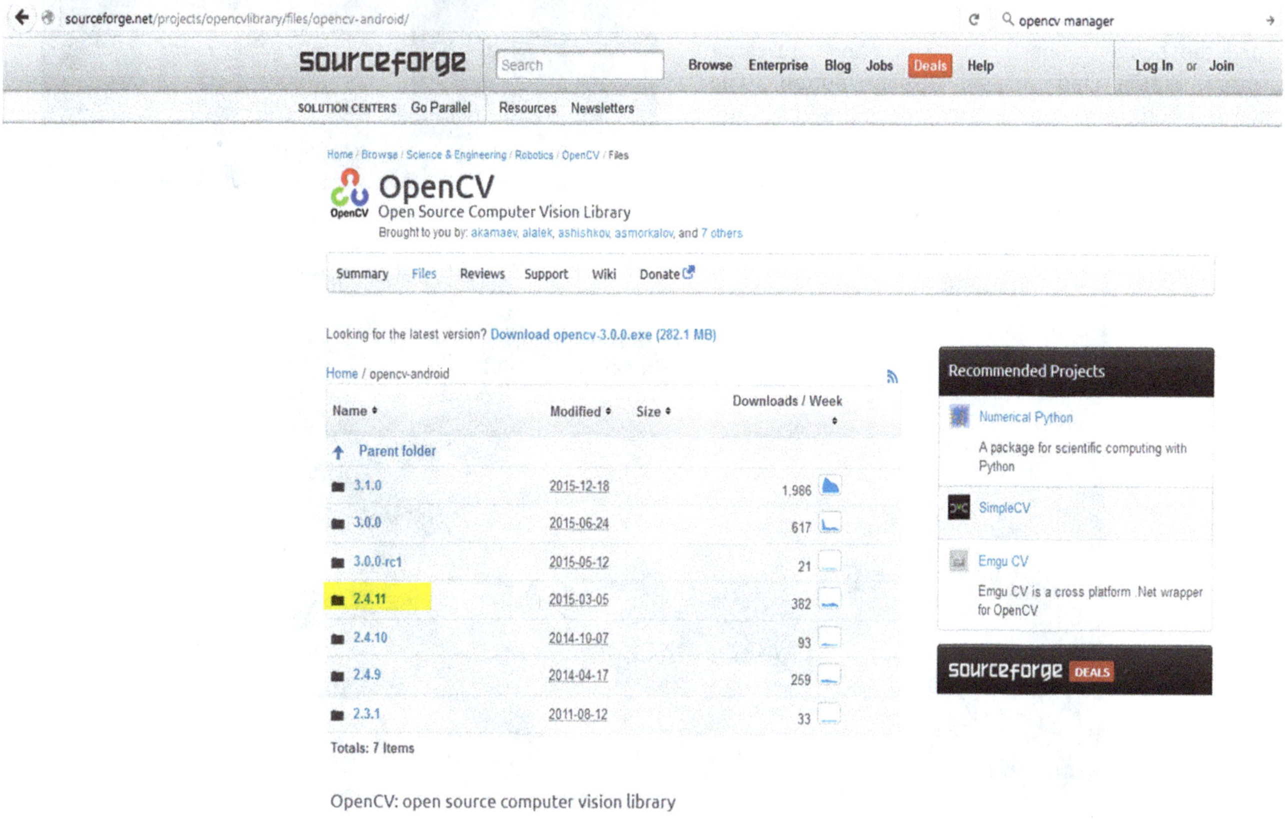

Figure 4. *Open CV 2.4.11.version.*

2.3. F.ID Process by Using Produce Application

After implementation of this Software, we are faced to primary page which is shown in fig. (6).

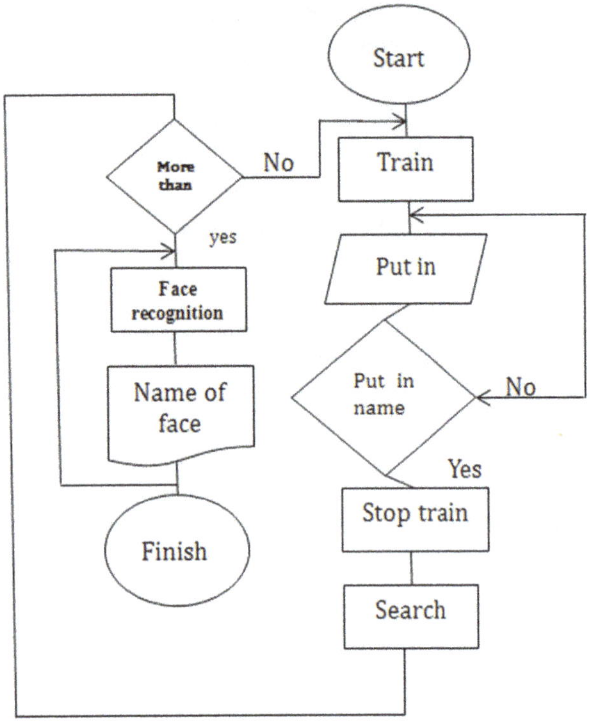

Figure 5. The overall process of implementation of face recognition application chart.

In this page (plane) we are faced to: camera deploys frame, button, and the behind camera change or phone, research button, train button and images display button. The application and buttons process is discussed in the following. At first we introduce the individual's images to the software by touching the train button. We write the name of the image and then touch the record button for server times and then save the names. At the end we touch the stop button.

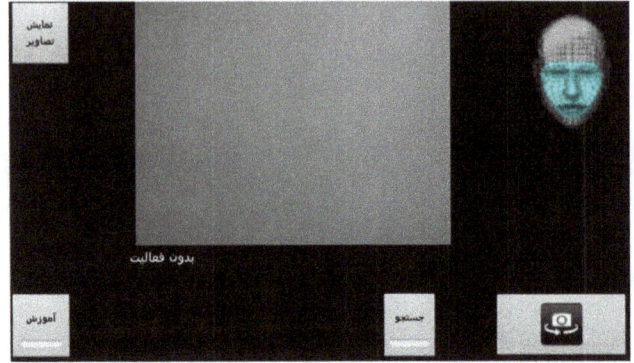

Figure 6. The initial (primary) program page after installing the application.

In this application we need more than one image for recognition. At least, we need two images to train and then by using search button, the application will start recognition. We used binary template histogram algorithm to exploit the characters. If the exploited characteristics from image are homologous with recognized face image, the name of that image will be displayed, infant the application will give us on unknown message (as shown below).

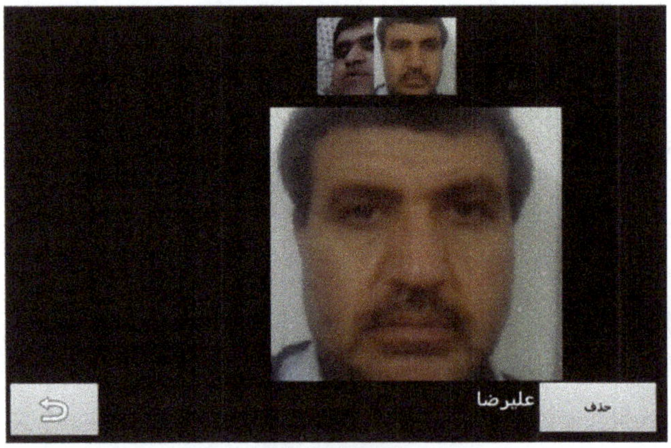

Figure 7. Saving data.

Figure 8. Individuals recognition and displaying their names.

It is also possible to recognize the face in little light due to using LBPH algorithm. LBPH algorithm has a good function against groovy (monotonous) light and can recognize the individual (as shown in fig. 9).

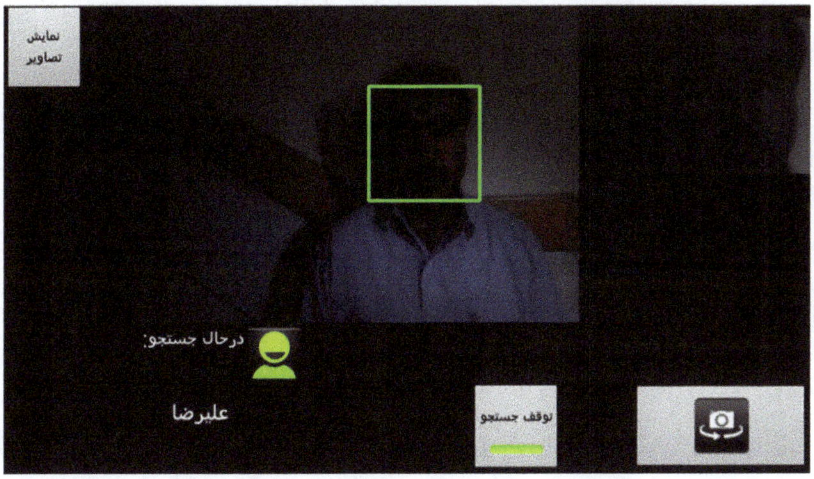

Figure 9. *Face ID in little light and displaying the name.*

3. Conclusion

Face ID is considered for researches connected to machine vision. It is also used in commercial and security applications. The applications in clouding:

Individual's security control, access control, criminal people recognition and interim are between human and computer. Recognition by F. ID is considered by lot of people, due to little necessary contribution of people and agreement. Today, the following device has provided a lot of possibilities for this target: intelligent phone, economical bases, application installation and open android operator system. In this study, we use the open CV as a processing kernel for F.ID, due to hardware limitation in android portable devices and low process ability in graphic processor or GPU. This technic has an appropriate function although there is limitation in today portable devices. The aim of this study is produce an application for F. ID in android platform by using Eclipse open CV and binary template histogram algorithm. Being info fixed against the light changes, we used a tissue powerful descriptor named LBPh algorithm. LBPh descriptor (delineator) is a powerful device to display the local structures and due to its simplicity of calculation, it is used for analyzing the F.ID.

References

[1] W. Zhao, R. Chellappa, P. J. Phillips and A. Rosenfeld, "*Face recognition: A literature survey Acm Computing Surveys (CSUR)*", Vol. 35, no. 4, pp. 399-458, 2003.

[2] P. Belhumeur, J. P. Lanfang, and D. Kriegman, "*Eigenfacesvs. fisherfaces: Recognition using class specific linear projection* ", Pattern Analysis and Machine Intelligence, IEEE Transactions on, Vol. 19, no. 7, pp. 711-720, March. 1997.

[3] M. Sharkas, and M. A. Elenien, *"Eigenfaces vs. Fisherfaces vs. ICA for facerecognition; a comparative study,"* Signal Processing, 2008. ICSP 2008.9thInternational Conference on. IEEE, pp. 914–919, Oct. 2008.

[4] A. Ozdil, and M. M. Ozbilen, "*A Survey on Comparison of Face Recognition Algorithms*," Application of Information and Communication Technologies (AICT), 2014 IEEE 8th International Conference on, pp. 1–3, Oct. 2014.

[5] X. Lu, "*Image Analysis for Face Recognition*," http://www.facerec.org/interest.ingpapers/General/ImAna4FacRcg_lu.pdf

[6] Gonzalez, Rafael C., and Richard E. Woods, "*Digital image processing*," (2002).

[7] W. W. Bledsoe. The model method in facial recognition. Technical report pri 15, Panoramic Research, Inc., Palo Alto, California, 1964.

[8] W. W. Bledsoe. Man-machine facial recognition: Report on a largescale experiment. Technical report pri 22, Panoramic Research, Inc,. Palo Alto, California, 1966.

[9] W. W. Bledsoe. Some results on multicategorypatten recognition. Journal of the Association for Computing Machinery, 13(2): 304–316, 1966.

[10] W. W. Bledsoe. Semiautomatic facial recognition. Technical rep ort sri project 6693, Stanford Research Institute, Menlo Park, California, 1968.

[11] W. W. Bledsoe and H. Chan. A man-machine facial recognition systemsome preliminary results. Technical report pri 19a, Panoramic Research, Inc., Palo Alto, California, 1965.

[12] M. Fischler and R. Elschlager. The representation and matching of pictorial structures. IEEE Transactions on Computers, C-22(1):67–92, 1973.

[13] T. J. Stonham. Practical face recognition and verification with wisard. In H. D. Ellis, editor, Aspects of face processing. Kluwer Academic Publishers, 1986.

[14] M. Turk and A. Pentland.Eigenfaces for recognition. Journal of Cognitive Neurosicence, 3(1): 71–86, 1991.

[15] M. McWhertor. "*sony spills more ps3 motion controller details to devs*". Kotaku. Gawker Media., June 19 2009. http://kotaku.com/5297265/sony-spills-more-ps3-motion-controllerdetails-to-devs

[16] Z. Pan, A. G. Rust, and H. Bolouri, "*Image Redundancy Reduction for Neural Network Classification Using Discrete Cosine Transforms*," Proc. Int. Joint Conf. on Neural Networks, Vol. 3, (Como, Italy), pp. 149-154, 2000.

[17] J. K. Sing, D. K. Basu, M. Nasipuri and M. Kundu, "*Face Recognition Using Point Symmetry Distance Based RBF Network,*" Jour. of Applied Soft Computing, 2005.

[18] M. J. Er, S. Wu, J. Lu and H. L. Toh, "*Face Recognition with Radial Basis Function (RBF) Neural Networks,*" IEEE Trans. on Neural Networks, Vol. 3, No. 3, pp. 697-710, 2002.

[19] Kirby, M., and Sirovich, L, "*Application of the Karhunen-Loeve procedure forthe characterization of human faces,*" IEEE PAMI, Vol. 12 ,pp. 103-108, (1990).

[20] Sirovich, L., and Kirby, M., "*Low-dimensional procedure for thecharacterization of human faces,*" J. pp. 519-524, (1987). Opt. Soc. Am. A, 4, 3.

[21] H. Anton, Elementary Linear Algebra 5e, John Wiley & Son Inc, 1987.

[22] M. Turk and A. Pentland, "*Face Recognition Using Eigenfaces*", Proc. IEEE Conf. on Computer Vision and Pattern Recognition, 1991, pp. 586-591.

[23] Raudys and A. K. Jain. Small sample size effects in statistical pattern recognition: Recommendations for practitioneers. - IEEE Transactions on Pattern Analysis and Machine Intelligence 13, 3 (1991), 252-264.

[24] T. Ojala, M. Pietikainen and T. T. Maenpaa .Multiresolution gray-scale and rotation invariant textureclassification with local binary pattern. IEEE Transactionson Pattern Analysis and Machine Intelligence. 24(7): 971-987, 2002.

[25] Gary, Bradski., and Adrian. Kaehler. "*Learning Open CV.*" (2008).

[26] J. F. Dimarzio "*Android a programmers guide.*" (2008).

Performance Evaluation of Hybrid Precoded Millimeter Wave Wireless Communication System on Color Image Transmission

Sk. Shifatul Islam[1], Shammi Farhana Islam[2], Mahmudul Haque Kafi[1], Shaikh Enayet Ullah[1]

[1]Department of Applied Physics and Electronic Engineering, University of Rajshahi, Rajshahi, Bangladesh

[2]Department of Material Science and Engineering, University of Rajshahi, Rajshahi, Bangladesh

Email address:

chisty56@gmail.com (Sk. S. Islam), shammi.farhana@gmail.com (S. F. Islam), mahmudkafi49@gmail.com (M. H. Kafi), enayet_apee@ru.ac.bd (S. E. Ullah)

Abstract: In this paper, a comprehensive study has been on the suitability of implementation of hybrid precoding scheme in performance analysis of the future generation wireless communication system. The 256-by-32 multi antenna supported simulated system incorporates various types of modern and classical channel coding schemes such as Low density parity check (LDPC), Repeat and Accumulate (RA), ½-rated convolutional and non-binary Bose-Chadhuri-Hocquenghem (BCH) and Zero-Forcing (ZF) signal detection technique. With consideration of ray path geometry based mmWave MIMO fading channel and properly designed precoders and combiners and their applicability in simulation works, it is seen from computer simulation results that the presently considered simulated system outperforms in retrieving color image in LDPC channel coding and 16QAM digital modulation and Zero-Forcing (ZF) signal detection schemes.

Keywords: Hybrid Precoding, Channel Coding, ZF, mmWave and Bit Error Rate (BER)

1. Introduction

The millimeter wave (mmWave) with frequency spectrum band ranging from 30GHzto300 GHz is expected to be a key component in the next generation 5G wireless communication systems. It enables the extensive use of allocated frequency spectrum to support greater data traffic for various multimedia services such as broadband mobile and backhaul services. The mmWave spectrum holds tremendous potential for providing multi-Gigabits-per-second data rates in upcoming cellular systems. One of the fundamental goals for 5G wireless mobile networks is to increase data rates through extreme densification of base stations with massive multiple-input-multiple-output (MIMO). In mmWave frequency bands, it has become a challenging task to execute cellular communication properly due to blockage, absorption, diffraction and penetration of mmWaves. However, advancement of CMOS radio-frequency technology along with the very small wavelength of mmWave signals allows for the packing of large–scale antenna arrays at both the transmit and receive ends and thus provides highly directional beam forming gains with reduced interference and acceptable signal-to-noise ratio (SNR) [1, 2]. In perspective of considering the potential of using of millimeter wave (mmWave) frequency for future 5G wireless cellular communication systems, an emphasis is being given on the study of large-scale antenna arrays for achieving highly directional beamforming. The conventional fully digital beamforming methods which require one radio frequency (RF) chain per antenna element is not viable for large-scale antenna arrays due to the high cost and high power consumption of RF chain components in high frequencies. To address the challenge of these hardware constraints, a hybrid beamforming architecture can be considered in which the overall beamformer consists of a low-dimensional digital beamformer followed by an RF beamformer implemented using analog phase shifters [3].

It is known from literature reviewing that Samsung

Electronics, an industry leader in exploring mmWave bands for mobile communications, has tested a technology that can achieve 2 Gbps data rate with 1 km range in an urban environment. Furthermore, Professor Theodore Rappaport and his research team at the Polytechnic Institute of New York University have demonstrated that mobile communications at 28 GHz in a dense urban environment such as Manhattan, NY, is feasible with a cell size of 200 m using two 25 dBi antennas, one at the BS and the other at the UE, which is readily achievable using array antennas and the beamforming technique [4].

The present study has been confined on the performance evaluative study of the simulated large-scale antenna mmWave system under consideration of hybrid beamforming structures.

2. Signal Processing Techniques

In this section, various signal processing techniques used for channel coding, geometry based fading channel estimation, hybrid precoding designing and signal detection have been outlined below.

2.1. Bose-Chadhuri-Hocquenghem (BCH) Channel Coding

BCH codes are a class of cyclic codes discovered in 1959 by Hocquenghem and independently in 1960 by Bose and Ray-Chaudhuri. The BCH codes are both binary and multi level The binary BCH code is parameterized by an integer $m\geq 3$. The t error correcting BCH code of length n is of varying nature depending on the value of m and is given by $n= 2^m-1$. Its roots include2t consecutive powers of α, the primitive element of $GF(2^m)$ [5]. In our study, a binary BCH code is of length 127 with a message is of length 64 have been used. In such [127, 64] BCH code, the value of error-correction capability, t is 10.

2.2. Low Density Parity-Check Matrix (LDPC) Channel Coding

LDPC code is a linear error correction code. Its parity check matrix H used in this paper is of 64 x 128 sized and this matrix is sparse containing less non zero elements irregularly in each row and column. The number of non zero element in each column ranges from 1 to 3and the number of non zero element in each row ranges from 5 to 6.The ½.-rated irregular LDPC code used here has a code length of 128bits.The parity-check matrix His formed from a concatenation of two matrices A and P, each with a dimension of64×64). The columns of the parity-check matrix H is rearranged to produce a modified form of parity-check matrix $\bar{H}$. With rearranged matrix elements, the matrix A becomes non-singular and it is further processed to undergo LU decomposition. The parity bits sequence p is considered to have been produced from a block based input binary data sequence $u=[u_1u_2u_3u_4.......u_{64}]^T$ and three matrices L,U and P(of $\bar{H}$) using the following Matlab notation:

p = mod (U\(L\z), 2);where, z = mod(P*u, 2);

The LDPC encoded 128×1 sized block based binary data sequencec is formulated from concatenation of parity check bit p and information bit u as: [c]=[p;u], The first 64bits of the codeword matrix [c]are the parity bits and the last 64bitsarethe information bits . The Log Domain Sum-Product LDPC is a soft decision decoding algorithm operating alternatively on the bit nodes and the check nodes through the Tanner graph. In such scheme, the received bits (0/1) are primarily converted into -1/+1 and assumed to be corrupted with AWGN channel noise of varianceσ^2 (=N0/2), N0 is the noise power spectral density. In processing, various required parameter values are compute diteratively with a view to finding out the mostly acceptable code words that satisfies the conditionc $\bar{H}^T=0$ [6, 7].

2.3. Repeat and Accumulate Channel Coding

The RA is a powerful modern error-correcting channel coding scheme. In such scheme, all the extracted binary bits from the color image has been arranged into a single block and the binary bits of the such block is repeated 2 times and rearranged into a single block containing binary data which is double of the number of input binary data[8].

2.4. Convolutional Channel Coding

In Convolutional Channel Coding, Convolutional codes are commonly specified by three parameters (n,k,m): n = number of output bits; k = number of input bits; m = number of memory registers. The quantity k/n called the code rate and it is a measure of the efficiency of the code.

The constraint length L(=k (m-1)) represents the number of bits in the encoder memory that affect the generation of the *n* output bits. Our presently considered Convolutional Channel scheme is specified with a coding rate of ½ and a constraint length of 7. The code generator polynomials G1and G2are171 and 133 in octal numbering system and can be written as [9]:

$$G1= x^0+x^2+ x^3+x^5+ x^6=1\ 0\ 1\ 1\ 0\ 1\ 1=133 \qquad (1)$$

$$G2= x^0+x^1+ x^2+x^3+ x^6=1\ 1\ 1\ 1\ 0\ 0\ 1=171$$

2.5. Zero-Forcing (ZF) Signal Detection

In the 256 x 32 MIMO hybrid precoded system, the transmitted and received signals are represented by $X=[X_1, X_2 ....X_{256}]^T$ and $Y=[Y_1,Y_2 ......Y_{32}]^T$ respectively. If $N=[N_1, N_2 ....N_{32}]^T$ denotes the white Gaussian noise with a variance σ_n^2 and the channel matrix is represented by $H=[H_1H_2.....H_{256}]$, we can write

$$Y = \mathbf{H}X + N \qquad (2)$$

As the interference signals from other transmitting antennas are minimized to detect the desired signal, the detected desired signal from the transmitting antenna with inverting channel effect by a weight matrix W is given by

$$\tilde{X} =[\tilde{X}_1,\tilde{X}_2,....\tilde{X}_{256}]^T =WY \qquad (3)$$

In Zero-Forcing (ZF) signal detection scheme, the ZF weight matrix is given by

$$W_{ZF} = (H^H H)^{-1} H^H \quad (4)$$

and the detected desired signal from the transmitting antenna is given by [10]

$$\tilde{X}_{ZF} = W_{ZF} Y \quad (5)$$

2.6. MIMO Fading Channel Estimation

In estimation of ray path geometry based 32×256 sized mmWave MIMO fading channel H, it is assumed that the $N_t(=256)$ transmitting and $N_r(=32)$ receiving antenna sare arranged in uniform linear array (ULA).Such MIMO channel has limited scattering with Lu(=6) scatterers. Each scatterer is assumed to contribute a single propagation path between the base station (BS) and mobile station(MS). The geometrical channel model $H \in C^{N_r \times N_t}$ can be written as:

$$H = \sqrt{\frac{N_t N_r}{L_u}} \sum_{l=1}^{L_u} \alpha_{u,l} a_{MS}(\theta_{u,l}) a^*_{BS}(\varphi_{u,l}) \quad (6)$$

where, $\alpha_{u,l}$ is the complex gain of the lth path including the path loss. The variable $\theta_{u,l}$ and $\varphi_{u,l} \in [0,2\pi]$ are the lth path's angle of arrival and departure(AoAS/AoDs) respectively. Finally, $a_{BS}(\varphi_{u,l})$ and $a_{MS}(\theta_{u,l})$ are the antenna array response vectors of the BS and MS respectively.

With available knowledge of the geometry of uniform linear antenna arrays, $a_{BS}(\varphi_{u,l})$ is defined as:

$$a_{BS}(\varphi_{u,l}) = \frac{1}{\sqrt{N_t}} [1, e^{j\frac{2\pi}{\lambda} d\sin(\varphi_{u,l})}, \ldots\ldots\ldots\ldots e^{j(N_t-1)\frac{2\pi}{\lambda} d\sin(\varphi_{u,l})}]^T \quad (7)$$

And

$$a_{MS}(\theta_{u,l}) = \frac{1}{\sqrt{N_r}} [1, e^{j\frac{2\pi}{\lambda} d\sin(\theta_{u,l})}, \ldots\ldots\ldots\ldots e^{j(N_r-1)\frac{2\pi}{\lambda} d\sin(\theta_{u,l})}]^T \quad (8)$$

where, λ is the signal wavelength and d is the distance between two consecutive antenna elements.

The MIMO channel H is further normalized to get its Frobenius norm value [11, 12]

$$E[\|H^2{}_F\|] = N_t N_r \quad (9)$$

2.7. Precoder and Combiner Designing

The optimal unconstrained precoder F^* and optimal unconstrained combiner W^* can be estimated from implementation of singular value decomposition (SVD) to MIMO channel H in normalized form. The unconstrained RF precoder F_{RF} at the transmitter side controls phases of the up converted RF signal. Its each (i,j) th element is given by

$$F_{RF}(i,j) = \frac{1}{\sqrt{N_t}} e^{j\varphi_{i,j}} \quad (10)$$

Where $\varphi_{i,j}$ is the unquantized phase of (i, j) th element of unconstrained RF precoder F_{RF} Each entry of F_{RF} are quantized up to B bits of precision, each quantized to its nearest neighbor based on closest Euclidean distance. The phase of each entry of F_{RF} can thus be written as:

$\hat{\varphi} = (2\pi\hat{n})/(2^B)$ where, $\hat{n}$ is chosen according to

$$\hat{n} = \arg_{n\in\{0,\ldots 2^{B-1}\}} \min \left| \varphi - \frac{2\pi n}{2^B} \right| \quad (11)$$

where, φ is the unquantized phase obtained from Equation (10). Then the unconstrained RF precoder F_{RF} is computed through substituting quantized phase $\hat{\varphi}$ in Equation (11).

The optimal unconstrained precoder F^* can be written in terms of unconstrained RF precoder FRF and the uncontained baseband precoder F_{BB} as:

$$F^* = F_{RF} F_{BB} \quad (12)$$

From equation (12), we can write,

$$F_{BB} = (F_{RF}^T F_{RF})^{-1} F_{RF}^T F^* \quad (13)$$

where, F_{RF}^T is conjugate transformed form of F_{RF}

With consideration of four RF chains, the constrained analog RF precoder F_{RF} and the constrained baseband precoder F_{BB} are estimated using the following relation based on Conjugate Gradient square method:

$$\min_{F_{RF}, F_{BB}} \left\| F^* - F_{RF} F_{BB} \right\|_F$$

$$\text{s.t.} \left\| F_{RF} F_{BB} \right\|^2 F = 4 \quad (14)$$

Equation (12) can be written in modified form as:

$$F_{RF}^T F_{RF} F_{BB} = F_{RF}^T F^* \quad (15)$$

In Equation (15), the unknown F_{BB} can be determined iteratively with minimization of residual $r^{(i)} = F_{RF}^T F^* - F_{RF}^T F_{RF} F_{BB}^{(i)}$ using Conjugate Gradient square method [13]. The iteration terminates when the estimated residual value is $\leq 1\times 10^{-10}$.

The iteratively re estimated F_{BB} value (F_{BB}) is substituted in equation (8) to get

$$F^* = F_{RF} F_{BB} \quad (16)$$

Equation (16) can be written in modified form as:

$$F_{BB}^T F_{RF} = F^{*T} \quad (17)$$

where, F_{BB}^T and F^{*T} are conjugate transformed form of F_{BB}

and F* respectively.

In Equation (17), the unknown F_{RF} can be determined iteratively with minimization of residual $r^{(i)}=F^{*T}-F_{BB}{}^{T}F_{RF}{}^{(i)}$ using Conjugate Gradient square method [13]. The iteration terminates when the estimated residual value is $\leq 1\times10^{-10}$.

The iteratively re estimated F_{RF} value (F_{RF}) would be such that

$$\left\|F^{*}-\mathbf{F_{RF}F_{BB}}\right\|_{F}=0 \; and \; \left\|\mathbf{F_{RF}F_{BB}}\right\|_{F}^{2}=4 \tag{18}$$

In case of unconstrained RF combiner W_{RF} at the receiver side, its each (i,j) th element is given by

$$W_{RF}(i,j)=\frac{1}{\sqrt{N_r}}e^{j\varphi_{i,j}} \tag{19}$$

Each entry of W_{RF} are quantized up to B bits of precision, each quantized to its nearest neighbor based on closest Euclidean distance. The phase of each entry of W_{RF} is estimated using equation (10). The unconstrained RF combiner W_{RF} is computed through substituting quantized phase $\hat{\varphi}$in Equation (19).

The optimal unconstrained combiner W^* can be written in terms of unconstrained RF combiner WRF and the uncontained baseband combiner W_{BB} as:

$$W^{*}=W_{RF}W_{BB} \tag{20}$$

From equation (20), we can write,

$$W_{BB}=(W_{RF}{}^{T}W_{RF})^{-1}W_{RF}{}^{T}W^{*} \tag{21}$$

where, $W_{RF}{}^{T}$ is conjugate transformed form of W_{RF} [12]

With consideration of four RF chains, the constrained analog RF combiner W_{RF} and the constrained baseband combiner W_{BB} are estimated using the following relation based on Conjugate Gradient square method:

$$\min_{W_{RF},\,W_{BB}}\left\|W^{*}-W_{RF}W_{BB}\right\|_{F}$$

$$\text{s.t.}\quad \left\|W_{RF}W_{BB}\right\|^{2}_{F}=4 \tag{22}$$

Equation (20) can be written in modified form as:

$$W_{RF}{}^{T}W_{RF}\,W_{BB}=W_{RF}{}^{T}\,W^{*} \tag{23}$$

In Equation (23), the unknown W_{BB} can be determined iteratively with minimization of residual $r^{(i)}=W_{RF}{}^{T}W^{*}-W_{RF}{}^{T}W_{RF}W_{BB}{}^{(i)}$ using Conjugate Gradient square method [13]. The iteration terminates for the estimated residual valueis $\leq 1\times10^{-10}$.

The iteratively re estimated W_{BB}value (W_{BB}) is substituted in equation (20) to get

$$W^{*}=W_{RF}W_{BB} \tag{24}$$

Equation (24) can be written in modified form as:

$$W_{BB}{}^{T}W_{RF}=W^{*T} \tag{25}$$

where, $W_{BB}{}^{T}$ and W^{*T} are conjugate transformed form of W_{BB} and W* respectively.

In Equation (25), the unknown W_{RF} can be determined iteratively with minimization of residual $r^{(i)}=W^{*T}-W_{BB}{}^{T}W_{RF}{}^{(i)}$ using Conjugate Gradient square method [13]. The iteration terminates when the estimated residual value is $\leq 1\times10^{-10}$.

The iteratively re estimated W_{RF} value(W_{RF}) would be such that

$$\left\|W^{*}-W_{RF}W_{BB}\right\|_{F}=0 \text{ and } \left\|W_{RF}W_{BB}\right\|_{F}^{2}=4 \tag{26}$$

In Appendix, our developed program for verifying $\left\|W^{*}-W_{RF}W_{BB}\right\|_{F}=0$ and $\left\|F^{*}-F_{RF}F_{BB}\right\|_{F}=0$ has been presented for developing idea for a typical assumed MIMO Rayleigh fading channel.

3. System Description

A simplified form of hybrid precoded millimetre wave wireless communication system is depicted in Figure 1. We consider that a color image of resolution 96 pixels (width) ×96pixels (height) will be processed in our simulated hybrid precoded millimeter wave wireless communication system.

The typically assumed color image is converted into three red, green and blue components with each component is of 96 pixels(width) ×96 pixels(height). The pixel integer values [0-255] are converted into 8 bits binary form. The binary converted signal vector $S\in(0,1)$ of dimention 1×221184 is channel encoded and subsequently interleaved to produce a signal vector $\tilde{s}$. The transformed signal vector $\tilde{s}$ is of size1×442368. In case of merely BCH channel coding, the signal vector $\tilde{s}$ would be of size1×438912 and after 16-arrayQAM/PSK/DPSK digital modulation [14], the number of digitally modulated symbols is 109728 and on adding additional 864 zeros in zero padding scheme. However, 110592×1 sized digitally modulated signal vector $\hat{s}$ is processed in serial to paralel converter to produce blocks with each block containing N(=1024) number of digitally modulated complex symbols [$\hat{X}_0,\hat{X}_1,\hat{X}_2..........\hat{X}_{N-1}$] prior to 1024 point IDFT implementation in OFDM modulation [10]. The number of OFDM block is 108. In each OFDM block, the samples are represented by [$\hat{X}_0,\hat{X}_1,\hat{X}_2..........\hat{X}_{N-1}$] and we can write,

$$\hat{X}_k=\frac{1}{N}\sum_{n=0}^{n=N-1}\hat{X}_n e^{j2\pi nk/N}$$

$$\text{for } k=0,1,2,3,\ \ldots\ldots N-1 \tag{27}$$

The IDFT implemented 1024×108 sized data vector $\ddot{X}$ is reshaped into single column data vector $\ddot{X}$ of dimension 110592×1in parallel to serial converter and passed through Spatial demultiplexer to produce4 data stream of a4×27648 sized signal vector X. The signal vector X is multiplied by a digital baseband precoder matrix FBB of dimension 4×4 and the digitally precoded signal is undergone in D/A conversion with execution of up sampling and filtering with raised cosine pulse shaping digital filter [15]. The D/A converted filtered 4 x 110632sized signal vector DA is multiplied with 256 x 4sizedanalog RF precoder FRF to produce 256 x 110632 sized signal vector FR. In RF up converter section, the signal is multiplied for each of 256 channels with multiplier ML=exp(1i*2*pi*Carrier_Freq.*t), where, Carrier_Freq is the carrier frequency in mmWave band (38GHz) and t is the sample time for each of the sample ranging from1 to 110632. A 256 x 110632 sized matrix MLL can be generated from ML using MATLAB notation MLL= repmat (ML, 256,1); The transmitted signal TX is given by TX=FR •MLL, where, •is the hadamard product which is indicative of element wise multiplication of two matrices.

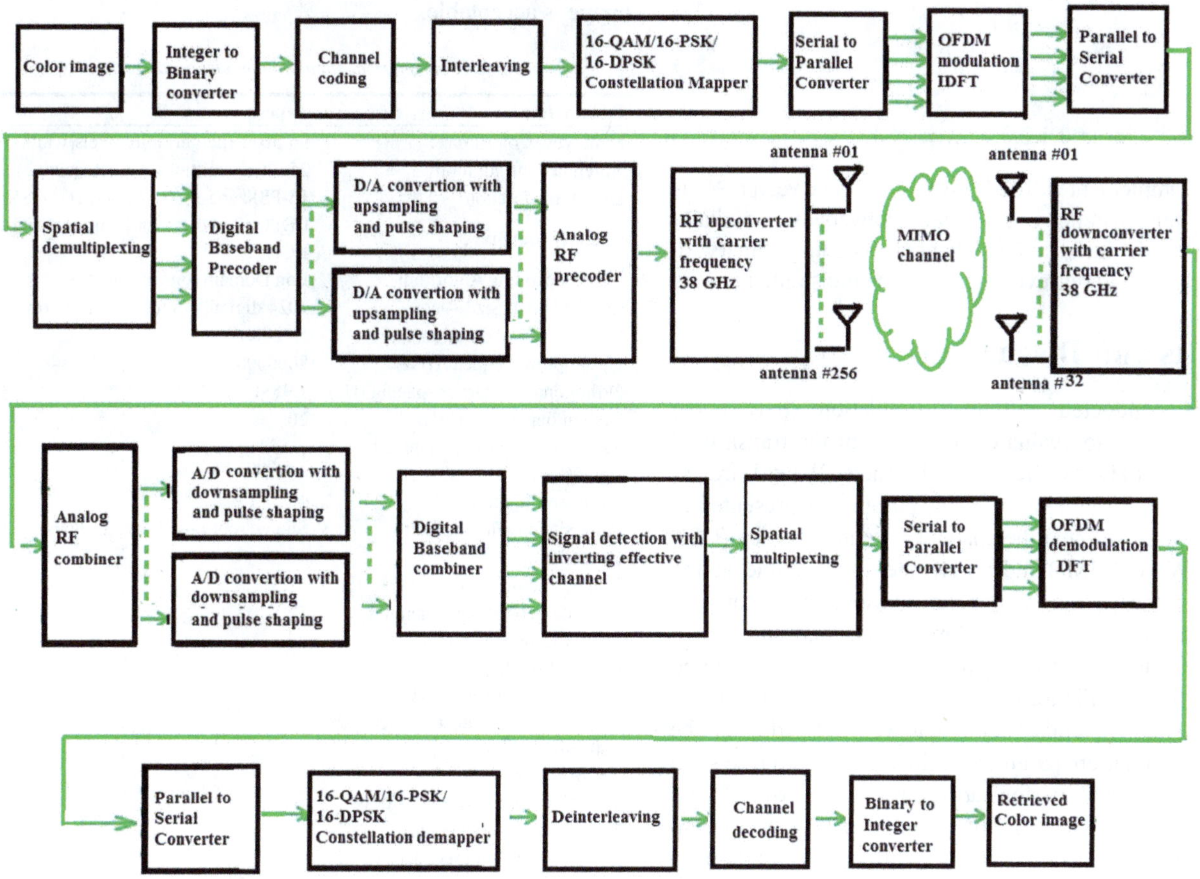

Figure 1. *Block diagram of Hybrid Precoded Millimeter Wave Wireless communication system.*

In receiving section, the 32x 110632 sized received signal Y with consideration of MIMO fading channel H is given by:

$$Y = HTX \tag{28}$$

The received signal is RF down converted through multiplication of 32x110632 sized matrix MLLL. The matrix MLLL can be generated from a matrix MLP using MATLAB notation

MLLL= repmat(MLP, 32,1) where, MLP=exp(-1i*2*pi*Carrier_Freq.*t). The received signal Y after multiplication with MLLL and in presence of addictive complex Gaussian noise N with i.i.d. CN(0, σ^2) is given by

$$\tilde{Y} = Y \bullet MLLL + N = HFRFDA + N \tag{29}$$

The noisy received signal is passed through 32 x 4 sized analog RF combiner WRF to produce modified form of received signal Y as:

$$\bar{Y} = WRF^T\tilde{Y} = WRF^T HFRFDA + WRF^T N \tag{30}$$

On executing A/D conversion theA/D converted 4 x 27648 sized received signal vector $\hat{\bar{Y}}$ is given by

$$\hat{\bar{Y}} = WRF^T HFRFFBB\bar{X} + WRF^T\bar{N} \tag{31}$$

After passing through 4 x 4 digital baseband combiner WBB, we can write,

$$\tilde{Y} = WBB^T\hat{\bar{Y}} = WBB^T WRF^T HFRFFBB\bar{X} + WBB^T WRF^T\bar{N} \tag{32}$$

Where, WBB^T is the complex conjugate transformed form of WBB.

In Equation 32, it is quite observable that the effective MIMO channel $\bar{H}$ is given by

$$\bar{H} = WBB^T WRF^T HFRFFBB \tag{33}$$

On applicability of ZF based signal detection technique, the spatially demultiplexed signal $\bar{X}$ is detected. The detected signal is spatially multiplexed to produce a single column data vector $\ddot{X}$ of dimension 110592 × 1. In serial to parallel converter, the signal vector $\ddot{X}$ is reshaped into 1024 × 108 sized data vector $\mathcal{X}$. Ineach of 108 blocksof $\mathcal{X}$, the samples are represented by [$\mathcal{X}_0, \mathcal{X}_1, \mathcal{X}_2 \ldots\ldots\ldots \mathcal{X}_{N-1}$] and after implementation of 1024 point DFT in OFDM demodulation section, the samples in each block is represented by

$$\bar{\mathcal{X}}_k = \sum_{n=0}^{n=N-1} \mathcal{X}_n e^{-j2\pi nk/N}$$

$$\text{for } k=0,1,2,3, \ldots\ldots\ldots N-1 \qquad (34)$$

The DFT implemented 1024× 108 sized data vector $\bar{\mathcal{X}}$ is processed for parallel to serial conversion, digitally demodulation, de interleaving, channel decoding, binary to integer conversion to retrieve eventually the transmitted image.

4. Results and Discussion

We have conducted computer simulation study using MATLAB R2014a to evaluate the quality of the transmitted color image in a Hybrid Precoded Millimeter Wave Wireless Communication System based on the parameters presented in Table 1 It is assumed that the channel state information (CSI) of the mmWave MIMO fading channel is available at the receiver and the fading process is approximately constant during the whole period of color I mage transmission.

On critical observation of grapical illustrations presented in Figure 2 through Figure 4, it is found that the performance of the simulated systemis very much well definedunder the considered simulation parameters. In all cases, the system shows satisfactory performance in 16-QAM digital modulation and comparitively worst performancein 16-DPSKdigital modulation. In Figure 2 with consideration of ½ -rated convolutional channel coding, the ber values at 1dB SNRarefound to have values of 0.1514, 0.2182 and 0.4733 in case of 16-QAM, 16-PSK and 16-DPSK which is in dicative of system performance improvement of1.59 dBin16-QAM as compared to16-PSK and 4.95 dB in 16-QAM as compared to 16-DPSK.At 10% BER, achieved system performance improvement in terms of signal to noise ratio (SNR) are 0.5 dB and 4.8 dB in 16-QAM relative to 16-PSK and16-DPSK respectively. In Figure 3 for LDPC channel coding, the ber values at 1dB SNRare 0.0457, 0.1879 and 0.3526 for 16-QAM, 16-PSK and 16-DPSK respectively which implies system performance improvement of 6.14dB in 16-QAMrelative to 16-PSK and 8.87 dB in16-QAM relative to16-DPSK. In Figure 4 for R and A channel coding, the system performance in all considered digital modulations shows almost linear response viz. improvement of system performance occurs linearly with increase in SNR values. At a typically assumed SNR value of 1 dB, the ber valuesare 0.1705, 0.2186 and 0.2472 for 16-QAM, 16-PSK and 16-DPSK respectively which implies system performance improvement of 1.08 dB in 16-QAMrelative to 16-PSK and 1.61 dB in 16-QAMrelative to16-DPSK. In Figure 5 for BCH channel coding, the estimated ber valuesat SNR value of 1 dB are 0.1278, 0.1809 and 0.2583 for 16-QAM, 16-PSK and 16-DPSK respectively which is indicative of system performance improvement of 1.51 dB in 16-QAMrelative to16-PSKand 3.06 dB in 16-QAMrelative to 16-DPSK. In Figure 6, it is observable that the simulated system shows satisfactory performance in retrieving color image. At reasonably low SNR value, the quality of the retrieved color image is acceptable.

Table 1. *Summary of simulation model parameters.*

Parameters	Types
Data type: Color image	96 pixels(height) ×96 pixels(width)
Antenna configuration	256(Transmitting) × 32(Receiving)
Digital modulation	16-PSK and 16-QAM and 16-DPSK
Channel coding	LDPC, R and A, Convolutional and BCH
LDPC decoding Algorithm	Log Domain Sum-Product
OFDM Block Size	1024 digitally modulated symbols
OFDM symbol duration(sec)	$5.1200\times^{-05}$
Upsampling frequency(Hz)	80000000
Orthogonal subcarrier spacing (Hz)	1.9531×10^{04}
System bandwidth (MHz)	20
Sampling time for transmitted signal(sec)	1.2500×10^{-8}
Oversampling rate	4
Pulse shaping digital filter	Square root raised cosine filter
Order of filter	40
Roll of Factor of filter	0.25
Filter delay(# of input samples)	5
Carrier frequency(GHz)	28
Path loss model(dB), λ=wavelength(m)of carrier frequency, d= distance(m) between transmitter and receiver	$-20\log_{10}(\lambda/(4\pi d))$
No. of RF chains in both Transmitter and Receiver sides	4
Number of channel paths	6
Size of Baseband Precoder F_{BB}	4 × 4
Size of RF Precoder F_{RF}	256 × 4
Size Baseband Combiner W_{BB}	4 × 4
Size RF Combiner W_{RF}	32 × 4
SNR	0-10 dB
Signal detection techniques used	Zero forcing (ZF)
Channel	AWGN and Ray path based geometrical MIMO

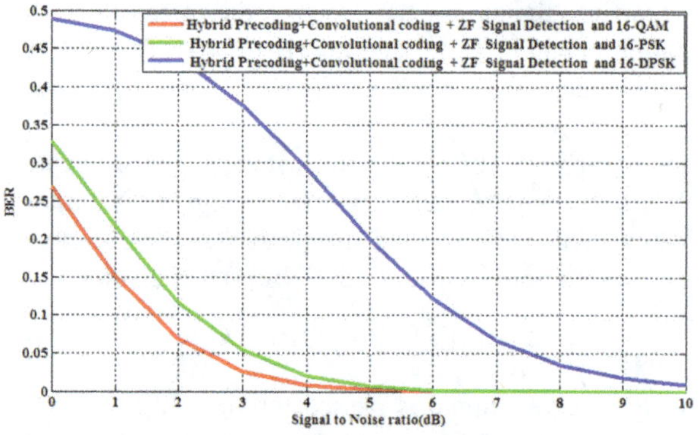

Figure 2. *BER performance comparison of Hybrid Precoded Millimeter Wave Wireless Communication system with implementation of ZF signal detection, ½ -rated convolutional channel coding and various digital modulation schemes.*

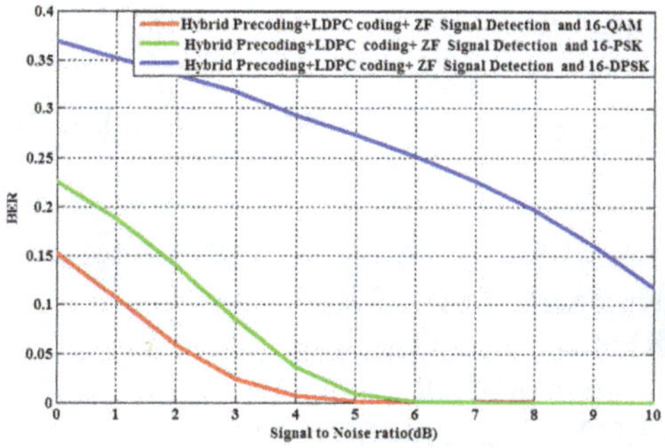

Figure 3. *BER performance comparison of HybridPrecoded Millimeter Wave Wireless Communication system with implementation of ZF signal detection, LDPC channel coding and various digital modulation schemes.*

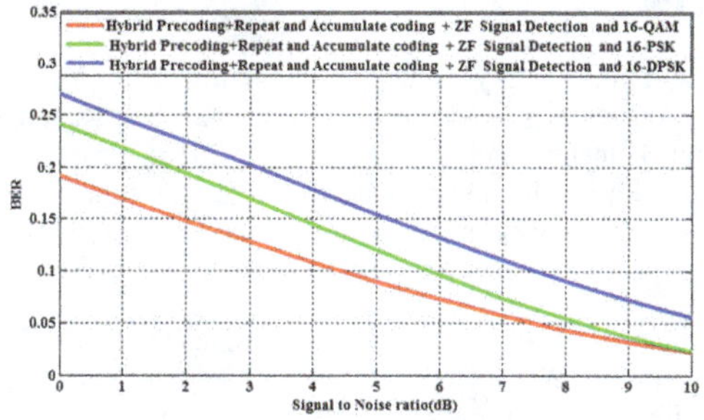

Figure 4. *BER performance comparison of Hybrid Precoded Millimeter Wave Wireless Communication system with implementation of ZF signal detection, Repeat and Accumulate channel coding and various digital modulation schemes.*

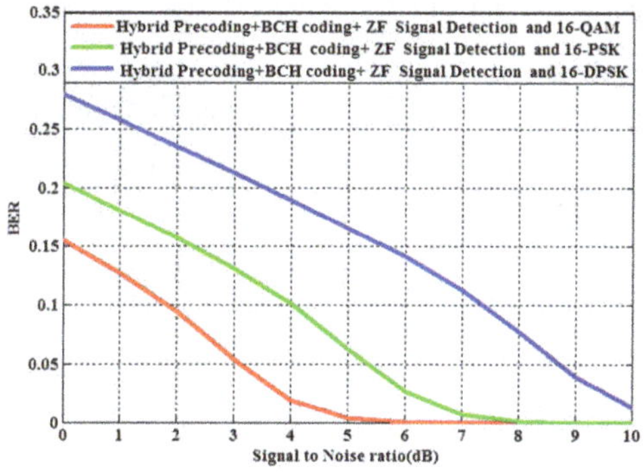

Figure 5. *BER performance comparison of Hybrid Precoded Millimeter Wave Wireless Communication system with implementation of ZF signal detection, BCH channel coding and various digital modulation schemes.*

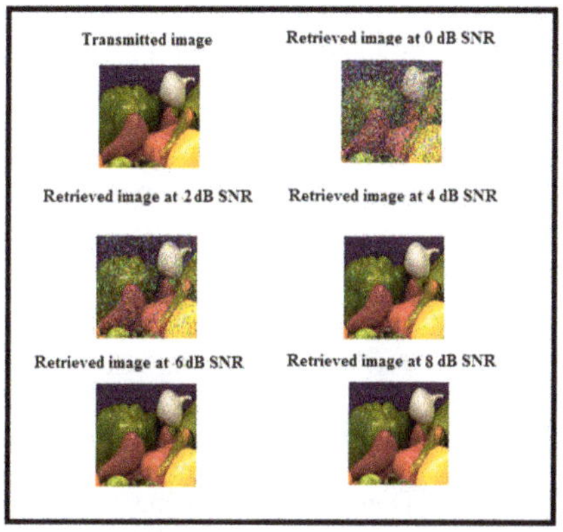

Figure 6. *Transmitted and retrieved images in Hybrid Precoded Millimeter Wave Wireless Communication system under implementation of ZF signal detection, LDPC channel coding and 16-QAM digital modulation scheme.*

5. Conclusions

In this present paper, we have made a comprehensive study on the performance analysis of millimeter wave (mmWave) wireless communication system under simultaneous implementation of both digital and analog precoding and combining schemes in hybrid form. Simulation results ratify that the Hybrid precoders and combiners have been designed satisfactorily with the application of effective iterative Conjugate Gradient Squared (CGS) Method and Singular value decomposition of ray path geometry based mmWave MIMO fading channel. Based on the results presented in this paper on color image transmission, it can be concluded that the presently considered Hybrid Precoded Millimeter Wave wireless communication System is undoubtedly a robust system in perspective of signal transmission in hostile fading channel under implementation of LDPC channel coding, 16 QAM digital modulation and Zero Forcing (ZF) signal detection schemes.

Appendix

```
clear all; close all;
%antenna configuration: 32 receiving × 256 transmitting
%MIMO fading channel generation
H=sqrt(1/2)*(randn(32,256)+sqrt(-1)*randn(32,256));
% MIMO channel
% Normalization of channel matrix
for kk=1:256
for kkk=1:32
H(kkk,kk)=H(kkk,kk)/ (abs(H(kkk,kk)));
end; end;
channel_normalization=(norm( H,'fro')).^2 ;
% its value would be 32 x 256=8192
%%%%%%%%%%%%%%%%%%%%%%%%%%%%%%
% Singular value decomposition (SVD) of channel matrix H
[U SIGMA VT] = svd(H); % U: 32 x 32
V=VT'; % 256 x 256
FSTAR=V(:,1:4); % 256 x 4, optimal unconstrained precoder
WSTAR=U(:,1:4); % 256 x 4 optimal unconstrained
combiner
%FSTAR=Baseband Precoder(FBB) X RF Precoder(FRF)
%WSTAR=Baseband Combiner(WBB) X RF
Combiner(WRF)
%%%%%%%%%%%%%%%%%%%%%%%%%%%%%%
%%
% Unconstrained FRF and FBB estimation
% no of stream =4, FRF= 256 x 4
quantized_phase=((0:2^7-1)*2*pi/2^7)';
unquantized_phase=2*pi*rand(256,4); % 256 rows x4 cols
for kk=1:4
for kkk=1:256
for kkkk=1:128
nhat(kkk,kk,kkkk)= abs((unquantized_phase(kkk,kk)-
quantized_phase(kkkk)));
end; end;end;
for kk=1:4
```

```
for kkk=1:256
[value(kkk,kk), integer(kkk,kk)]= min(nhat(kkk,kk,:));
phihat(kkk,kk)=2*pi*integer(kkk,kk)/(2^7);
end; end;
%%%%%%%%%%%%%%%%%%%%%%%
FRF= (1/sqrt(256))*exp(j*phihat); % 256 x 4 RF Precoder
%%%%%%%%%%%%%%%%%%%%%%%%%%%%
FBB=inv(FRF'*FRF)*FRF'*FSTAR;
%%%%%%%%%%%%%%%%%%%%%%%%%%%%
% Reestimate FRF and FBB such that FRF*FBB=FSTAR
modified_FRF=FRF'*FRF;
modified_FSTAR=FRF'*FSTAR;
%Conjugate Gradient  square Method
for kk=1:4
Estimated_FBB(:,kk)=
cgs(modified_FRF,modified_FSTAR(:,kk), 1e-10);
end;
%FBB'*FRF' =FSTAR', FSTAR'= 8 rows x 256 cols FRF' =
8 rows x 256 cols
FSTART=FSTAR';
FBBT=Estimated_FBB';
for kk=1:256
Estimated_FRF(:,kk)= cgs(FBBT,FSTART(:,kk), 1e-10);
end;
Estimated_FRF=Estimated_FRF';
% Restimate WRF and WBB such that WRF*WBB=WSTAR
unquantized_phase1=2*pi*rand(32,4); % 32 rows x 8 cols
% From Low-Complexity Hybrid Precoding equation 6,
quantized phase
for kk=1:4
for kkk=1:32
for kkkk=1:128
nhat1(kkk,kk,kkkk)= abs((unquantized_phase1(kkk,kk)-
quantized_phase(kkkk)));
end;end;end;
%%%%%%%%%%%%%%%%%%%%%%%%%%
for kk=1:4
for kkk=1:32
[value1(kkk,kk), integer1(kkk,kk)]= min(nhat1(kkk,kk,:));
phihat1(kkk,kk)=2*pi*integer1(kkk,kk)/(2^7);
end;end;
%%%%%%%%%%%%%%%%%%%%%%%%
WRF= (1/sqrt(32))*exp(j*phihat1); % 32 x 4 RF Precoder
WBB=inv(WRF'*WRF)*WRF'*WSTAR;
%%%%%%%%%%%%%%%%%%%%%%%%%%%%%%
%
modified_WRF=WRF'*WRF;
modified_WSTAR=WRF'*WSTAR;
% Equation modified_WRF*WBB=modified_WSTAR
%Conjugate Gradient square Method
for kk=1:4
Estimated_WBB(:,kk)=
cgs(modified_WRF,modified_WSTAR(:,kk), 1e-10);
end;
%WBB'*WRF' =WSTAR', WSTAR'= 8 rows x 256 cols
WRF' = 8 rows x 256 cols
WSTART=WSTAR';
WBBT=Estimated_WBB';
for kk=1:32
Estimated_WRF(:,kk)= cgs(WBBT,WSTART(:,kk), 1e-10);
end;
Estimated_WRF=Estimated_WRF';
%%%%%%%%%%%%%%%%%%%%%%%%%%%%%%%
result1= round(norm(FSTAR-
Estimated_FRF*Estimated_FBB,'fro'))
result2= round(norm(WSTAR-
Estimated_WRF*Estimated_WBB,'fro'))
%%%%%%%%%%%%%%%%%%%%%%%%%%%%%%%
```

References

[1] Sooyoung Hur, Sangkyu Baek, Byungchul Kim, Youngbin Chang, Andreas F. Molisch, Theodore S. Rappaport, Katsuyuki Haneda, and Jeongho Park, 2016: Proposal on Millimeter-Wave Channel Modeling for 5G Cellular System, Transactions on Wireless Communications, vol. PP, issue 99, pp. 1-16.

[2] Diana Maamari, Natasha Devroye, Daniela Tuninetti, 2016: Coverage in mmWave Cellular Networks with Base Station Cooperation, Transactions on Wireless Communications, vol. PP, issue 99, pp. 1-14.

[3] M. N. Kulkarni; A. Ghosh; J. G. Andrews, 2016:A Comparison of MIMO Techniques in Downlink Millimeter Wave Cellular Networks with Hybrid Beamforming, IEEE Transactions on Communications, vol. PP, issue 99, pp. 1-18.

[4] Jonathan Rodriguez, 2015: Fundamentals of 5G Mobile Networks, John Wiley and Sons, Ltd, United Kingdom.

[5] Peter Sweeney, 2002: Error Control Coding, From Theory to Practice John Wiley and Sons limited, England.

[6] Yuan Jiang, 2010: A Practical Guide to Error-Control Coding Using MATLAB, Artech House, Boston, USA.

[7] Bagawan Sewu Nugroho, https://sites.google.com/site/bsnugroho/ldpc.

[8] Giorgio M. Vitetta, Desmond P. Taylor, Giulio Colavolpe, Fabrizio Pancaldi andPhilippa A. Martin,2013:Wireless Communications Algorithmic Techniques.John Wiley and Sons Ltd, United Kingdom.

[9] Sneha Bawane and V. V. Gohokar, 2014: Simulation of Convolutional Encoder, International Journal of Research in Engineering and Technology (IJRET), India, 3(3), 557-561.

[10] Yong Soo Cho, Jackson Kim, Won Young Yang, Chung G. Kang, 2010: MIMO-OFDM Wireless Communications with MATLAB, John Wiley and Sons (Asia) PTE Limited, Singapore.

[11] Ahmed Alkhateeb, Geert Leus, and Robert W. Heath Jr., 2015: Limited Feedback Hybrid Precoding for Multi-User Millimeter Wave Systems, IEEE Transactions on Wireless Communications, vol. 14, issue 11, pp. 6481–6494.

[12] Weiheng Ni, Xiaodai Dong, and Wu-Sheng Lu, 2015: Near-Optimal Hybrid Processing for Massive MIMO Systems via Matrix Decomposition, pp. 1-9, http://arxiv.org/abs/1504.03777.

[13] Richard Barrett, Michael Berry, Tony F. Chan, James Demmel, J une M. Donato, Jack Dongarra, Victor Eijkhout, Roldan Pozo, Charles Romine and Henk Van der Vorst, 1994: Templates for the Solution of Linear Systems: Building Blocks for Iterative Methods, SIAM, Philadelphia, USA.

[14] Theodore. S. Rappaport, 2004: Wireless communications: Principles and Practices, Second Edition, Prentice Hall Inc., New Jersey, USA.

[15] Mathuranathan Viswanathan, 2013: Simulation of Digital Communication systems using Matlab, second edition.

Infrastructure Sharing Among Ghana's Mobile Telecommunication Networks: Benefits and Challenges

Sylvester Hatsu[1], Ujakpa Martin Mabeifam[2], Philip Carlis Paitoo[3]

[1]Computer Science Department, Accra Polytechnic, Accra, Ghana

[2]Faculty of Information Technology and Systems Development, International University of Management (IUM), Dorado Park Campus, Windhoek, Namibia

[3]Graduate School, Ghana Technology University College, Takoradi Campus, Accra, Ghana

Email address:

shatsu@apoly.edu.gh (S. Hatsu), ujakpamabeifam@gmail.com (U. M. Mabeifam), ellycarly14@gmail.com (P. C. Paitoo)

Abstract: In Ghana network companies such as Vodafone, Mobile Telecommunication Network, Airtel, Tigo, Expresso and Globacom (GLO) are experiencing increasing subscribers for voice calls, internet and video services. Competition in the industry has been intensified making service providers searching for innovative strategies to survive the competition. Strategies adopted to survive the stiff competition include rebranding, infrastructure sharing and mergers and acquisitions. This study focuses on infrastructure sharing as a strategy to reducing cost for these telecommunication service providers in Ghana. Mobile telecommunication industry in developing countries has players a remarkable role in providing services to large portion of the population. Despite the achievement in reaching large numbers, extra efforts are needed to increase the mobile service penetration. In increasing the penetration, attention needs to be focused on the rural areas. High network infrastructure cost has been the major problem. As operators strive to recoup investment cost associated with building the expensive infrastructure, customers tends to suffer from high network charges/ prices. Infrastructure sharing presents itself as a means of lowering network deployment cost, especially in rural and marginalized areas. Sharing has an advantage to stimulate migration to new technologies and mobile broadband deployment. Arguably, another advantage is the stirring up of competition between mobile operators and service providers, when safeguards are used to prevent anti-competitive behavior.

Keywords: Telecommunication, Competition, Mobile, Telegraph

1. Introduction

1.1. Background of the Study

The current liberalization, privatization and globalization concept which Ghana embraced during the economic recession period of the 1980s has resulted in the complete removal of Government's hands in active business and economic activities in the country. It is assumed that through liberalization, economic and allocative efficiency will be achieved in the country which will result in total national development. This approach is also to correct the balance of payment deficit and result in higher Gross Domestic Product (GDP), which is a necessary indicator for economic development of any country. Goods and services will be allocated by "invisible hands" in the words of Adam Smith. One clear result of liberalization, privatization and globalization, embarked by the government of Ghana in the 1980s, was the upsurge of telecommunication industry operators in Ghana, after the deregulation of the telecommunication industry.

In Ghana Government's corporation, known as the Ghana Post, Telephone and Telegraph dominated and monopolized the telecommunication industry of Ghana. Liberalization, privatization and globalization policies of the Bretton Woods organizations in Ghana led to the relaxation of deliberate controls and policies which seeks to protect and monopolized state organizations to favour the encouragement and/or engagement of private ownership in business. The period from1994 to 2000 saw a dramatic competition within the telecommunication industry of Ghana. This period, characterized by liberalization and privatization, experienced a shift from government's controlled Ghana Post, Telephone and Telegraph to a privatization, technically termed deregulation of the telecommunication sector. This resulted in a competitive telecommunication industry, which allowed the operation of private internet and/or mobile telecom network providers in Ghana, after 1994.

To [1] Ghana's adoption of liberalization, privatization and globalization policy of the World Bank within its telecommunications sector, led to the birth of a development and implementation of a policy document known as the Accelerated Development Program (ADP). This programme was initiated in the year 1994 and spanned to the year 2000, meaning a six years deregulation programme of the telecommunication Industry. The Accelerated Development Programme was a six years comprehensive development plan which was meant to restructure the telecommunication industry of Ghana. Drawn together with the assistance from the World Bank consultants, the policy objectives of the Accelerated Development Programe, according to [1] were to achieve a density between 1.5 and 2.5 lines per 100 people, to provide payphone facilities in the whole country (both rural and urban) to improve public access to communication and improve the coverage of mobile services in the country. The other objectives were to encourage the ownership and operation of Ghanaian telecommunication companies and finally the creation of a national body called the National Communication Authority (NCA), to regulate and control the activities of the telecommunication networks. Telecommunication industry has made a remarkable improvement in Africa and has become one of the robust industries. The service has expanded to include innovative packages such as accessibility to internet, banking and retail transactions [2] Through mobile telecommunication, Africa has reached 400 million mobile subscribers, many of which resides in remote areas with no fixed telephone connectivity. [3] on a telecommunication industry growth survey, reports a growth of 9.3 percent (from the year 2002 to 2007), making the industry one of the fastest growing in the world.

In Ghana, mobile telecommunication market has experienced a geometric growth making it one of the fastest on the African continent. Sustaining such a market carries for itself a huge cost burden. Mobile telecommunication investors and operators have no option than to spend huge capital expenditures on assets/infrastructure to become competitive on the market. Thus, telecommunication industry invests heavily in infrastructure, equipment, logistics and software to improve service delivery services for its clients. Factors such as customer sophistication, increased advancements in technology and globalization makes it imperative for mobile telecommunication service providers to invest in such items, which comes with huge cost. Normally, these investments take a longer time for these mobile telecommunication service providers to recoup their investment. However, they have no option than to acquire the equipment to survive the competition. Management of such companies spends time to think about strategies to reduce cost and increase revenue and profit margins. Some of the areas of focus to achieve cost reduction and increase revenue and profit margins are:

1. To continue with the acquisition of state of the art technology (thus increase assets- transmission, equipment).
2. Introduction of new products and services and strengthening of marketing of service and products.
3. Efficient management of debt and the practice of stiff internal controls.

Acquisition of infrastructure has become an inevitable part of the operations of these mobile telecommunication companies. These infrastructures are expensive to acquire. Mobile telecommunications regulators, together with policy makers in the industry are encouraging the sharing of infrastructure globally.

A benefit to the mobile telecommunication is seen from enhanced services.

Sharing induces and intensifies competition among the mobile telecommunication operators, thus also reducing tariffs on the consumer. Resource conscious advocates argue in support of infrastructure sharing as an environmental friendly approach as it saves the depletion rate of the environment. Also, savings can be made on energy consumption of infrastructure. Aesthetics wise, infrastructure sharing is a plus on street furniture. Mobile telecommunication companies in Ghana have started practicing infrastructure sharing. The idea is to reduce cost and provide quality services which meet the pocket of every Ghanaian. This sharing has introduced special advantages and challenges of which this study is interested to unearth.

1.2. Problem Statement

Growth and survival of any company, of which mobile telecommunications companies are included, is intrinsically linked to globalization, productivity, quality, and satisfaction of customers. The needs to attract and retain customers are necessary conditions for growth and survival of these mobile telecommunication companies. Thus, making the customer the centre of attraction on which business revolves and therefore the major focus for business success. So sharing is like a compulsory acquisition of infrastructure from Econet who has invested a lot in infrastructure building for the benefit of other companies who have not spent on building infrastructure.

1.3. Objectives of the Study

The general objective of the study is to investigate infrastructure sharing among mobile telecommunication networks in Ghana. Specifically, the study will achieve the following objectives. These are to: Assess the types of infrastructure mobile telecommunication networks share among themselves; examine the benefits of mobile telecommunication infrastructure sharing;

1.4. Research Questions

The following research questions guided the study. These are:

What types of infrastructure are mainly shared among the telecommunication networks?

What are the benefits of infrastructure sharing?

1.5. Significance of the Study

A study of this nature makes the government and the mobile telecommunication network operators, academia and the general public. The government is one of the primary drivers of infrastructure sharing, through its regulatory body called the National Communication Authority (NCA). The government seeks to ensure that mobile telecommunication operators share infrastructure to reduce their operational cost. The savings made can be channeled into other areas such as improving quality of service for customers. Reduction in operational cost can also manifest in less service charges, thus saving customers from exorbitant charges from network operators. It will aid the National Communication Authority in making appropriate policies in the areas of infrastructure sharing.

2. Literature

2.1. Generic Strategies: Theory

- It is important for a firm to choose a position [4]. These positions, in the words of Porter, called variables are central to position; Porter names the two variables as: competitive advantage and competitive scope.
- Porter argues that a firm's success in the long run, relative its competitors, is its sustainable advantage. Two types of competitive advantage were identified: lower cost and differentiation. Porter explains lower costs as the ability to design, produce and market a comparable product at more cost-efficiently than its competitors. In the long run, lower costs translate into higher returns. Higher returns are attained when prices are less or near the competitors' prices. The mobile telecommunication industry in Ghana is highly competitive and technology intensive. Customers are always looking for cheaper means of enjoying services and tend to accept companies which provide better services at lower cost. However, for a mobile telecommunication industry to provide less costly and improved services, it must invest more into infrastructure. Acquiring infrastructure is expensive. Those who are able to afford these expensive usually charges higher service cost in a view of recouping the cost of infrastructure. However, infrastructure sharing presents one of the possible means for mobile telecommunication companies to achieve Porter's competitive advantage and serve in the industry.
- The ability of the firm to provide unique and superior value to customers, in the areas of product quality, special features, after-sale service is referred to as differentiation. Achieving competitive advantage, as explained by Porter, helps to achieve higher productivity compared to the firm's competitors. Competitive advantage makes a firm productive using fewer inputs, with higher revenues per unit. Porter cautions on the impossibility to achieve quality service at lower cost, since inherent in good performance, quality service is more cost. A successful strategy, such as infrastructure sharing, pays close attention to both.
- Competitive scope variable looks into the breadth of the firm's target within its industry. Questions on what to produce (products), distribution channels to use, targeted buyers, geographic areas to sell products and the related competitive industries it will operate. The segmented nature of industries makes competitive advantage an important concept. Another fact that asserts the importance of competitive scope is that firms can sometimes gain competitive advantage from breadth through competing globally or from exploiting interrelationships by competing in related industries. Though different strategies exist, competitive advantage is inherently central to any strategy a firm adopts. Achieving competitive advantage requires choice making. Making a choice of competitive advantage and scope spells out where the firm can strategically achieve the advantage. Infrastructure sharing among mobile telecommunication network provides the best strategy for the achievement of such an advantage to the companies on board.

2.2. Defining Infrastructure Sharing

- Resource sharing here is equated to mean same as infrastructure sharing. Infrastructure sharing, according to [5] is the same use of part or parts of infrastructure by more than one mobile telecommunication operators. Resource sharing is country specific and thus its implementation varies from country to country. Due to variations, different legal backing documents exits on support of infrastructure sharing. Coordination, is one of the forms through which sharing can be made possible. Coordination involves exchange of information on infrastructure use. This result in more efficient online or offline use of resource(s) [5].
- Infrastructure sharing gained attention during 3G development in Europe through [6] spectrum licensing and agreements on sharing network.

2.3. Ghana's Mobile Telecommunication Industry

The telecommunication industry in Ghana has a long history. In 2003, there were less than one million telephone lines in Ghana, the number has increased to more than 17 million at the end of 2010 (75 percent mobile penetration). Mobile voice subscriptions from the month of January to

August 2015 have been presented here. Increases have been recorded in January 2015 (30,629,604) to August 2015 (32,826,405), according to the National Communication Authority in 2015.Mobile communications sector's growth has been attributable to the deregulation policy.

2.4. Drivers of Infrastructure Sharing

Globally, trends towards voluntary, unlike mandatory, sharing is been pictured and championed. Intensive growth together with competition from other operators and rising trend in the cost of capital expenditure, associated with expending network to reach more operators [7]. Infrastructure sharing has been the way out. Infrastructure sharing reduces risk [8] through sharing of site building risk among operators. Sharing has been championed on commercial lines as it serves the following purposes. These are:

a It allows Mobile Network Operators to avoid permit securing challenges in building new sites.

b Merges existing networks. Sharing of risk in areas with low population densities among the mobile network operators.

c Encourages reduction in cost, whiles increasing the efficiency through pooling of spectrum.

d Cuts down on duplication. Saved resources are channeled to other areas such as product innovation (likeLTE, mobile broadband).

e Drivers of infrastructure sharing are mainly classified into two [7]. These are government's regulatory and competition perspective.

2.5. Government's Regulations

- Regulations of the National Communication Authority affect decisions of infrastructure sharing. National Communication Authority of Ghana usually comes in with such policies to save the country and its citizens. For example, the country is currently facing power challenges. Sharing infrastructure is one of the strategies to reduce power consumption of multiple infrastructure belong to different mobile telecommunication operators. Also, there have been public out-cry on the damaging and health hazards that mobile telecommunication infrastructure, such as mast, poses to the health of the immediate environments it is constructed. There was a mass protest in Accra in 2009 after the collapse of a mast killing one person and injuring others. [9] Estimates that 40 per cent of the public complains of telecommunication mast were health related concerns. 33 percent proximity to residential property and schools and 27 per cent were lack of informed consent, meaning the communities were not sufficiently consulted before erecting the mast. Fumes and noise from generators were also reported by community.
- Infrastructure sharing becomes one of the options for reducing, as much as possible, the number of such masts. Also, through infrastructure sharing the operational cost of mobile telecommunications operators are reduced by an average margin of 60 percent. This shaving can be channeled into other areas of production, which are equally important for the provision of better services to mobile subscribers. The cost of constructing a mast is estimated to be US$ 250,000. After constructing such an infrastructure, mobile telecommunication operators have no option than to shift the cost of constructing such an infrastructure to the customer, who is the weakest in this relationship. Customers are charged higher for mobile services. So infrastructure sharing tends to reverse such a trend. These are the major reasons for the government to champion infrastructure sharing in Ghana.
- In Ghana, health, aesthetics and safety were main argument for the involvement of government to encourage infrastructure sharing in the year 2009. An Inter-Ministerial Committee was commissioned to develop a framework to enhance infrastructure sharing. The commission involved the Ghana Civil Aviation Authority (GCAA), National Communications Authority (head), Ghana Atomic Energy Commission (GAEC), Environmental Protection Agency (EPA), Metropolitan, Municipal and District Assemblies. The task of the commission was to develop a guiding document for the deployment of communication towers. The head was to collaborate with industry and other stakeholders.
- Additionally, Sharing prevents duplication of infrastructure. It enables investments in financially underserved areas. [10] Argue that sharing results in improved customer service and product innovation. They further positions four important regulatory dimensions for government's interventions. These are: pricing, policy, safeguards and policy enforcement (Table 1). National Communication Authority encourages the sharing of infrastructure that carry with no risk of lessening of competition. The infrastructure includes: Antenna, Masts, tower structures and Rights of Way. Others include Space in buildings, Trenches, Electric power and Ducts.

***Table 1.** Regulators' Intervention areas in Infrastructure Sharing.*

Category	Description
Regulatory Policy	Regulatory authorities should create in collaboration with local authorities and municipalities, policy that encourages sharing. The policy should be such that encourages incumbents and new entrants to balance their shared network rollout. Telecom laws facilitate, mandate or empower the regulator to enforce infrastructure sharing.
Pricing Regulation	Cost-based prices should be applied so as to allow operators to recoup their investments. Each network element should be priced separately so as to allow the requesting operators to pay only the elements that they need. In case of new networks and increase of existing capacity, costs should be shared. Late entrants should compensate existing partners for any shared investment before being given access to the network.

Category	Description
Regulatory Safeguards	Regulators should ensure that infrastructure sharing is in line with the general regulatory standards of telecom sector: transparency, efficiency, independence and non-discrimination. Capacity should be sold on first come first served basis and the regulator should intervene to distribute scarce resources when it is necessary. Unused capacity should be returned and operators should not order excess capacity. The operators could get penalties if their ordered capacity exceeds the utilized capacity by a certain percentage. Regulators should constantly monitor the sharing activities. Physical separation of shared network components can be used as long as it does not affect efficient sharing.
Policy Enforcement	Sharing should be encouraged by creating incentives. Penalties should be given to the operators that fail to comply with the adopted regulation. Regulators should intervene to solve disputes and clearly define the dispute-resolution procedures. Compliance should be monitored by regulators.

Source: Adopted from [11]

2.6. Quality of Service

Besides lower charges for service, quality of service provided by the network operators have be of prime focus. Quality of service is the main factor distinguishing among the competing mobile telecommunication companies. With the power in the hands of the subscriber to shift to other mobile telecommunication service with better service, it behooves on the mobile network operators to improve service provision by the acquisition of infrastructure; which has become possible under the jacket of infrastructure sharing.

2.7. Enhancement of Profitability

Infrastructure sharing is one of the strategies for reducing operational cost of mobile telecommunication companies. Capital expenditure is also reduced with opportunity for higher Return on Investment. It has positively impact on Profit and Loss account.

2.8. Demands from New Technologies

The introduction of new technologies such as 3G and its resultant use of Wi-MAX (Worldwide Interoperability for Microwave Access) have made it necessary to share network to improve services. WI-Max will increase the demand for sharing.

2.9. Benefits of Infrastructure Sharing

Improving quality of service of network operators and expenditure reduction has been seen as the immediate benefit of infrastructure sharing [12]. Earning of extra income from rentals of non-core assets, new entrants enjoys faster market entry and the focus it gives operators to focus on other important services such as customer service were the further benefits (ibid). Through sharing, customers, who are at the receiving end, enjoy better charges for mobile phone services and areas considered not to be economically feasible to receive mobile networks, get the chance to network. It also serves to add up esthetics to the city. From the perspective of network operators, benefits of infrastructure sharing manifests in cost savings [10] 30% savings on capital and operational expenditure from passive sharing. Further, Ghandhi, an Indian telecommunication expert, opines passive infrastructure sharing allows operators to focus on marketing and sales areas. Leasing of towers generates additional income. Entry of new operators into the market becomes easy. To the environmentalist, negative environment and health hazards perceived to be carried in the form of emissions are reduced through infrastructure sharing.

2.10. Models of Infrastructure Sharing

Five different sharing models have been explained. Choice and/or combination of sharing models depend on some factors. The primary among them is the operator. Through theoretically, it is possible to share almost everything, regulatory authorities, sometimes, prohibits the sharing of certain configuration. Possibly, theory makes it easy to, say level one and level three, without sharing level two. Operators can share sites and masts. Additionally, they share the radio network. Another sharing is to geographically divide the country. They can then build in different areas. Typical mobile network infrastructure has been presented in. Each operator has then its individual network but offers extended coverage, known as national roaming, to other operators and their customers. Operators, finally, deploy a common shared network. On this they all elements are shared on levels: from one to five, Detail explanation of the model is provided in Figure 1.

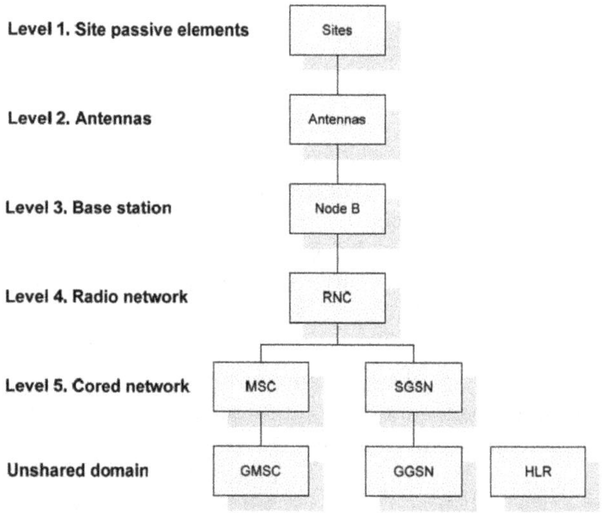

Source: Adopted from [13]

Figure 1. *Different Technical level Infrastructure Sharing Models*

2.11. Sharing Site and Mast

This is the simplest form of network sharing. Under this operators have common facilities and site. Passive elements are shared on level one. All passive and active elements on site are involved. Examples of passive and active elements are: antennas, sites, cooling, masts, civil works, power and towers [14]. Areas with challenges such as delayed or impossibility in getting building permits. Also municipality with limited space and restrict the installation of separate antenna, are good for this type. With this, challenges arises from the charge one operator can impose on the other operator. Employees of one operator can damage the equipment of the other. Lastly, when to plan for new site is also a challenge. The advantages here are the simplicity and the cost savings are huge [15]. There is a positive effect of sharing and overall expenses, as overall expenses reduces from sharing cost associated with sites. Civil works constitutes30% and site rental 20% of network CAPEX and OPEX respectively. Reduction of visual pollution of mast is an additional advantage [14]. Infrastructure sharing does not minimize competition to any appreciable level.

2.12. Geographical Sharing

Operators can decide to build in different geographic locations of the country, and strategically, enhance nationwide coverage through national roaming advantage. This strategy promotes the fast growing and cost reduction method for the network operators. For instance, one operator can build in the northern part of the country, while another builds in the southern part. This is an efficient way of providing coverage while each operator keeps its individual network. The sharing is not on any certain technological level, but each operator holds the entire infrastructure (on all levels) in its own part of the network.

Regulatory authorities in some countries have restrictions that can affect this model of sharing. As stated earlier, each operator in Sweden must own radio infrastructure that covers at least 30 percent of the population. Consequently, it is not possible for one operator to only cover a small area and use national roaming to get access to another operator's network in other parts of the country. However, when the operators split the total coverage in two equally large parts, this restriction is fulfilled. Further, the network operators are often willing to roll out separate networks in urban areas, as the income in these areas justifies separate networks. In such areas, the own network is the preferred one. One problem with national roaming is that many operators have the same target groups, and thus identify the same target areas within a country for their network deployment.

This can reduce the possibility of agreeing on larger national roaming agreements [14]. The operators probably solve problems of this kind by constructing contracts in a way that prevents one part from taking extensive advantage of the situation. Another problem is that the operators cannot be certain that their counterpart does not favour his own clients. There is a reason for one operator to favour large corporate customers. This can e.g., be done by tweaking the antennas in a way that the corporate customers get excellent quality while people on the street below the corporate customer's building get lower quality [14]. There are also complex business issues. Operators must agree on roaming prices, quality and capacity guarantees, and how the interfaces toward the customers are dealt with. Furthermore, agreements on how to deal with customers of third party network operators must be reached. This last problem becomes even more complex if the user visits urban areas where there is overlap in coverage [14]. Geographically split networks have certain effects on competition. National roaming prevents operators from competing against each other on the network level. Factors such as coverage, capacity, quality of service (QoS), and reliability in the shared areas are the same for both operators [14].

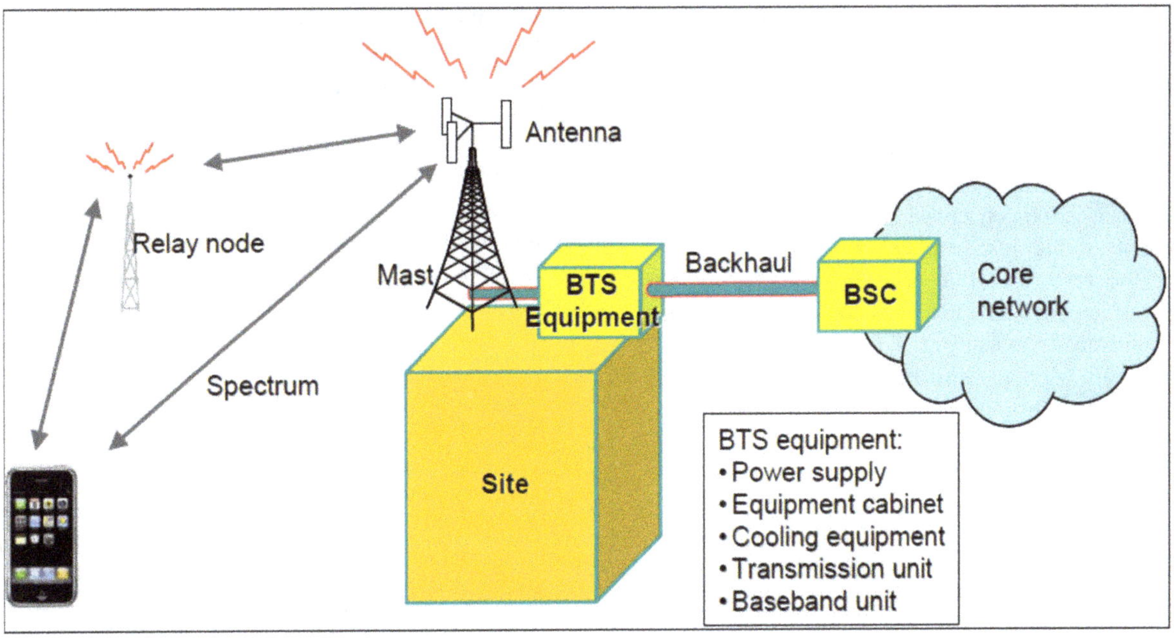

Source: Adopted from [5]

Figure 2. *Infrastructure Elements of Mobile Telecommunication Industry.*

3. Philosophical Underpinnings of the Study

The study was qualitative in nature and draws its philosophical perspectives from the epistemological and ontological foundations of research. The general argument here is that social reality (in the case of infrastructure sharing) is socially constructed, through government's policies and discussions which involve the agreement of the companies. Infrastructure sharing did not emanate by it, however, the interactions of people and phenomenon, which is socially constructed brought, gave birth to it. Therefore it is influenced by society. Therefore any effort to study such a phenomenon must involve the active presence of the society.

This argument is in line with the assumptions of both subjective ontology and subjective/constructive epistemology. The subjective ontology and constructive epistemology both agree that the participants and the researchers influence the knowledge to be created ([16]; [17]. Their experience, perception and beliefs have impact on the kind of knowledge that is created [18]. Another assumption from the camp of the subjective ontology and constructive epistemology is that knowledge is constructed rather that it being discovered ([17]; [18]. The active involvement of the researcher and the researched in their natural setting will have influence of the kind of knowledge that is constructed. The subjective ontology and the subjective/constructive epistemology also influence the interpretive position of the research. The interpretive assumption believes that knowledge is constructed through the experience, perception and beliefs of the researcher and the participants in their natural setting ([18]; [16]; [17]. Data collected in such is full of words rather than numbers ([18; [19].As such, the best way to conduct such research is to adopt a qualitative approach and design to research.

3.1. Study Design

From the camp of [20] research design is an arrangement of conditions. These conditions capture data collection and analysis. Further, it is the conceptual structure which defines the environment of the research. Other authors call it a blueprint for the research. Blue print in it sense that is provides the framework and guidelines for data collection, measurement and finally to analysis. To the man on the street guideline outlining what the researcher will in his study, sweeping from hypothesis setting, operationalization and final analysis constitutes research design.

The research design adopted for this study is the qualitative and quantitative design. Qualitative design describes the characteristics of the study subject as put forth by the research objects (individual, group). It also involves diagnosis which determines the frequency of occurrence of an event or association with other things. Further diagnosis looks into the association between variables. Summary, descriptive studies makes predictions, narrate facts and characteristics of individual, group or phenomenon. This study is involved in narrating and description of infrastructure sharing practices among mobile telecommunication networks and to bring out their characteristics. On the hand quantitative design involves the use of statistical or mathematical formulas such as mean rankings, chi square, relative importance index, etc.

Requirements for the use of descriptive studies, according to [20] are: what the study seeks to measure and methods for measuring it. There should be a clear cut definition of study population. This study use the mobile telecommunication networks and focused on the management of these networks that are responsible for infrastructure; thus a clear cut population. The study has clearly defined specific study objectives with operationally defined variables within the objectives, making it easy to measure. The design made adequate provision against bias and the use of triangulation to ensure reliability of results.

3.2. Population and Sample Size of the Study

The population of the study includes all the mobile telecommunication networks operating in Ghana, together with their employees. All the six networks were used due to the small number for the study. Now for the employees, since all sharing issues are handled from the top management in the head office in Accra, only the population of the top management personnel from the six mobile telecommunication networks in Accra was used. Therefore all the mobile telecommunication networks operating in Ghana automatically became part of the population for the study. Staff like middle and low level staff were not considered due to their les role in the infrastructure sharing arrangement for the company.

The total number of all the top management staff for the six mobile operating companies was 30. From the statistical table for determining sample sizes for study by [21], a population of 30 requires a sample size of 28. However the study, adds an additional two of the sample size to make it 30. Similar [9] (which is on Ghana), [7] [13] (which is on Nigeria) and [11] (which is on Europe) (all discussed in literature review, under the sub-heading empirical review) made use of 30 as a sample size, making the sample size justified within empirical research circles within the topic.

3.3. Sampling Procedure

Non-probability sampling procedure was employed for the study. This was considered suitable for the study due to the qualitative nature of the study. Non-probability sampling procedure is that type of sampling which does not follow the law of probability; which gives each member in the population equal, non-zero and non-calculable chance of being included in the sample [22].

Purposive sampling method of non-probability sampling was employed to select respondents for the study. This method dwells on the expertise and judgments of the researcher to select respondents who best suit the objectives and purpose of the study. Respondents who deal directly with infrastructure management of mobile telecommunication networks, with in-depth knowledge on infrastructure sharing were considered and involved in the study. Five each was collected from each of the six mobile telecommunication operators in Ghana, focusing on the Head Office in Accra. This made a total sample of 30 respondents, selected from the top management personnel.

3.4. Reliability and Validity

Reliability and validity are very important aspect of research. They can be statistically calculated using quantitative approach and design. In the case of qualitative research, reliability and validity are done using different approach. Validity is defined as the ability and probability that a measurement is measuring what it is intended to measure [20]. Validity in a research can be content validity which is the extent to which an instrument or a research has adequate coverage of the subject under study, criterion related validity which comprise of predictive and concurrent validity and construct validity [20].

Validity in qualitative research can best be achieved using triangulation [18].Triangulation in qualitative research refers to the use of different data collection instruments to collect data ([18]; [20]. In such case, two or more different data collection is used in the collection of data in qualitative research. The use of observation and interview or questionnaire and interview in qualitative research promotes validity of the data that is collected. This by way helps to enhance the validity of the data that is collected and analyzed. The stability and equivalency of the data that is collected and analyzed ensure reliability of research. The conformity of data collected by a researcher to that collected by a different researcher on the same study and participants is stability. Stable data over a period of time promote the reliability of research [20].

3.5. Data Collection Instrument

Primary data was collected using one set of semi-structured interview guides for all the respondents selected from the various mobile telecommunication networks operating in Ghana. Semi- structured interview schedules were designed and used for the study. The instrument facilitated the conduct of interview of respondents on one-on-one basis with the various respondents selected from the mobile telecommunication networks operating in Ghana.

3.5.1. Sources of Data

Primary and secondary sources of data were used for the study. Primary data were collected from the field with the use of semi-structured interview guides prepared by the researcher to seek responses that best answers the research questions set out for the study. The questions on the instruments were sectioned according to the various research questions set out for the study. Therefore each section answers a research question. Secondary data were collected from the records of the mobile telecommunication networks. Also, information on the official websites of the mobile telecommunication networks was also used for the study.

3.5.2. Pre-test

Instruments were pre-tested using Vodafone regional office in the Takoradi Metropolis. This was done to assess the reliability and validity. Mistakes identified during the pre-testing exercise were noted down. Questions with ambiguous meaning were also corrected to ensure easy understanding by all respondents. Validity and reliability of the instruments of the data collection instruments used were confirmed during this exercise.

3.5.3. Field Work

Actual fieldwork was done a day after the pre-testing was done. Data collection on field took three days. Two research assistants aided the process. The researcher was the team leader and also part of the data collection team. By role, tem leader supervised all the field work and brought the field strategy to be used.

The final data collected were edited by the filed data collection supervisor. Incomplete instruments were sent back to be fully completed. Serial numbers were provided for all the instruments. This aided the keying of results into the data analysis software.

3.5.4. Data Analysis

Descriptive statistics and content analysis were used for data analysis. The statistical package employed for the data analysis was the Statistical Product and Service Solutions (SPSS) software, 19. Microsoft Excel (2013 version) was also used for the study. Likert Scale was used to assess the responses from the respondents. Respondents were asked to rank on a scale of 1 to 3, the significance of the issues presented on the data collection instruments. 1: represent Disagree, 2: represent neutral, 3: represent Agree. In the words of [23], [2].

Relative importance index was employed to rank the responses.

With relative importance index, frequencies attached to responses were multiplied by their corresponding ranking values (measured from 1 to 3) for each factor (W). These were respectively added up ($\sum W$) and subsequently, divided by the product of the total number of respondents (N) and the highest figure or integer on the five point Likert Scale (A= 5) [24]. For a three-point response item, RII produces a value ranging from 0.0 to 1.0. Appendix B explains RII formulae as in [25].

4. Introduction

Results of data collected are analyzed and presented in this chapter. Further, the findings of the study, which answers the study research questions, are presented and discussed. Respondents were purposively selected based on the judgment of the researcher to select five respondents from each of the six mobile telecommunication companies operating in Ghana. A sample of thirty (30) respondents was used for the study. The study begins with the analysis of the socio-demographic characteristics of the respondents. The analysis has been sectioned to correspond to the substantive objectives/ research questions set for the study.

4.1. Socio-demographic Characteristics of Respondents

The study looks into the social background of the respondents. The study reveals that all the respondents were married and were ranging between 35 to 45 years of age. All the respondents had more than five years of work experience on the job. The sex distribution of respondents stood at 75 percent for males and 25 percent for females. Making the profession, male dominated. All the respondents had tertiary level education and were all on the rank of management level in their respective organizations. The rest of the analysis focuses on the substantive objectives of the study and have been done under sub-headings.

4.2. Benefits of Infrastructure Sharing

The study considered the benefits of sharing infrastructure. Respondents were asked to report the benefits they have received from sharing infrastructure with other networks. The benefits of sharing have been enumerated in this section. These are: Infrastructure sharing speeds up the process of network installation. Cost of building a site is expensive, so with sharing, all the operators come on board and share the cost. Infrastructure sharing is one of the effective strategies of encouraging and promoting cooperation among the different operators in the industry.

Respondents further note benefits such as reduction in operational and capital cost. Reduction in site visits, wide area coverage and ease/ quick site building site. Lastly, sharing of site maintenance is a responsibility.

In a similar study conducted in Nigeria for infrastructure sharing between Mobile Telecommunication Network and Zain Nigeria limited reveals the following benefits [7], which conforms to this study. The benefits includes, but not limited to, low set up cost, wide area coverage, 30 percent reduction in capital expenditure savings and low operational cost (e.g. fuel, electricity etc.). Others include reduction in site visits due to site maintenance and quick to put up site on air.

4.3. Challenges of Infrastructure Sharing

Accessibility to the cell site was reported to pose a major problem since each sharing partner gives a different name to the same cell site. Security at the site was also reported to be lows most of the appliance and the devices used at the cell site often gets stolen at the site due to weak security measures from the original owners of the tower.

Timing wasting was reported to be another factor in sharing infrastructure and this result in the agreement processes. Some operators sometimes refuses co-location agreement due to weak towers or that a loaded. Another reason for refusal in share is to gain monopoly in a particular area to boost revenue and increase the number of subscribers. It was also reported that other players use inferior and other infrastructure which makes it incompatible to other sharing operators. Table 2 presents the details on the challenges.

Table 2. *Challenges faced by Operators in Infrastructure Sharing.*

Challenges	Percent
Inferior equipment	30
Timing delays	30
Monopolistic behavior	22
Security	15
Accessibility to site	3
Total	100

Source: Field data, 2016

Table 2 presents the challenges operators of network face, arranged logically from the highest to the lowest, reflecting the responses from the respondents. These are the regular use of inferior and incompatible equipment (30%), delays in completing sharing agreements (30%) and monopolistic behavior (22%) of some operators on the field. These so called "giant operators" refuses to share their infrastructure in order for them to continue to enjoy and maintain monopoly over certain services, thus becomes their competitive advantage tool.

The rest of the challenges, enumerated by the respondents, were security (15%) and site accessibility (3%). Respondents complained of frequent loss of equipment and tools on cell sites. With shared sites, each operator has her own engineers. Frequent use and changes with engineers goes with losses of equipment and tools. Lack of uniformity in cell site naming, because of the differences in the names each operator gives to her cell site. This creates problems locating the cell site.

In a similar study on infrastructure sharing between Zain and Mobile Telecommunication Network Nigeria limited reveals [7] the following challenges. These challenges are similar to that of this study. These are: incompatible equipment and inferior equipment. The rest were refusal to share by some of the operators.

5. Conclusions

5.1. Summary of Findings

The study investigated infrastructure sharing among Ghana's mobile Telecommunication Networks and focused on the benefits and challenges that operators face in the process of sharing infrastructure. The study used all the head offices of the six mobile telecommunication operators in Ghana- viz Globacom (GLO), Mobile Telecommunication Network, Tigo, Vodafone, Airtel and Expresso. Five (5) respondents were purposively selected from the head office of each of the operators, making the total sample size of 30. All the 30 were in top management position for the firms. Qualitative sampling procedure was used. Interview guide were used on all the respondents. Primary and secondary sources of data were used for the study. Scaling and percentages techniques were used for the data analysis. Key findings from the study have been presented here. With respect to the first objectives of the study, the types of infrastructure shared were:

Must is the most shared infrastructure

Space in building is the second most shared infrastructure

Electricity power was the third most shared infrastructure

The least shared infrastructure was: antenna, microwave equipment.

To support the first objective, the study assessed the drivers of sharing among the operators in Ghana. The network operators in Ghana were compelled to share infrastructure because of:

High cost involved in building a cell site. Sharing will allow for sharing of the cost;

Reduction in operational and capital cost of operators;

Promotion of access to areas of strategic importance and also areas considered as under or not served. This helps operators to capture more subscribers.

Promotion is ease in the acquisition of cell site.

With respect to the second objective of the study, the benefit of sharing was:

Speed- up process of network installation, also known as quick in building cell site

Reduced cost of building sells sites

Promotion of corporation among the operators

With respect to the third and the final objective, the challenges linked with sharing infrastructure were:

Inferior equipment used by some of the operators

Delays in the preparation of the sharing agreement

Monopolistic behavior of some of the operators to enjoy competitive advantage

Low security of equipment at cell sites

Lastly, difficulty in accessing cell sites due to the different naming by different operators.

5.2. Conclusion

In this current age where information is a key for development, the role of mobile telecommunication industry cannot be downplayed. It is therefore an urgent need to promote and sustain the mobile telecommunication industry, since aside providing a means for communication has also created vast job opportunities to masses of unemployed persons in Ghana and Africa at large. Infrastructure sharing is one of the strategies to adopt to enhance, promote and sustain the mobile telecommunication industry. This study was embarked within this mindset.

The first objective made it real that infrastructure sharing is going on practically among the mobile operators in Ghana. However, the stage is at its lowest levels since only few infrastructures are shared. It can be said that the sharing of infrastructure is still at its infant level.

It was revealed from the second objective that benefits are being derived from the current infrastructure sharing going on among the network operators in Ghana. These benefits have had a positive effect on customers, who now stand the chance to enjoy comparatively cheaper calls and internet service rates, compared with the situation before. Customers are also enjoying better services from the operators. Operators now can focus on core business such as customer service and service quality matters and therefore making lots of profits.

The last objective makes it clear that despite the efforts of promoting infrastructure sharing, the process is bedeviled with problems which hinder its growth. In all, the study and its findings were in line with the conceptual framework developed for the study, which combines the various concepts and theory used for study and also the variables within the empirical reviews. The conceptual framework depicts that drivers of infrastructure sharing gives ways for sharing, epitomized in the various types of infrastructure sharing. And the final result of the sharing is profit making and service improvement. All these were achieved in the study, with respect to its findings.

References

[1] Addy-Nayo, C. (2001). 3G mobile policy: The case of Ghana.ITU: Geneva, retrieved from www.itu.int/osg/spu/ni/3G/casestudies/ghana/ghanafinal.doc on Jan, 1, 2016.

[2] WSJ (2012). Telecom Giants Battle For Kenya. Retrieved from: http://online.wsj.com/article/SB10001424052748704514504575612012681373530.html, (Accessed on: January 1, 2016).

[3] Ernst & Young (n.d) as cited in Ernest IT News Africa (2009). Africa's telecom industry is fastest growing. Retrieved from, http://www.itnewsafrica.com/2009/02/africas-telecom-industry-is-fastest-growing/

[4] Porter, M. E. (1998). The competitive advantage of nations – with a new *introduction by the author.* London: MacMillan Press Ltd.

[5] SAPHYRE. (2010). D5.1a. Contract No.FP7-ICT-248001.

[6] Maitland, C., Bauer, J., & Westerveld, R. (2002). The European market for mobile data: Evolving value chains and industry structures. *Telecommunications Policy,* 26(9-10). 485-504.

[7] Onuzuruike E. (2009). Telecom Infrastructure Sharing as a Strategy for Cost Optimization and Revenue Generation: A Case Study of MTN Nigeria/Zain Nigeria Collocation. A thesis submitted to the Blekinge Institute of Technology, for the award of Master degree in Business Administration.

[8] Chanab, L. A. et. al. (2007). Telecom Infrastructure Sharing: Regulatory Enablers and Economic Benefits. *Booz Allen Hamilton Consulting*, December 2007, pp. 1–12.

[9] Dernab, S. (2011). Telecom Infrastructure Sharing as a Strategy for Cost Optimisation. Thesis submitted to the Kwame Nkrumah University of Science and Technology for the award of Commonwealth Executive Master in Business Administration.

[10] Chanab, L., El-Darwich, B., Hasbani, G. and Mourad, M. (2007). Telecom Infrastructure Sharing: Regulatory Enablers and Economic Benefits. Booz Allen Hamilton Inc.: Available from http://www.boozallen.com/media/file/Telecom_Infrastructure_Sharing.pdf (Retrieved July 20, 2014).

[11] Chatzicharistou, L. (2010). *Infrastructure Sharing in Mobile Service Market: Investigating the final decisions of the network operators*. Thesis submitted to the Delft University of Technology for the award of Master of Science in Engineering and Policy Analysis.

[12] Bala-Gbogbo, E. (2009). Telecom Industry Operators Opt for Infrastructure Sharing. Access on the 25th day of January, 2016 from the website http://www.234next.com/csp/cms/sites/Next/Money/Business/5418647/story.csp

[13] Loizillon, F. et al. (2002). Final Results on Seamless Mobile IP Service Provision. *Economics Deliverable number 11.* Information Society Technologies.

[14] Northstream. (2001). *Network sharing: Savings and Competitive effects*. The Swedish Post and Telecom Agency.

[15] Björkdahl, J. & Bohlin, E. (2003). *3G Network Investments in Sweden*. The Swedish Post and Telecom Agency.

[16] Creswell, W. J. (2003). *Research design: Qualitative, quantitative and mixed methods approach* (2nd edn.). London: Sage Publication.

[17] Creswell, W. J. (2004). *Research design: Qualitative, quantitative and mixed methods approach* (3nd edn.). London: Sage Publication.

[18] Kusi, H. (2012). *Doing Qualitative Research: a Guide for Researchers*. Accra: Emmpong Press.

[19] Kuhn, Thomas S. 1964. The structure of scientific revolutions. 3rd edition. Chicago: University of Chicago Press.

[20] Krejcie, R. V., & Morgan, D. W. (1970). Determining sample size for research activities. Educational and Psychological Measurement, 30 (3), 607-610.

[21] Sarantakos, S. (2005). *Social Science Research* (3rd edn.). New York: Palgrave Macmillan.

[22] Ayarkwa, J., Dansoh, A. & Amoah, P. (2010). Barriers to Implementation of EMS in Construction Industry in Ghana. *International Journal of Engineering Science,* 2(4), pp. 37-45.

[23] Fugar, F. D. & Agyakwah-Baah, A. B. (2010). Delays in Building Construction Projects in Ghana. *Australasian Journal of Construction Economics and Building,* 10 (1/2), p. 103-116.

[24] Kothari, R. C. (2004). *Research Methodology: Methods and Techniques*, (2ndEdn.). New Delhi: New Age International Publishers.

[25] Muhwezi, L., Acai,. J. & Otim, G. (2014). An Assessment of the Factors Causing Delays on Building Construction Projects in Uganda. *International Journal of Construction Engineering and Management,* 3(1), pp. 13-23.

A slotted circular monopole antenna for wireless applications

Gopal M. Dandime*, **Veeresh G. Kasabegoudar**

Post Graduate Department., Mahatma Basveshwar Education Society's, College of engineering, Ambajogai, India

Email address:
gopaldandime@gmail.com (G. M. Dandime), veereshgk2002@rediffmail.com (V. G. Kasabegoudar)

Abstract: In this paper, a compact slotted circular monopole antenna with spike shaped slots embedded in it is presented. Also, the proposed antenna has ground length limited to 33% of substrate length. This ground is provided on the back side of the antenna geometry to excite the antenna by microstrip line feed. A slotted circular patch element will be achieved by subtracting 45 degrees rotated square patch of 12mm x 12mm, and then by proper scaling. In this study the geometry is scaled separately by 60%, 40%, 20%, and finally the resultant is obtained by uniting them. This final geometry offers an ultra wide band operation. The overall size of the antenna is 30mm×32.4mm×1.6mm including finite ground feeding mechanism. The antenna operates in the frequency range from 2.5-15GHz covering FCC defined UWB band with more than 130% impedance bandwidth. Stable omni-directional radiation patterns in the desired frequency band have been obtained. Measured data fairly agree with the simulated results.

Keywords: Ultra Wideband, Printed Monopole Antenna, Slot Antenna, Defected Ground Structure (DGS)

---

1. Introduction

As we know that in patch antenna's narrow bandwidth is a major disadvantage and current wireless applications demand for wide bandwidth to integrate several applications. Therefore, it is necessary to design antenna with compact size, wide bandwidth, and low profile. In the recent years printed monopole antennas are popular due to some advantages like low profile, etched on a dielectric substrate and can provide feature of broadband and multiband. Novel printed monopole antenna presented in [1] has cross slotted shaped slot with tapered shaped microstrip to achieve broadband matching. Planar monopole antennas can be used for UWB applications due to their wideband characteristics [2].

The Ultra Wideband (UWB) refers to the frequency range of 3.1 to 10.6 GHz [3]. As this band is unlicensed by FCC which encourages the designers to go for UWB antennas [1-13]. Microstrip line feed is more suitable to excite a patch antenna as it facilitates proper impedance matching. Hence, these microstrip feeds are preferred for wireless applications and also the similar feed is used in this work. In this paper we propose a printed monopole antenna with operating frequency ranges from 2.8 GHz to 12 GHz which is close to 124% (impedance bandwidth). There are several bandwidth enhancement techniques used in monopole antennas to achieve such a huge bandwidth. In proposed antenna we have used the slots in circular shaped patch in which square is subtracted and then scaled & united together. A microstrip feed with dimensions of length equal to 11.9mm & width of 2.9mm is used for excitation. In this work, a compact microstrip fed slotted circular patch monopole antenna is presented. Although, the antenna demonstrated in this paper is similar to the geometry presented in [12], has several changes like much smaller dimensions, microstrip feed rather than CPW feed to acheive better gain. Even though the microstrip feed is used, it offers impedance bandwidth similar to [12]. The antenna is simulated using Ansoft's High Frequency Structure Simulator (HFSS) v.13. More details on this geometry are discussed in subsequent sections. Section 2 presents the geometry structure of the proposed antenna. Design and optimization procedure of the proposed antenna is presented in Section 3. Section 4 presents the validation of the fabricated prototype and discussions on the measured results are also presented there. Finally, conclusions of this study are presented in Section 5.

2. Antenna Geometry

Figure 1(a) shows the basic geometry of proposed circular monopole antenna for UWB operation. The antenna is symmetrical with respect to the longitudinal direction. Substrate used for the design is FR4 with dielectric constant of 4.4, and thickness of 1.6mm. Finite ground planes, with dimensions of length L_g and width W_g are placed symmetrically on either side of the microstrip feed-line. The proposed geometry is fed by the microstrip line. The detailed optimization procedure of the proposed antenna and its optimum dimensions, and characteristics are presented in Section 3. All the parameters of the geometry are indicated in Figure 1 (a) and optimized dimensions are listed in Table 1.

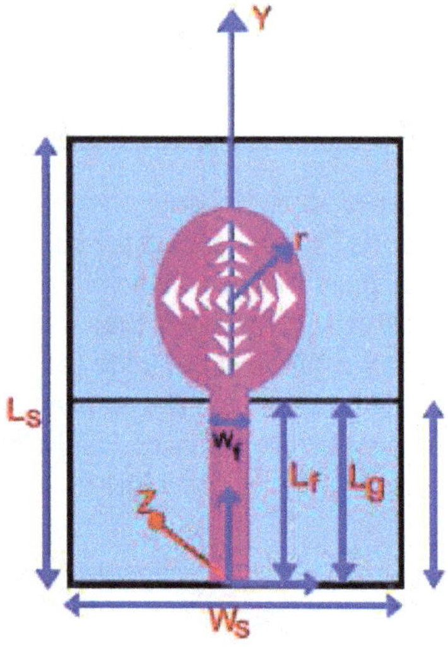

(a) Top view

(b) Bottom view (ground)

Figure 1. *Geometry of proposed microstrip fed monopole antenna.*

Table 1. *Optimized dimension of the proposed antenna*

Geometry Parameter	Notation	Dimensions of geometry in mm
Radius of Patch	r	9.50
Substrate Length	L_s	32.4
Substrate Width	W_s	30.0
Feed Length	L_f	11.9
Feed Width	W_f	2.90
Ground Length	L_g	10.9

3. Geometry Optimization and Discussions

In this section parametric study is conducted to optimize the proposed antenna. The key design parameters used for the optimization are radius of circular patch, square slot, ground plane length, and length & width of the feed. The detailed analysis of these parameters is investigated in the following subsections.

3.1. Effect of Square Slot and Scaling of Patch Geometry

As shown in Figure 2, circular patch geometry is subtracted by a square slot of dimensions 12mm by 12mm & then whole geometry is scaled to 60%, 40% & finally 20% to see scaling effect on the performance of antenna. Return loss characteristics of this study are presented in Figure 3. From Figure 3 it may be noted that the fourth iteration offers wide impedance bandwidth. This is mainly due to the excitation of closely spaced multi resonance which merges and offer wide impedance bandwidth.

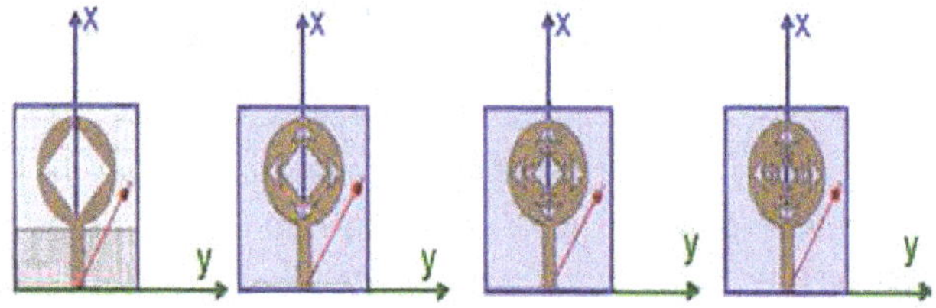

(a) 1st Iteration (b) 2nd Iteration (c) 3rd Iteration (d) 4th Iteration

Figure 2. *Geometry modification of antenna by scaling process of patch*

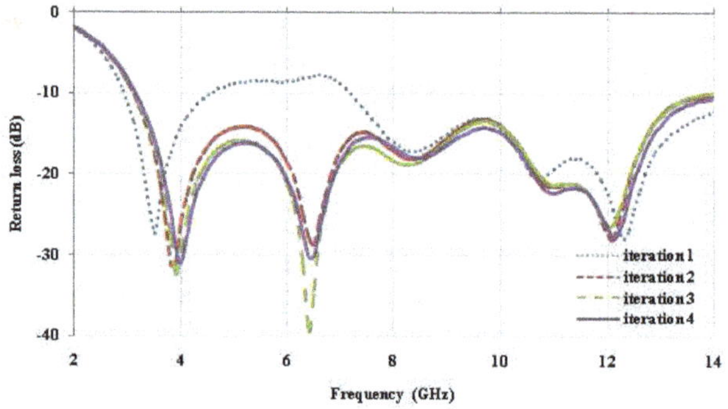

Figure 3. *Return loss vs. frequency plot for variation in geometry of patch as indicated in Figure 2.*

Table 2. *Effect of variation of geometry on bandwidth of proposed antenna*

Iteration	1st (No Scaling)	2nd (60% Scaling)	3rd (40% Scaling)	4th (20% Scaling)
Frequency range (GHz)	7.5-15	3-14.8	3-15	3-15
Bandwidth (%)	66.6	132.5	133.3	133.3
Positive Gain Freq. Range (GHz)	2.8-6.8	2.8-7.6	2.8-7.8	2.8-8.0

3.2. *Effect of Radius & Ground Dimensions*

To study the effect of circular patch dimensions on the antenna performance, its radius (r) is varied. Initially, the radius of the patch (r) is varied from 6.5 mm to 12.5 mm in steps of 1mm keeping ground dimensions & feed dimensions constant (L_f=11.9mm, W_f=2.9mm). From the simulations we found an optimum radius (r) of 9.5mm. Return loss characteristics of this study are depicted in Figure 4 and calculated bandwidth values are presented in Table 3. From Figure 4, it may be noted that the lower cut-off frequency remains nearly constant whereas upper cut-off frequency varies slightly i.e., bandwidth varies with respect to radius parameter. From Table 3 it is found that the geometry offers the maximum impedance bandwidth of 124%. Also, the defected ground behind the monopole geometries offer a wide impedance bandwidth [14-17], in this work we have included a finite ground of size equal to L_g x W_s. Width of the ground is chosen equal to the substrate width and length is chosen such that it covers only feed line of the monopole antenna.

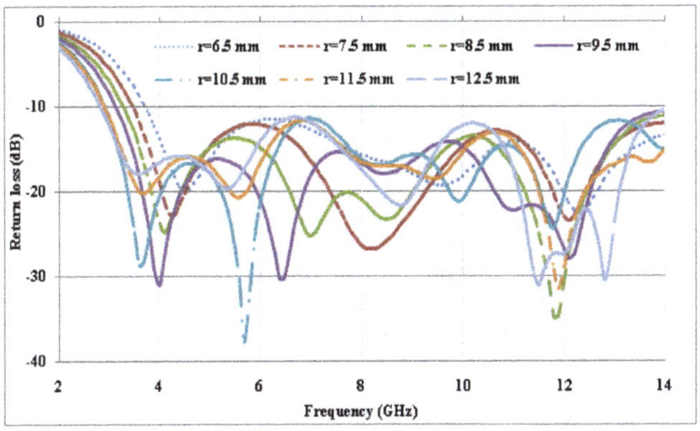

Figure 4. *Return loss vs. frequency plot with variable radius of the circular patch (r).*

Table 3. *Effect of variation of radius of patch on bandwidth of proposed antenna*

Radius (*r*) (mm)	6.5	7.5	8.5	9.5	10.5	11.5	12.5
Frequency Range (GHz)	3.76- 15	3.5- 15	3.26-15	3.12-15	3.22-14.56	3.21-14.72	3.72-15
Bandwidth (%)	119.8	124.3	128.5	131.2	127.5	128.3	120.5

3.3. *Effect of Length & Width of Feed Line*

In this investigation, first by keeping feed width constant (W_f=2.9mm) the length of feed line was varied from 7.9mm to 14.9mm in steps of 2mm and its effects are presented in Figure 5(a) and Table 4. From Figure 5 (a) it may be noted that the optimized results are obtained for feed length L_f =11.9mm. In the further study, feed length L_f=11.9mm was kept constant and feed width (W_f) was varied from 2.3mm to 2.9mm in steps of 0.2mm The S_{11} characteristics of this study are presented in Figure 6 and Table 5. From this study, optimized value of feed width equal to 2.9mm is obtained. It may be noted that these dimension vary the input impedance of the antenna and optimized as explained above to yield maximum impedance bandwidth.

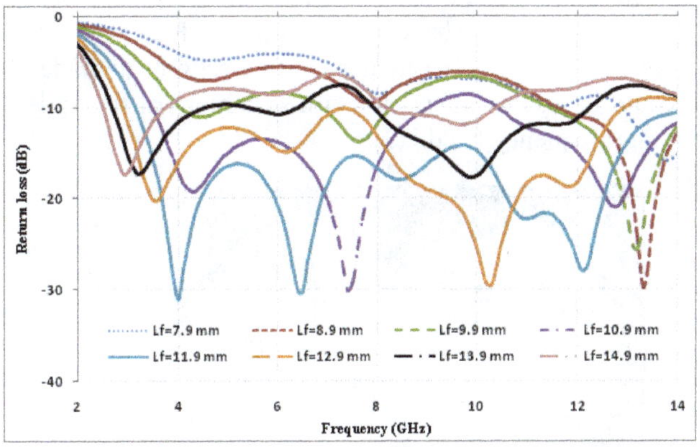

Figure 5. *Return loss vs. frequency plot for feed length variation.*

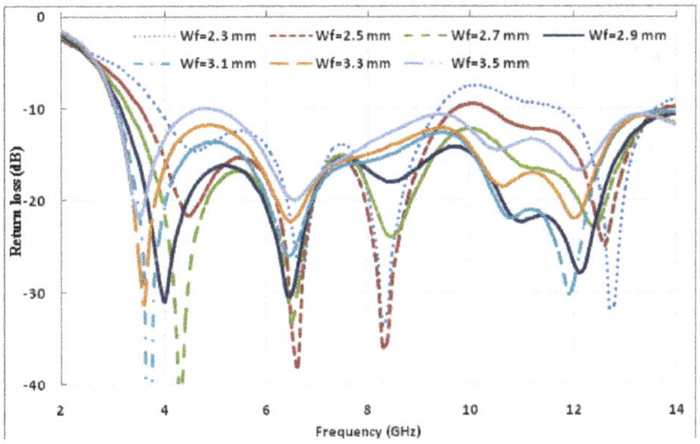

Figure 6. *Return loss vs. frequency plot for feed width variation.*

Table 4. *Effect of feed length on bandwidth of proposed antenna*

Feed Length(mm)	9.9	10.9	11.9	12.9	13.9	14.9
Bandwidth(%)	66.6	88	131.25	129.11	126.53	26.087

Table 5. *Effect of feed width on bandwidth of proposed antenna*

Feed Width(mm)	2.3	2.5	2.7	2.9	3.1	3.3	3.5
Bandwidth (%)	83.73	118.6	128.76	131.25	124.9	127.9	124.45

4. Experimental Results and Discussions

The geometry shown in Figure 1 with its optimized dimensions using HFSS EM software [18] presented in Table 1 was fabricated and tested. The substrate used for the fabrication is the FR4 glass epoxy with dielectric constant of 4.4, and thickness of 1.6mm. A photograph of the fabricated prototype is shown in Figure 7. Return loss comparisons of measured and simulated values are compared in Figure 8. The measured results are fairly agreed with the simulated values.

From Figure 8 it may be noted that the proposed antenna is having measured operating frequency range from 4.5 GHz to 12 GHz. This corresponds to an impedance bandwidth (measured) close to 100%. Radiation patterns of the geometry are presented at various frequencies in the band of operation (Figure 9). These patterns are symmetrical at the start and middle frequencies of the band of operation with nearly omni-directional shape in the H-plane. However, at the end of operating band of frequencies the patterns degrade which could be due complex current distributions at the end of operating frequency band [13].

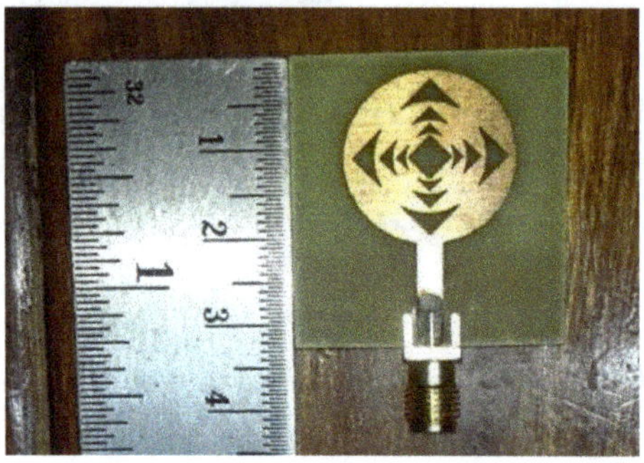

Figure 7. *Fabricated prototype of antenna shown in Figure 1.*

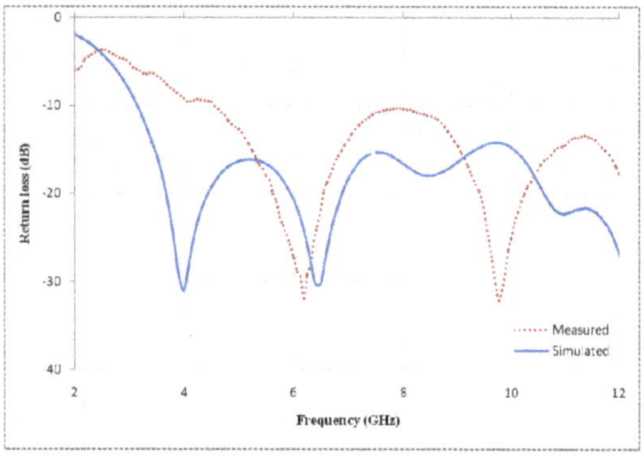

Figure 8. *Return loss characteristics comparisons of the antenna geometry shown in Figure 1.*

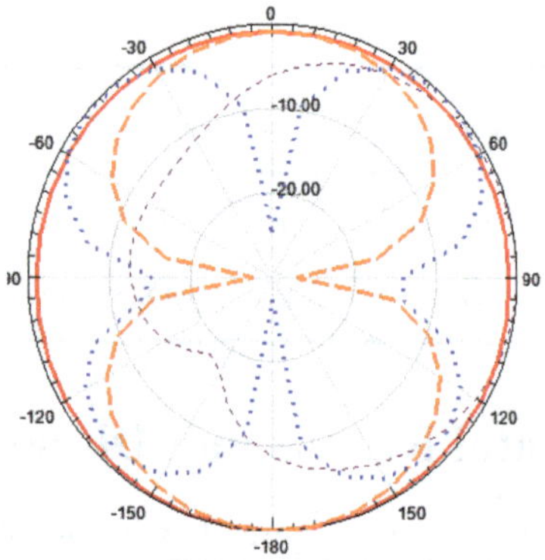

(a) 3.2GHz

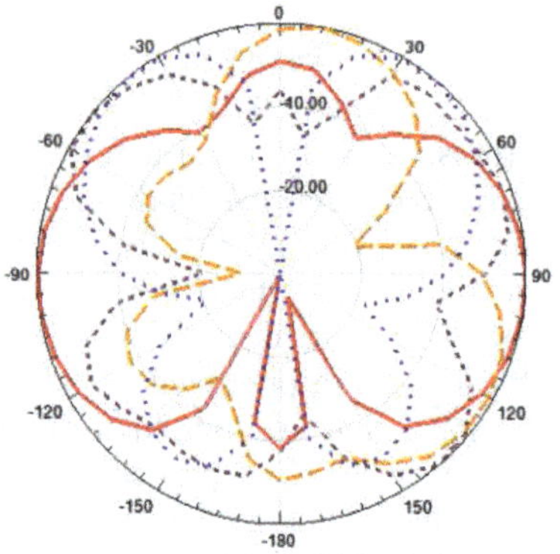

(b) 9.1GHz

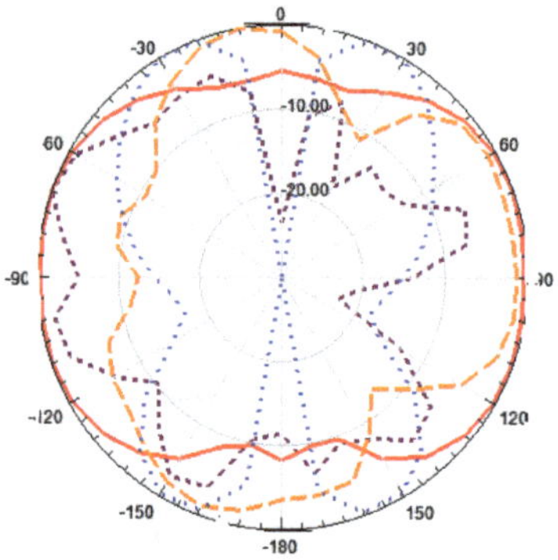

(c) 12 GHz

Figure 9. *Radiation patterns at various frequencies of the antenna shown in Figure 1. Red (solid): H-Co. Poln.; Orange (Long-Dash): E-Co. Poln.; Blue (Doted): E-Cross Poln.; Brown (Short-Dash): H-Cross Poln.*

5. Conclusions

A compact circular shaped slotted monopole antenna mounted on rectangular base with finite ground has been presented. Ground geometry was varied & fixed with optimum size to 33% to offer wide impedance bandwidth and good gain. Also, by varying feed length & width a perfect impedance matching can be obtained. The antenna presented here offers an impedance bandwidth of more than 124% in the frequency range of 2.8GHz to 12GHz which essentially covers the FCC defined ultra wideband (UWB) frequency range. Omni-directional radiation patterns were obtained throughout the band of operation. Therefore, the antenna presented here is a suitable for the FCC defined UWB band of operation.

References

[1] R. S. Kshetrimayum and R. Pillalamarri, "Novel UWB printed monopole antenna with tapered feed lines," Indian Institute of Technology, Guwahati, India 2009.

[2] G. Kumar and K. P. Ray, Broadband Microstrip Antennas, Artech House Boston, London, 2002.

[3] J. R. Panda and R. S. Kshetrimayum, "An F-shaped printed monopole antenna for dual-band RFID and WLAN applications," Indian Institute of Technology, Guwahati, India 2010.

[4] First Report and Order, Revision of part 15 of the commission's rule regarding ultra- wideband transmission systems FCC 02-48, Federal Communications Commission, 2002.

[5] W.-J. Liu, Q.-X. Chu, and L.-H. Ye, "A low-profile monopole antenna embedded with a resonant slot," Progress In Electromagnetic Research Letters, vol. 14, pp. 59-67, 2010.

[6] K. C. gupta, R. Gerg, I. Bahl and P. Bhartia, Microstrip Lines and Slotline, Artech House, Boston, London.

[7] R. S. Kshetrimayum, J. R. Panda, and R. Pillalamarri "UWB printed monopole antenna with a notch frequency for coexistence with IEEE 802.11a WLAN devices," Indian Institute of Technology, Guwahati, India, 2009.

[8] Z.-A. Zheng and Q.-X. Chu, "Compact CPW-fed UWB antenna with dual band notched characteristics," Progress In Electromagnetics Research Letters, vol. 11, pp. 83-91, 2009.

[9] N. Kushwaha and R. Kumar, "An UWB fractal antenna with defected ground structure and swastika shape electromagnetic band gap," Progress In Electromagnetics Research B, vol. 52, pp. 383-403, 2013.

[10] S. Sadat, M. Fardis, F. Geran, and G. Dadashzadeh, "A compact microstrip square-ring slot atenna for UWB applications," Progress In Electromagnetics Research, vol. 67, pp. 173-179, 2007.

[11] C.A. Balanis, Antenna Theory Analysis & Design, 3rd Edition, John Wiley & Sons, New York, 2011.

[12] R. Kumar and K. K. Sawant, "Design of CPW-fed fourth iterative UWB fractal antenna," International Journal of Microwave & Optical Technology, vol. 5, no. 6, pp. 320-327, 2010.

[13] M. K. Kulkarni and V. G. Kasabegoudar, A CPW-fed triangular monopole antenna with staircase ground for UWB applications," Int. J. Wireless Communications and Mobile Computing, vol. 1, no. 4, pp. 129-135, 2013.

[14] Y.S. Li, W.X. Li and Q.B. Ye, " Compact reconfigurable UWB antenna integrated with stepped impedance stub loaded resonators and switches," Progress In Electromagnetics Research, vol. 27, pp. 239-252, 2007

[15] J. K. Ali, A.J. Salim, A.I. Hammoodi and H. Alsaedi, "An Ultra-wideband printed monopole antenna with a fractal based reduced ground plane," Progress In Electromagnetics Research Symposium Proceedings, vol. 19, pp. 613-617, 2012

[16] K. R. Dharani and D. Pavithra, " a simple miniature U-shaped slot antenna for WIMAX applications," International Journal of Advance in Engineering & Technology, pp. 1256-1262, 2013

[17] R. Ghatak, A. Karmakar, and D.R. Poddar, "A circular shaped sierpinsik carpet fractal UWB monopole antenna with band rejection capability," Progress In Electromagnetics Research, vol. 24, pp. 221-234, 2011.

[18] HFSS13.0 User's Manual, Ansoft Corporation, Pittsburgh.

Design and Implementation of a Testbed for Mobile Adhoc Network Protocols

Akhtar Hussain, Aimel Khan, Abdul Rehman Qaiser, Muhammad Mohsin Akhtar, Obaidullah Khalid, Muhammad Faisal Khan*

Dept. of Electrical Engineering, National University of Sciences and Technology, Islamabad, Pakistan

Email address:
mfaisalkhan@mcs.edu.pk (M. F. Khan)

Abstract: Due to the significant growth in the area of wireless communication in the last few years, Quality of Service (QoS) has become an important consideration for supporting variety of applications. Elaborate testing of newly developed Mobile Adhoc Network's routing protocols in real world scenarios is a key step for providing QoS to the users. In order to test the developed Adhoc Routing Protocols, an IEEE 802.11 based testbed has been developed in the paper. Some well-known Adhoc network protocols were implemented and tested in user space daemon in Linux. All the key parameters necessary for assessing an Adhoc Network Protocol have been analyzed. A user application was developed in Java NetBeans Integrated Development Environment (IDE) to ensure repeatability and to provide a mobility model to the users in an efficient way. Customized and special purpose nodes have been developed by integrating various hardware components, in order to improve the efficiency and robustness of the testbed. Some of the well-known protocols have been tested by exposing the configured nodes to an outdoor atmosphere to cater all the unforeseen environmental factors, which affect the performance of Adhoc Protocols. Different performance metrics like overhead, throughput, end to end delay, average jitter and packet loss were evaluated by varying mobility, number of hops, packet size and pause time.

Keywords: Ad-hoc Routing Protocols, AODV, B.A.T.M.A.N, File Transfer Protocol, Mobile Ad-hoc Networks, Test-Bed, Network Time Protocol, OLSR

1. Introduction

Mobile Ad hoc Network (MANET) is an infrastructure-less network and can be deployed instantly to serve temporary or urgent purposes. MANETs promises to be useful in disaster management, for military in combat zones, for spontaneous meetings and for all those scenarios where the existing infrastructure is damaged or difficult to deploy a trivial communication infrastructure. Several protocols have been proposed and many of them are being practiced in real world scenarios. Bluetooth and Wi-Fi are among the common communication links used for mobile adhoc networks. Whenever new protocols are developed, their viability and efficiency is being evaluated, to assess their key features, conquer their shortcomings and add more appealing features.

There are several methods for testing the developed protocols. Theoretical evaluation provides elementary insight into the features of the investigated approach. Software simulations can be used for initial design and for an estimation of the results, but they do not realistically duplicate the physical layer. What emulator testbeds have in common is that they try to address the problems of scaling, management and test repeatability; but emulator testbeds cannot always be a substitute for real world scenarios. Testbeds with indoor experiments by using MAC filtering are also common. Indoor testing of protocols doesn't cater to the environmental conditions and open-air interruptions. MANETs are usually used for emergency purposes, and they find their major usages in open-air scenarios.

An emulated testbed was proposed in [1] which compresses the network and emulates the mobility without actually moving the nodes. An End to end delay analysis model was proposed in [2] with special consideration to MAC delay contention time while [3] compares the end to end delay in pervasive multimedia networks. A testbed for evaluating various performance parameters of Adhoc protocols was suggested by [4] with varying mobility and node congestion. Jitter of different Adhoc protocols was compared by [5]. Some security issues related to Adhoc networks were discussed in [6] and a quantitative analysis of a full-scale multi-hop Adhoc network testbed was carried out by [7].

The requirements for a testbed to be reliable and successful for testing the newly developed protocols are numerous. Some of them are scalability, cost-effectiveness, reproducibility, reliability and management [2]. A testbed with a user application and dedicated hardware will serve this purpose in an efficient way.

This paper proposes an IEEE 802.11 based testbed for adhoc network protocols. Some well-known Adhoc network Protocols i.e AODV, OLSR and B.A.T.M.A.N were implemented in user space daemon in Ubuntu. All the tests have been carried out in open air to consider all the possible interferences and unseen environmental factors, which influence the performance of the routing protocols. The developed testbed has been equipped with an application which was developed in Java NetBeans IDE. Testing nodes have been developed by integrating various hardware components to make a customized hardware to facilitate the users in testing and to ensure reproducibility in an efficient and robust way.

The rest of the paper is organized as follows. In section 2, a brief introduction of the tested protocols and performance metrics for assessing Adhoc Protocols has been presented. In section 3, the developed testbed has been discussed in detail; while in section 4, the results of the tested protocols and discussions regarding each result have been presented and are followed by conclusions.

2. Adhoc Network Protocols and Performance Metrics

IEEE has standardized the 802.11 protocol for Wireless Local Area Networks. It works in two modes: DCF (Distributed Coordination Function) and PCF (Point Coordination Function). 802.11 DCF can be used as a MAC scheme for multi-hop Wireless Adhoc Networks. It is CSMA/CA with binary slotted exponential bakeoffs. Data transmission by a mobile node in DCF mode using RTS and CTS is shown in figure 1.

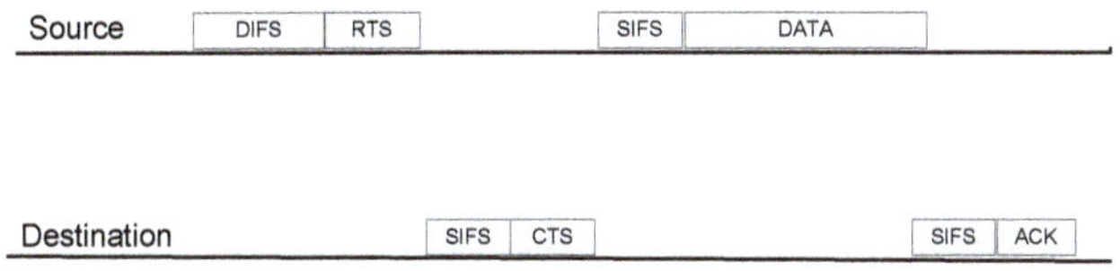

Fig. 1. *Data Transmission in IEEE 802.11in DCF mode*

Protocols governing the routing of packets within a MANET's scenario can be categorized as Reactive, Proactive and Hybrid Protocols. In proactive protocols, the route information is kept up to date by exchanging the control packets between the nodes periodically. While proactive protocols initiate a route discovery only when a node requires a route to a certain node. Hybrid protocols are the result of an intelligent combination of reactive and proactive protocols, in order to increase scalability and reduce the overhead.

2.1. Tested Adhoc Network Protocols

The protocols which have been used for testing the viability of the developed testbed in this paper are Adhoc On demand Distance routing Vector (AODV), Optimized Link State Routing (OLSR) and Better Approach To Mobile Adhoc Networking (B.A.T.M.A.N). The basic working principle of each protocol is discussed in the following section.

2.1.1. Adhoc On Demand Distance routing Vector

AODV is a reactive protocol and its working is based on the Distance Vector algorithm, which tells the information about its neighbors and number of hops required to reach them. Figure 2 shows the different types of messages used by the AODV for finding a route to a destination [7]. Where Hello messages are used to detect and monitor links to the neighbors.

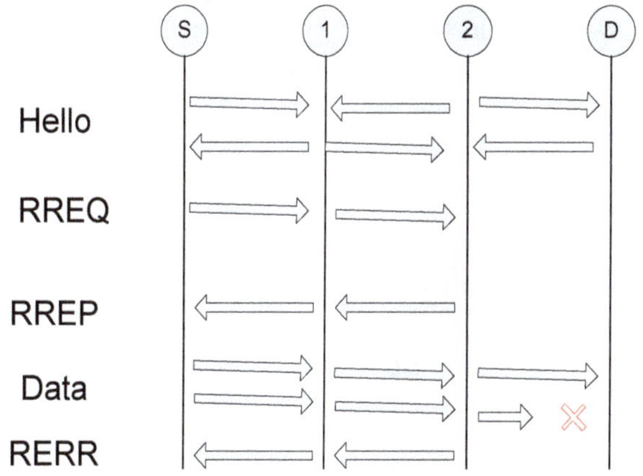

Fig. 2. *AODV Protocol Messaging*

When the source has to send data to an unknown destination, it broadcasts a Route Request (RREQ) message to that destination. The destination generates a Route Reply (RREP) and unicasts to the source. If the data is flowing and a link failure is detected, Route Error (RERR) is sent to the source.

2.1.2. Optimized Link State Routing

OLSR is an optimized version of Link State Routing Protocol. The protocol inherits the stability of Link State Algorithm and has the routes readily available pertaining to its proactive nature. In order to reduce the overhead, only the nodes designated as MPRs (Multipoint Relays) are responsible for forwarding the control information. Figure 3 shows the selection of the MPRs [8]. The rule for selecting an MPR is "For all 2 hops neighbors' n there must exist an MPR m so that n can be connected via m".

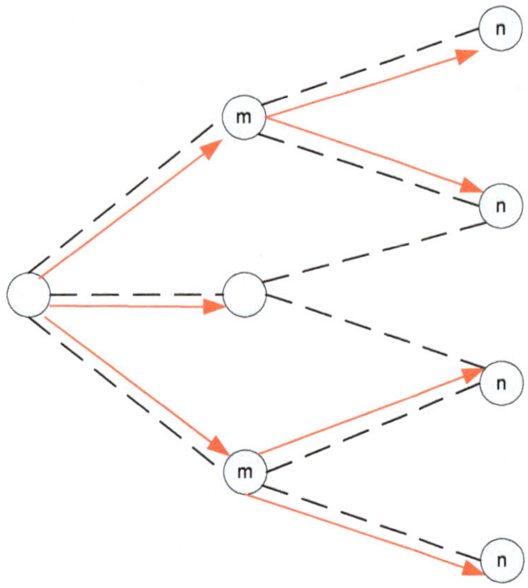

Fig. 3. MPR Selection in OLSR.

2.1.3. Better Approach to Mobile Adhoc Networking

B.A.T.M.A.N is also a proactive protocol and is especially designed for networks where the view of the topology is ambiguous and altering constantly. It finds the optimal gateway node to the destination instead of a complete route. It uses OMGs (Originator Messages) to calculate link quality and to identify the path. Figure 4 shows the OMGs and its contents.

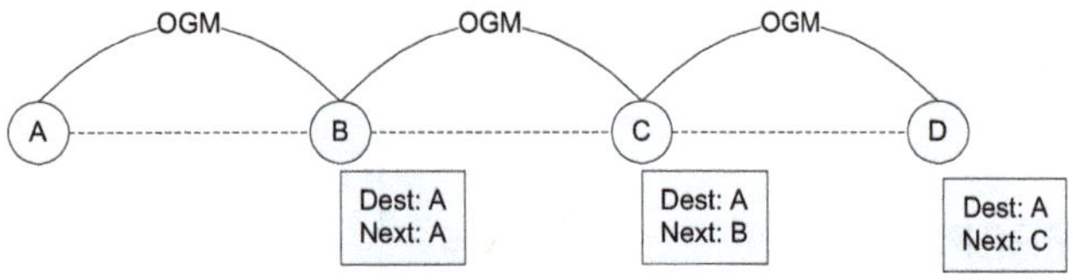

Fig. 4. B.A.T.M.A.N Originator Messages.

The approach used by BATMAN is to generate an Originator Message periodically. The source name and sequence number is used to uniquely identify a packet and is used to detect duplicates. On receiving an OMG, the routing tables are updated [9].

2.2. Performance Metrics

Performance Metrics encompasses QoS (Quality of service) to the end users in terms of several general parameters [10]. The perceived Quality of Service can be measured in terms of several parameters. The key parameters for evaluating an Adhoc Network Protocol are Overhead, Throughput, Jitter, End to End Delay and Packet Loss. Several papers have been written on these performance metrics [10], [11], [12], [13]. All of these five parameters have been analyzed by the developed testbed in this paper. Details of each parameter are in the following sections.

2.2.1. Overhead

Control traffic overhead mainly depends upon the topology and the data traffic in addition to the protocol being used. AODV generate overhead only when a new route is needed while OLSR and BATMAN continuously generate control traffic. The formula for calculating control traffic in both cases is shown in table 1[14].

Table 1. *Equations for Overhead Computation*

	Reactive Protocols	
	Packets	**Bandwidth**
Fixed	ßôN^2+ÞN	ÞPN+ßôRQrN^2
Mobile	ôµaLN^2	ôpµaRQrN^2
	Proactive Protocols	
	Packets	**Bandwidth**
Fixed	ßN+ôptpN^2	ÞPN+ôptpTpN^2
Mobile	ôµANpN^2	ôpµANpTpN^2

Given the parameters shown in table 2, overhead of both types of protocols can be calculated. Table 1 contains different equation for computing overhead for fixed and mobile nodes. Overhead is a key parameter in analyzing adhoc routing protocols and has been discussed in detail in different papers [14], [16].

Table 2. *Parameters for equation in table 1.*

Reactive Protocol Parameters	
ß	Route creation rate per node
ô	Route request optimization factor
N	Number of Nodes
Þ	Hello rate
RQr	Average size of route request
µ	Link breakage rate
L	Number of route reply messages
µa	Active node link breakage rate
ôp	Broadcast optimization factor
Proactive Protocol Parameters	
ß	Route creation rate per node
Ôp	Broadcast optimization factor
N	Number of Nodes
Þ	Hello rate
ANp	Active next hop ratio
Tp	Average size of topology broadcast packet
P	Number of route reply messages
P	Active node link breakage rate
tp	Topology broadcast rate

2.2.2. Throughput

A networks end to end throughput is a measure of the network's successful transmission rate and is defined as bytes per second [20]. End to end throughput can be calculated by using equation (1) [15].

$$TP = \sum_{n=1}^{N} \left(\frac{PS(n)}{PAT(n) - PST(n)} \right) \quad (1)$$

Where TP: Throughput, PS: Packet Size, PAT: Packet Arrival Time, PST: Packet Sending Time and N is the number of packets transmitted. In order to calculate the throughput, the size of each data packet was added. This gives the total data transferred [15]. Throughput is an important metric for evaluation of any Adhoc network protocols. Depending on the type of data used for transmission, the acceptable thresholds of throughput will be different. Throughput is an index for Quality of Service in all types of data.

2.2.3. End to End Delay

End to end delay is the average time between generation and successful delivery of the packets for all nodes in the network [17], [18]. The major sources of delay could be processing delay, network delay, propagation delay and destination processing delay. Equation for calculating end to end delay is given as (2) [15].

$$E2ED = \sum_{n=1}^{N} \left(\frac{PAT(n) - PST(n)}{N} \right) \quad (2)$$

Where E2ED is end to end delay and remaining parameters are the same as in the equation (1).

Packet arrival is the time when a packet reaches the destination. End to end delay may be larger in open air due to unforeseen environmental effects.

2.2.4. Average Jitter

Jitter is defined as the variation of data communication packets in the network. It is the variation in time between each of the packets arriving [19]. Jitter is an index for consistency and stability of the network. Jitter can be calculated by equation (3) [15].

$$PPD(n) = \sum_{n=1}^{N} (PAT(n+1) - PST(n+1)) \quad (3)$$

$$CPD(n) = \sum_{n=1}^{N} (PAT(n) - PST(n)) \quad (4)$$

Where PPD is Previous Packet Difference, CPD is Current Packet Difference and AJ is Average Jitter. Average jitter can be calculated by using equation (3) and equation (4), as shown in equation (5).

$$AJ = \sum_{n=1}^{N} \left(\frac{PPD(n) - CPD(n)}{N-1} \right) \quad (5)$$

Average jitter is a critical element in determining the performance of the network and QoS offered by the network.

2.2.5. Packet Loss

Packet loss or corruption of packets indicates the packets which have been sent by the sender but not received by the destination node. It affects the perceived quality of the application. Packet loss could be due to unstable wireless connection, overflowing of the queuing buffer or congestion in the network. The formula formulated for computing lost packets is equation (6) [15].

$$PL = \frac{\sum_{l=1}^{L} (LP(l) * Size(l))}{\sum_{n=1}^{N} (SP(n) * Size(n))} * 100 \quad (6)$$

Where PL is Packet Loss, LP is Lost Packet, SP is Sent Packet, L is number of lost packets. All of the above mentioned performance metrics are evaluated by the developed testbed. The development process of the complete testbed is explained in the following section.

3. MANETs Testbed

The developed Mobile Adhoc Networks testbed is capable of providing a generic platform for the testing of Adhoc protocols in real world scenarios, equipped with a suitable software environment for the facilitation of user's utility. Real world testing is being carried out to cater for deficiencies of simulations, which are based on a significant level of abstractions.

All the protocols mentioned in section 2 were implemented in user space daemon. A daemon is a program that runs in the background, rather than under the direct control of a user. It communicates with the kernel module to discover and maintain the routes. User space daemon implementation is advantageous to kernel modification in many aspects. The user space communicates with the kernel space through sockets as shown if figure.5.

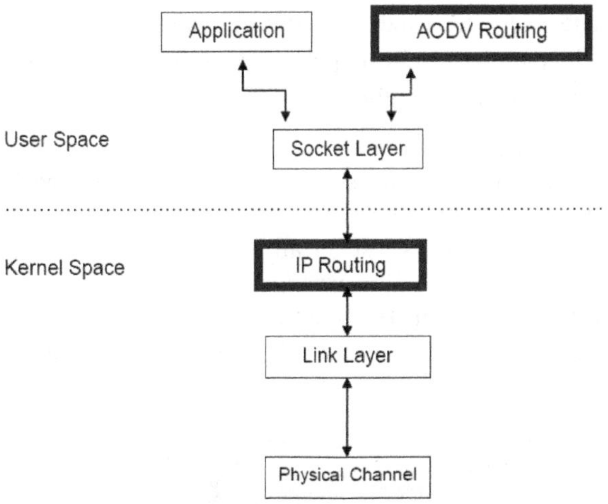

Fig. 5. User Space Daemon

The developed testbed is composed of a hardware implementation and a software application. The software application was developed in Java NetBeans IDE and testing nodes were developed by integrating several hardware components. Details about tested data, software application and hardware are in the following sections.

3.1. Tested Data

Different types of data have been used in the proposed testbed to analyze the robustness of the testing protocol. Test data includes file transfer through FTP and video streaming through VLC. All the nodes have been time synchronized with an NTP server. Precise time synchronization is required to ensure the viability of the test results resulting in a reliable analysis. NTP is used for synchronizing the clocks of the computer systems over packet-switched, variable-latency data networks. It utilizes UDP on port 123.

FTP is based on a client-server architecture which makes

use of separate control and data connections between the client and server. The client is allowed to access to the server's database after entering the correct username and password.

VLC has been utilized to stream videos from one to multiple nodes over the adhoc network. Before streaming is initialized, various parameters such as transcoding rate and the video format together with the IP address of the destination are required to be specified.

3.2. Software Application Development

In order to facilitate user interaction, our MANET's testbed is equipped with an intelligent application. The software application was developed in Java NetBeans IDE and several features for testing Adhoc network protocols were incorporated in it. This utility enables the configuration of a multitude of mobility and data models. In order to ensure repeatability, this interactive utility allows the arrangement of various scenarios with specific parameters.

The developed application is broadly categorized into a profile manager and a client application. Profile manager is a centralized configuration tool which allows the user to manipulate every test node in terms of mobility and data model. The purpose of client application is to implement the mode as defined by the profile manager.

3.2.1. Profile Manager

Profile manager is composed of different modules, each having a particular function. The profile manager's function is to build the MANETs test for all the participating nodes rather than being a part of the experiment itself.

On startup, the Parameter Selection Window (PSW) pops up and it requires information regarding the number of nodes, experiment time and testing protocol.

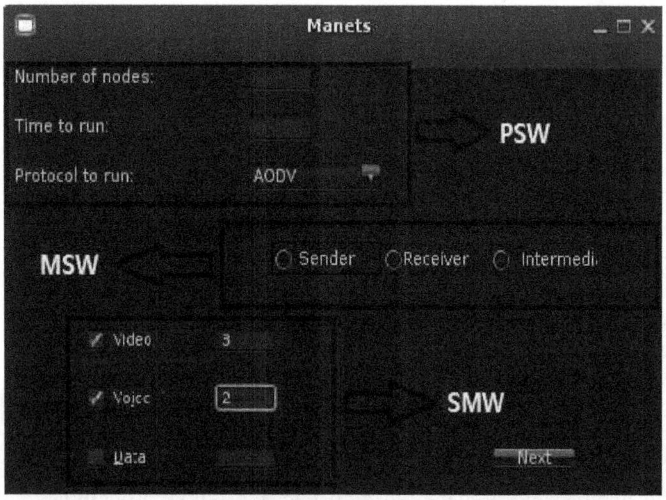

Fig. 6. Parameter, Mode and Model Window in Profile Manager.

The top of figure 6 shows the parameter section window. After filling the selection window fields, the next step is to select which node will act as a sender, receiver or an intermediate node. The next window is named Mode Selection Window (MSW) and is shown below the parameter selection window in figure 6. Multiple senders and multiple receivers can also be selected in the mode selection window.

If a node is selected as a sender node, a new window named Sender's Model Window (SMW) will appear. The user is restricted to select at least one type of data in the model window. The model window for sender nodes is shown at the bottom of figure 6. Multiple data types can be selected at a time. The number of data files for each data type should also be entered in each data field. Parameter Selection Window, Mode Selection Window and Model Window are three separate windows; however, for simplicity, all have been combined in figure 6.

Once a specific data model is selected along with specifying the number of files to be transmitted, the path for each file has to be specified. This feature is implemented by the Data Source Window (DSW). The data source window is shown at the top of figure 7.

After specifying the paths of all the data files, which were specified in the previous steps, the next step is to load the Map and Instruction menu. The bottom of figure 7 shows the Map and Instruction Menu Window (MIMW). Data Source Window and Map & Instruction Window are two separate windows in the actual application. For simplicity, both of these windows have been combined in figure 7.

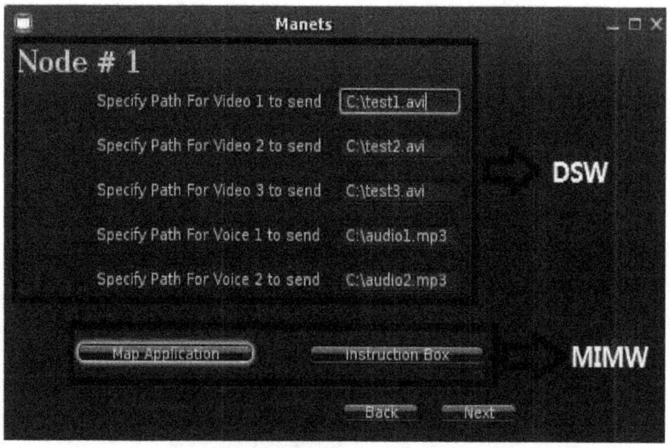

Fig. 7. Data Source and Map and Instruction Menu Window..

To specify a particular mobility model for the participating nodes, the map allocation utility is employed. The purpose of this utility is to create a route to be followed by the mobile nodes. A particular testing region is selected, in our case, the Military College of Signals was chosen as the test area. Different control points were marked on the map and users were given the liberty to join specific points to define the path to be traversed by a particular test node.

Users can select an Offline or an Online Map application. The instruction box is a manual guiding for the nodes about the actions which have to be executed together with the paths they have to traverse in a sequential order. It is in the form of text file and is made available to each and every test node. The instruction box comprises of two components: actions to be performed and movements

Online Map application has been used for testing in the developed testbed and marked points are shown in figure 8.

Fig. 8. Online Map Application in Profile Manager.

3.2.2. Client Application

The client Application is running on the nodes, which are part of the Mobile Adhoc Network Testbed. This application provides a generic platform for the clients to execute the profile created by the profile manager. It demonstrates the mobility model in the form of a map, which represents the physical path to be traversed by the node, displays the instructions from the configuration file and provides the necessary options to execute the actions listed. The major components of this application are protocol selection box, actions to be performed, map application and instruction box. Figure 9 shows the layout of the client application developed for the testbed.

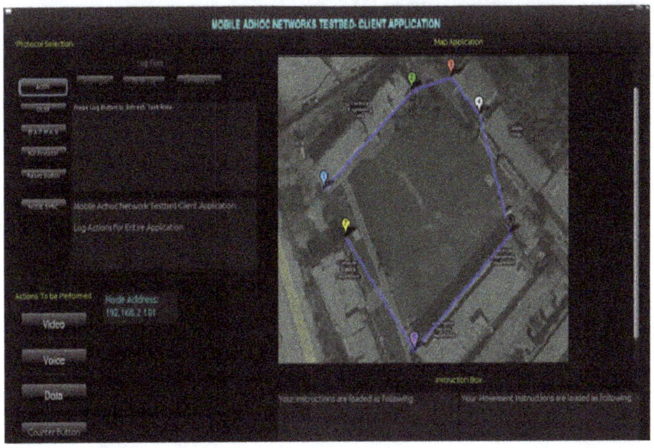

Fig. 9. Client Application in MANETs User Application.

3.3. Hardware Implementation

Several customized portable nodes were developed by integrating different hardware components at the hardware implementation phase of the testbed. It comprises of an Intel Atom Development board, DDR3 2GB RAM, Wifi card, Wifi Extender Antenna, SATA 160GB Hard Disk, NiCad battery Pack, Pico PSU-160 ITX power supply, Miniature keyboard, VGA enabled LCD and power supply design. As opposed to standard laptops our device has been manually configured specially for a Mobile Adhoc Networks environment. This configuration consists of the implementation of adhoc routing protocols namely AODV, OLSR and B.A.T.M.A.N. Some of the major hardware components are shown in figure 10.

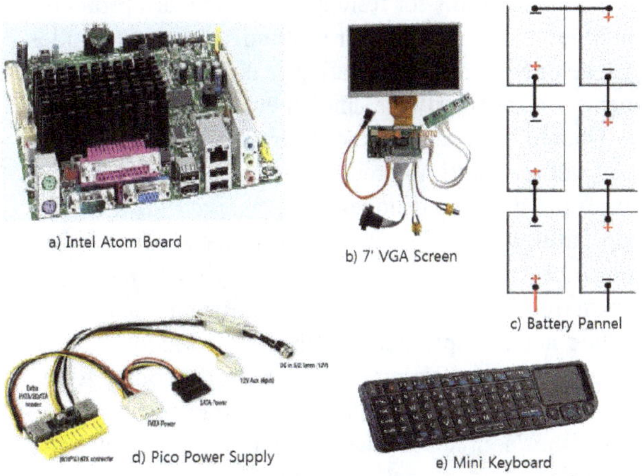

Fig. 10. Major Hardware Components of the Testbed.

The Intel Desktop Board D525MW is designed to support internet-centric computing, delivering incredible capabilities in the new flexible Mini-ITX form factor, featuring the integrated 45nm Intel Atom processor 330 and the Intel 945GC Express Chipset. This board is energy efficient. In order to design a miniature, portable and customized device, DDR3 RAM has been used. D-Link wireless adapter has been used for the purpose of communication between mobile computing nodes. Wifi extender antenna acts as a range extender. It effectively increases the operating distance of the wireless network and conventionally avoids the additional need for power cables or the device clutter.

The Nickel-Cadmium battery is a type of rechargeable battery. The major advantages of NiCad battery are high charge density, durability and enhanced charging discharging cycles. PicoPSU-160-XT High power, 24pin mini-ITX power supply is small efficient package. The developed board operates at 12V; in order to provide this voltage PiscoPSU is being used. It is highly reliable, contains lesser cables and is durable. A miniature keyboard has also been incorporated, in order to make the node portable and so the users could easily carry it.

In order to run Intel the atom board and the accessories like a hard disk, Wifi card, VGA card, and control circuitry power supply has been designed to operate on a battery. A battery pack of 16V and 5600mAH is made to power up all these components.

A regulator circuit was used to get 12V, which acts as an input to PicoPSU160. Micrels are high current, high accuracy, low drop out voltage regulators. A Micrel regulator is used. The Mic29302 that is an adjustable regulator with a maximum current of 3A has been used. It is fully protected against high current faults, reversed input polarity, over temperature operations and negative and positive transient voltage spikes.

4. Results and Discussion

4.1. Experimental Setup

A software application was developed in Java NetBeans IDE and several features for testing Adhoc network protocols were incorporated in it. Several customized portable nodes were developed by integrating different hardware components at the hardware implementation phase of the testbed. Ubuntu was set as the operating system for the developed nodes. Multi-hop topologies have been configured by planting the developed nodes at the marked locations by the map allocation. All the nodes have been time synchronized by using the NTP. Different types of data have been transferred between the nodes by utilizing each of the three Adhoc network protocols explained in section 2. FTP has been used for transferring the files from the source to the destination. The VLC has been used for transferring video data from the source to the destination by using the application described in section 3. Wireshark has been used to analyze the data and to calculate the different performance parameters, explained in the following section. Information window of Wireshark is shown in figure 11.

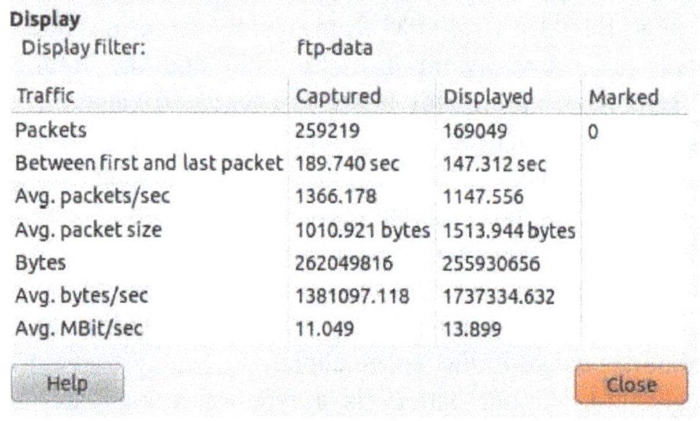

Fig. 11. *Wireshark Display Window.*

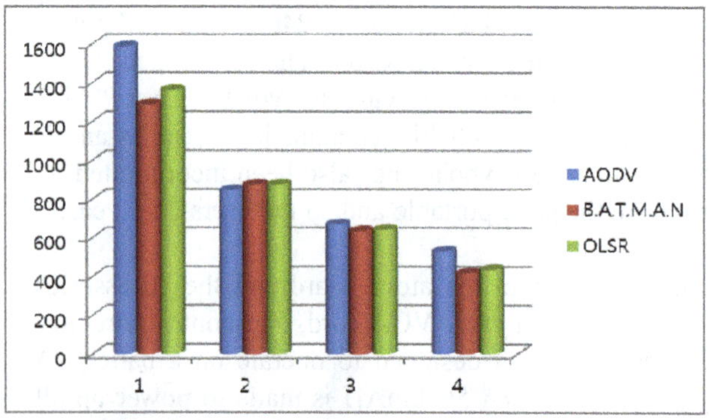

Fig. 12. *Throughput Vs Number of Hops*

4.2. Throughput

Throughput and Packet Loss has been computed by the changing the number of hops. Table 3 shows the performance of each of the three protocols in terms of throughput and percent packet loss as a function of the number of hops from the source to the destination. X-axis shows the number of hops while throughput and packet loss is on the Y-axis of figure 12 and figure 13 respectively.

Figure 12 shows the throughput versus the number of hops for AODV, OLSR and BATMAN protocol. It is evident from the graph that the performance of AODV degrades as the number of nodes increases, while OLSR and BATMAN show a gradual degradation. Being reactive protocol AODV consumes large part of the bandwidth for transferring control information, while OLSR and BATMAN only transfers the control messages once they are requested.

Table 3. *Throughput and Error Rate with variable Hops*

Number Of Hops	Protocol	Parameter	Values
1	AODV	Throughput	1586.6
		Error Rate	0.333
	OLSR	Throughput	1361.75
		Error Rate	0
	BATMAN	Throughput	1288.5
		Error Rate	0.333
2	AODV	Throughput	845.95
		Error Rate	4.33
	OLSR	Throughput	876.49
		Error Rate	1
	BATMAN	Throughput	875.2
		Error Rate	1.678
3	AODV	Throughput	668.65
		Error Rate	7
	OLSR	Throughput	640.35
		Error Rate	2.33
	BATMAN	Throughput	34.15
		Error Rate	2
4	AODV	Throughput	530
		Error Rate	8.677
	OLSR	Throughput	436.05
		Error Rate	2.678
	BATMAN	Throughput	417.9
		Error Rate	2.678

4.3. Packet Loss Rate

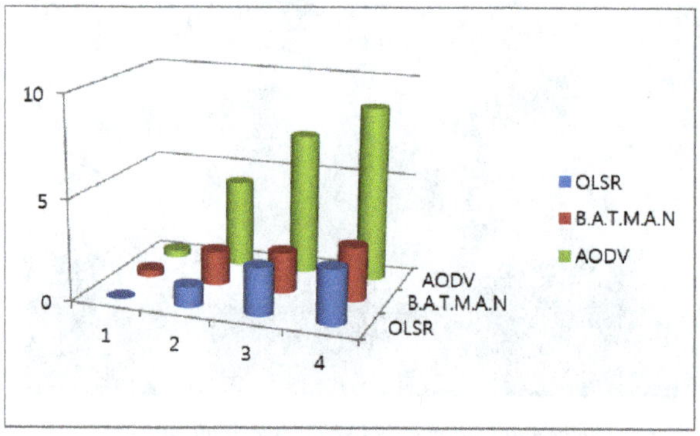

Fig. 13. *Packet Loss Rate Vs Number of Hops*

Figure 13 shows the packet loss rate, calculated by using equation (6) versus the number of nodes. It can be observed in figure 13 that the percent error rate propagates exponentially for AODV, while it increases on a linear fashion for OLSR and BATMAN with an increase in the number of nodes. Due to the reactive nature of AODV with the increase in the number of nodes, the quantity of control packets also increases and the path get congested. While in the case of OLSR and BATMAN, the control information only exchanges upon request.

4.4. Overhead Packets

Number of overhead packets and end to end delay has been compared against pause time. Pause time is equal to the time waited by a particular node after arriving at the destination provided as a mobility plan by the developed application. Figure 14 shows the number of overhead packets transferred verses the pause time in seconds. Pause time is on the X-axis and number of overhead packets is on the Y-axis.

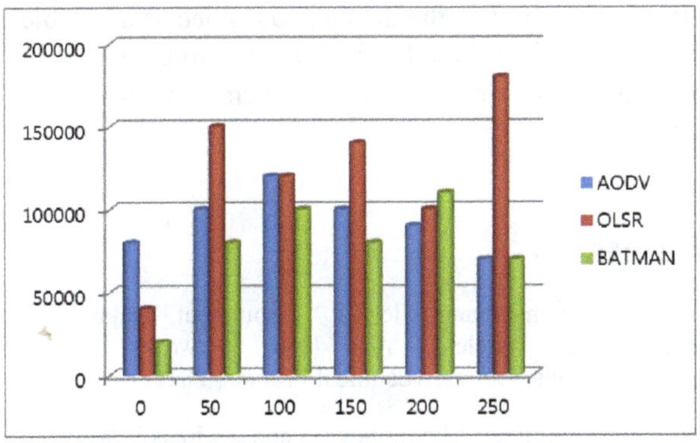

Fig. 14. Number of Overhead Packets Vs Pause Time (sec)

It can be deduced from the graph that BATMAN shows the minimum number of control packets. AODV shows a gradual increase initially and then a decrease. OLSR shows a higher number of overhead packets compared to other two protocols. Decrease in control packets after 50 seconds for AODV indicates that the AODV settles after 50 seconds. BATMAN has a built-in capability to control the number of overhead packets. In the case of OLSR, it can be concluded that overhead doesn't only depend on the pause time, but also depends on several factors like a change in the configuration of the network, the speed of moving nodes and many other factors. X-axis in Figure 15 shows the pause time while Y-axis represents the End to End delay. AODV shows a zigzag behavior while BATMAN and OLSR show a gradual decrease with increase in pause time.

4.5. End to End Delay

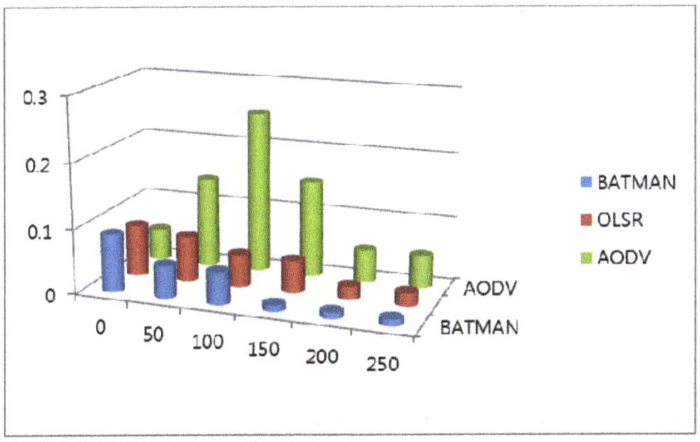

Fig. 15. End to End Delay Vs Pause Time (sec)

The zigzag behavior of AODV is because of the overhead load due to the proactive nature of the protocol. In the case of OLSR, when pause time is increased, the link between the source and destination becomes stable. Due to the stable connection, the end to end delay decreases. Same is the case for BATMAN. Due to the inherit overhead reduction capability of BATMAN, reduction in end to end delay is more rapid as compared to OLSR.

Table 4. *Overhead and End to End Delay with variable Pause Time*

Protocol	Parameter	Pause Time	Values
AODV	Over Head	0	80000
		50	100000
		100	120000
		150	100000
		200	90000
		250	70000
	End to End Delay	0	0.05
		50	0.14
		100	0.25
		150	0.15
		200	0.05
		250	0.05
OLSR	Over Head	0	40000
		50	150000
		100	120000
		150	140000
		200	100000
		250	180000
	End to End Delay	0	0.08
		50	0.07
		100	0.05
		150	0.05
		200	0.02
		250	0.02
BATMAN	Over Head	0	20000
		50	80000
		100	100000
		150	80000
		200	110000
		250	70000
	End to End Delay	0	0.09
		50	0.05
		100	0.05
		150	0.01
		200	0.01
		250	0.01

4.6. Average Jitter

Average Jitter is shown in the Y-axis while packet size is on the X-axis of figure 16. By analyzing figure 16, it can be concluded that AODV is more prone to jitter as compared to OLSR and BATMAN. Table 5 shows the performance of each protocol against the size of the packet.

OLSR shows little variation in performance in terms of jitter, while AODV and BATMAN show almost steady performance.

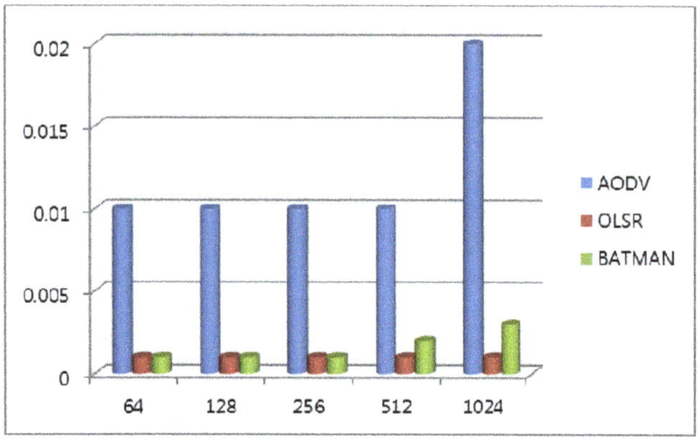

Fig. 16. *Average Jitter Vs Packet Size*

These types of deductions can be valid for all other protocols, which are tested by using the developed testbed in this paper.

Table 5. *Average Jitter with variable packet size*

Protocol	Packet Size	Average Jitter
	64	0.01
	128	0.01
AODV	256	0.01
	512	0.01
	1024	0.002
	64	0.001
	128	0.001
OLSR	256	0.001
	512	0.001
	1024	0.001
	64	0.001
	128	0.001
BATMAN	256	0.001
	512	0.002
	1024	0.003

By analyzing all the results of all the five performance metrics, it can be deduced that AODV performs much better in networks with static traffic. OLSR and BATMAN function appropriately in conditions where traffic is highly dense and sporadic, but scalability acts as a limiting constraint as the network size increases. Heavy flooding of control packets is observed in the case of AODV upon route discovery. In the case of OLSR, there is a non-linear increase in the size of the route table and control messages may block the actual data transmission.

5. Conclusion

The developed testbed provides a generic platform for the testing of Adhoc routing protocols in real world scenarios. The developed testbed is capable of implementing and testing Adhoc routing protocols in real life scenarios in order to test the viability of the newly developed protocols. A software tool has been developed in Java NetBeans IDE to facilitate the users carrying out different tests under same constraints, thus ensuring the repeatability and helping out in finding the exact pros and cons of the protocols being tested. The testbed is equipped with a customized, flexible and dedicated hardware, which is specially designed by integrating several hardware components for Adhoc Networks in contrast to the standard laptops or testing devices. All necessary parameters for evaluating the performance of a newly developed protocol are included in the testbed. The developed testbed is capable of testing newly developed Adhoc network protocols' viability, robustness and efficiency, to assess their key features, conquer their shortcomings and add more appealing features.

Test results have proven the developed testbed as a reliable, efficient and robust testbed for real time testing of newly developed protocols before using them in real life applications.

References

[1] Pan Li, Yuguang Fang, Jie Li "Throughput, Delay, and Mobility in Wireless Ad Hoc Networks" IEEE Communications Society, Technical Program 2010.

[2] Matthew Brand, Petar Maymounkov, and Andreas F. Molisch "Routing with probabilistic delay guarantees in wireless ad-hoc networks".

[3] Humayun Bakht "Survey of Routing Protocols for Mobile Ad-hoc Network" International Journal of Information and Communication Technology Research Volume 1 No. 6, October 2011.

[4] Erik Nordstrom, Per Gunningberg, Henrik Lundgren "A Testbed and Methodology for Experimental Evaluation of Wireless Mobile Ad hoc Networks".

[5] Yongguang Zhang, Wei Li "An Integrated Environment for Testing Mobile Ad-Hoc Networks" EPFL Lausanne, Switzerland 2002.

[6] Chandra Kanta Samal "TCP Performance through Simulation and Testbed in Multi-Hop Mobile Ad hoc Network" International Journal of Computer Networks & Communications (IJCNC), Vol.2, No.4, July 2010.

[7] Ian D. Chakeres, Elizabeth M. Belding-Royer "AODV Routing Protocol Implementation Design".

[8] P. Jacquet, P Muhlethaler, T Clausen, A Laouiti, A Qayyum, A Viennot "Optimized Link State Routing Protocol for Adhoc networks".

[9] Laurent Delosi`eres and Simin Nadjm-Tehrani "BATMAN Store-and-Forward: the Best of the Two Worlds "Second International conference on Pervasive networks for emergency managements 2012.

[10] Sagar Sanghani, Timothy X Brown, Shweta Bhandare, Sheetalkumar Doshi "EWANT: The Emulated Wireless Ad Hoc Network Testbed" IEEE, 2003.

[11] Virendra Singh Kushwah1, Kamal Kumar Chauhan and Amit Kumar Singh Sanger "A Comparative Study of Mobile Ad Hoc Network Protocols for Throughput, Average End-to-End Delay and Jitter" International Journal of Computational Intelligence Research Volume 6, Number 3, 2010.

[12] David A. Maltz Josh Broch David B. Johnson"Quantitative Lessons From a Full-scale Multi-Hop Wireless Ad Hoc Network Testbed" IEEE, 2000.

[13] Joni Birla, Basant Sah, "Performance Metrics in Ad-hoc Network". International Journal of Latest Trends in Engineering and Technology. Vol. 1 Issue 1 May 2012.

[14] Laurent Viennot, Philippe Jacquet and Thomas Heide Clausen"Analyzing Control Traffic Overhead versus Mobility and Data Traffic Activity in Mobile Ad-hoc Network Protocols".

[15] Ahmed A. Radwan, Tarek M. Mahmoud and Essam H. Houssein "Performance Measurement of Some Mobile Ad Hoc Network Routing Protocols" IJCSI International Journal of Computer Science Issues, Vol. 8, Issue 1, January 2011.

[16] Soumendra Nanda, Zhenhui Jiang, David Kotz "A Combined Routing Method for Wireless Ad Hoc Networks" Dartmouth College Technical Report TR2007-588, June 2007.

[17] Rakesh Kumar, Manoj Misra, Anil K. Sarje "A Simplified Analytical Model for End-To-End Delay Analysis in MANET" IJCA Special Issue on "Mobile Ad-hoc Networks" MANETs, 2010.

[18] Kamal Kumar Sharma, Hemant Sharma and A. K. Ramani "Modeling and Analysis of End-To-End Delay for Ad Hoc Pervasive Multimedia Network" Proceeding of the international multi conference of engineers on computer scientists, Hong Kong , 2010.

[19] Swati Bhasin, Ankur Gupta, Puneet Mehta "Comparison of AODV, OLSR and ZRP Protocols in Mobile Ad-hoc Network on the basis of Jitter" International Journal of Applied Engineering Research Vol.7 No.11 2012.

[20] R. Marutha Veni, R. Latha"Mobile Ad hoc Network" International Journal of Science and Research, Volume 2 Issue 4, April 2013.

Recent Update on One-Minute Rainfall Rate Measurements for Microwave Applications in Nigeria

Obiyemi Obiseye O.[1], Adetan Oluwumi[2], Ibiyemi Tunji S.[3]

[1]Department of Electrical and Electronic Engineering, Osun State University, Osogbo, Nigeria
[2]Department of Electrical and Electronics Engineering, Ekiti State University, Ado Ekiti, Nigeria
[3]Department of Electrical and Electronics Engineering, University of Ilorin, Ilorin, Nigeria

Email address:
obiseye.obiyemi@uniosun.edu.ng (Obiyemi O. O.), oadetan@gmail.com (Adetan O.), tibiyemi@unilorin.edu.ng (Ibiyemi T. S.)

Abstract: Rain rate statistics is required for planning both satellite and terrestrial links, especially in the microwave and millimeter wave bands. Presented in this work is the one-minute rain rate statistics observed over seventeen months using an electronic weather station - Davis Vantage Vue. The installation is at the main campus of Osun State University, Osogbo (7° 76' N, 4° 60' E), Nigeria. The cumulative rain rate distribution from the measured rain rate is presented alongside predictions by other prominent models. The $R_{0.01}$ estimate as high as ~ 120 mm/h was obtained from the surface data and subsequently employed in estimating the fade margin over a hypothetical DTH link for the reception of digital television content at 12.245 GHz from EUTELSAT W4/W7. Estimates presented over time percentages ranging between 0.001% and 1% are dissimilar. However, their suitability for predicting fade margins over this location could be ascertained via a performance analysis, based on experimental attenuation estimates over the link. The first point rain rate estimate from surface data over Osogbo is reported here and will be very useful for modeling rain attenuation and for planning both terrestrial and earth-space microwave links.

Keywords: One-Minute, Rain Rate Statistics, Rain Attenuation, Earth-Space Microwave Links, Point Rain Rate

1. Introduction

Rainfall is a major climatic factor and its variability plays several disturbing roles in a number of activities ranging from the transmission of radiowave signals to malaria epidemiology[1]. Among other similar disturbing factors is the hail [2], which is one of the most terrible natural disaster with severe impacts on mankind. For the propagation of radiowave signals however, attenuation due to rain remains a major loss factor required for planning communication systems and it affects applications operating within the microwave (3-30 GHz) and millimeter wave (30-300 GHz) bands, particularly for frequencies beyond 10 GHz [3-5].

Although rain is not the only form of precipitation affecting signal propagation in this band, it accounts for the most severe impairment on transmitted signals than other atmospheric components such as snow, fog, cloud, water vapour, fog, etc. [6-10]. The propagation path, to and from the satellites is actually well-defined through the earth's atmosphere, where the transmitted signal attenuates, scatters, refracts and depolarizes [11]. Most of these propagation mechanisms are however rain induced and more pronounced in tropical climates where rainfall is characterized with high intensity events.

Rainfall data logged in one-minute integration time is considered suitable for monitoring the rapid attenuation fluctuation on the radio path. Earlier studies have presented useful tools for planning satellite and terrestrial microwave links over Nigeria [12, 13], the one-minute rain rate estimates presented were deduced mainly from satellite data. Interestingly, a recent comparison by Ojo and Omotosho in [14] however reveals that one-minute rain rate estimates derived from ground data generally performs better than those obtained from satellite data. Since the dearth of ground data, particularly in the required one-minute integration time is still a concern to radio scientists and engineers, the cumulative rain rate distribution derived from ground-based observations in Osogbo will therefore provide a useful update.

Moreover, several rain rate models have been developed from long term precipitation statistics over different locations. Prominent among others are those developed from data collected in Brazzaville, Congo [15], those developed from a data bank containing 290 data sets for 30 countries [16], those predicting rain rain rate based on local climatological data available over any location [17], and the global ITU-R rain rate model as recommended in [18]. These models have been useful for the prediction of the cumulative rain rate distribution for selected locations in Nigeria [13, 19-21]. The radio climatological data required to ascertain their suitability for predictions over Nigeria is still sparse. Investigations on the performance of such models over some tropical sites however reveals that the rainfall rates derived using such models are either overestimated or underestimated [19, 21]. However, the implication is as stated in [22], where over-prediction is considered to account for designs that are less effective in cost, while under-prediction limits reliability of the link. There is the need to update radio-climatological database for accurate propagation design purposes. This can be achieved through a deliberate and aggressive precipitation measurement campaign, using a network of rain gauge installations across Nigeria.

Besides, the knowledge of surface precipitation data would enhance reliable link planning, required to meet the increasing demand for bandwidth and multimedia services. Likewise, communication satellites retain backhaul obligations for broadband internet service provision and digital DTH distribution. The detailed estimate of the fade margins required to maintain the high quality content also depends on the local rain rate date. These satellites offer trans-border services through its large footprint (typically delivering contents or services across rural, sun-urban and urban settings). The apparent digital divide can therefore be minimized through a deliberate effort to provide reliable services using satellite. This also dictates the need for improved availability, especially for mission critical applications, where little or no downtime can be tolerated.

Generally, the attenuation induced by rain A (dB) is estimated from the knowledge of the rainfall rate R (mm/h), which is expressed as [6]:

$$A = aR^b l \quad (1)$$

where a and b are coefficients depending on the Drop Size Distribution (DSD), frequency, elevation angle, temperature of the raindrops and polarization of the radiowave, while l is the equivalent path length of the rainy region, and is usually set to be about 5 km on high elevation angle on the earth-space link [23, 24].

Since the accurate prediction of rainfall attenuation depends on the knowledge of the long-term local precipitation data over a particular area, the rain gauge installation at Osun State University, Osogbo is an addition to the South-western precipitation measurement network [4, 24-27] for propagation studies in Nigeria. Due to the short integration time requirement for radio propagation purposes, the daily precipitation measurement at the Osogbo Aerodrome has not been directly useful in this regard.

The analysis of the ground-based one-minute rain rate measurement at Osun State University, Osogbo, South-West Nigeria is presented in this work. The cumulative rain rate estimates derived from the seventeen-month observation is a useful update to the precipitation database required for radio-climatic studies and for estimating rain fade in the design of satellite and terrestrial microwave links.

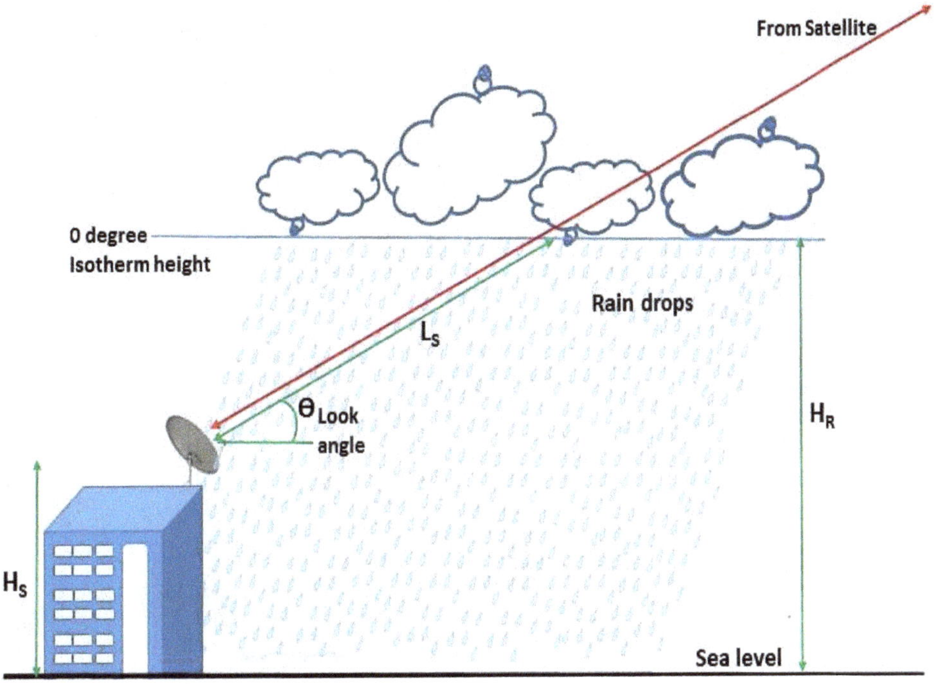

Figure 1. *Geometry of a typical DTH link.*

2. Rain Rate and Rain Attenuation Prediction Techniques

Several techniques have been employed for the prediction of rain rate over the years. They become very useful, particularly in locations where precipitation data are not available in the preferred short integration time format. Prominent among others are techniques developed by Moupfouma and Martins [15], Rice and Holmberg[17], Ito and Hosoya [16] and the ITU-R [18]. Estimates based on these techniques are compared with the measured rain rate statistics using the cumulative distributions predicted for Osogbo, Osun State.

The point rain rate estimate derived for Osogbo will remain useful for planning over terrestrial and satellite microwave and millimeter wave links, especially in other locations within the P zone of the rain climatic region, where such data do not exist. This knowledge is also employed for estimating the attenuation induced by rain on a typical Direct-To-Home (DTH) link for reception of digital television (TV) content via EUTELSAT W4/W7, geostationary at longitude 360 East. The selected models are the ITU-R model [6], Garcia model, Svjatogor model, Bryant model and the Australian model. Details of these models can be found in [28]. It is also important to investigate the suitability of these models. However, this will require an experimental data obtained over a period during – typically a year or more, at the earth station of a practical DTH link. The geometry of a typical DTH link is shown in Figure 1.

3. Experimental Site and Measurement

Rain gauge represents the most common equipment for the acquisition of surface precipitation data over a site. The data used in this work was obtained from a Davis Vantage Vue weather station, manufactured by Davis Instruments, Hayward, California, USA. The installation is at the rooftop of the College of Science, Engineering and Technology (CSET) building (about 360 m above sea level) at the main campus of Osun state University, Osogbo, Nigeria (7.760 N and 4.600 E). The weather station is a tipping spoon type and also measures and records data for other quantities such as temperature, relative humidity, wind speed and direction, all combined in the Integrated Sensor Suite (ISS). A sensor interface module (SIM) collects outside data from the ISS and transmits the data to a Vantage Vue console via low power radio at a frequency range of 868.0 – 868.6 MHz with a maximum line of sight distance of 300 m for reliable communication between the outdoor ISS and the indoor console. Figure 2 shows the measurement site in Osun State, South-Western Nigeria.

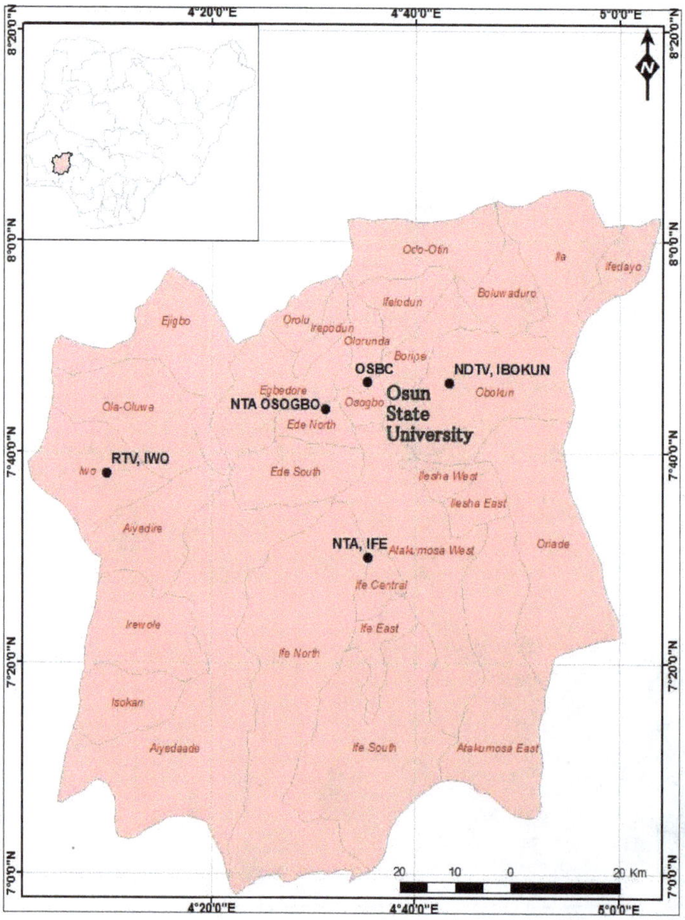

Figure 2. Map of Osun state showing the experimental site.

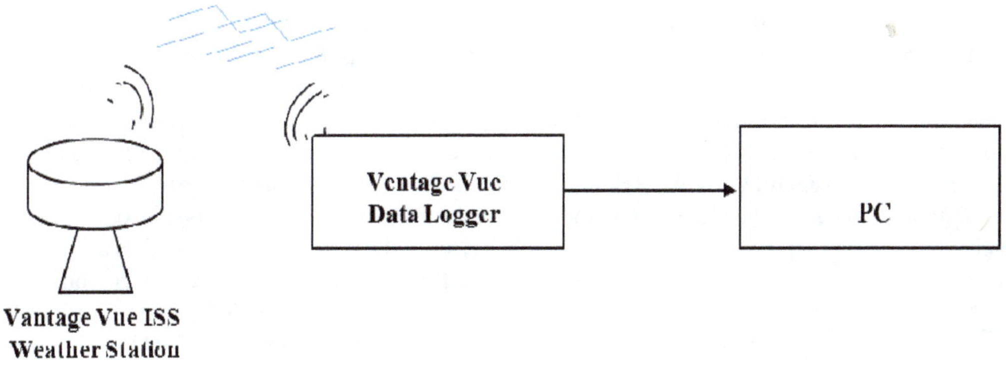

Figure 3. Block diagram of the measurement setup at Osun State University, Osogbo.

The accuracy of the gauge is ±1% at 1 litter/h with a measuring range of a minimum of 2 mm/h to a 400 mm/h. The gauge is accurate to within 2% up to 250 mm/h. Rainfall up to 400 mm/h is measured with the resolution of 0.2 mm. The data logger scans the data at every one second intervals and integrated over one-minute interval. The availability of the gauge is about 99.2 %. And the 0.8% unavailability is due to system maintenance and system shutdown as a result of power drain. The console incorporates a Weatherlink software and USB data logger that is synchronized with the personal computer, hence facilitating improved weather monitoring capabilities and a continuous data logging. The block diagram of the measurement setup is as shown in Figure 3.

4. Results and Discussion

The one-minute rain rate data collected using the Davis Vantage Vue weather station for the period between March 2013 and July 2014 was sorted and analyzed to provide the precipitation data required for rain attenuation prediction over Osogbo. Statistical analysis of the 17-month precipitation data is also presented here. The parameters extracted from the weather station include; rainfall rate for each month, number of one-minute rain rate recorded throughout the period and a number of other parameters.

Table 1. Precipitation statistics for the 17-month observation.

Month	Minimum rain rate (mm/h)	Maximum rain rate (mm/h)	Total rain rate (mm/h)	Number of rain duration	Number of one-minute rain rate recorded per month
March, 13	0.8	57.4	569.8	72	4914
April, 13	0.8	256	11759	940	18013
May, 13	0.8	122.9	5601.8	406	16768
June, 13	0.2	94.5	1618.7	303	20498
July, 13	0.8	286.6	4066.5	509	31470
August, 13	1.0	27.9	317.7	90	15633
September,13	1.0	87.6	7666	1041	28778
October, 13	0.8	281	5062	450	27894
November, 13	1.0	115.3	816	33	25994
April, 14	1.0	18	240.1	71	10520
May, 14	1.0	124	5401.3	1000	18820
June, 14	1.0	98.3	4434.2	283	11732
July, 14	1.0	190	5109.1	429	8860
Total	11.2	1759.5	52662.2	5627	239894

Table 1 shows the precipitation statistics for the observation period. The highest rainfall intensity was observed in the month of July, 2013 with an intensity of 286.6 mm/h, followed by October, 2013 with 281 mm/h. Usually, Rainfall in Nigeria is usually observed between the months of March and October. The minimum total rainfall intensity was recorded in November of 2013, while the highest was recorded in September, 2013. The total rain rate in August is low due to the "August break", which typically lasts for about 3 weeks.

The total number of one-minute rain rate recorded per month varies from 4914 to 31470. The lowest number was observed in March, 2013 with a total record of 4914, while July 2013 is said to the have the highest number with a total number of 31470 one-minute rain rate recordings. The total record taken for the whole 17-month observation is 239894.

Table 2 shows the number of days when rain accumulation is either equal to or less than 0.2mm. Stratiform rain type of longer duration is witnessed when rainfall accumulation is less than or equal to 0.2mm, much of these are observed in July and October 2013. On the other hand, when rainfall accumulation is greater than 0.2 mm, the convective rain type is said to be in occurrence. The month of April, 2013 and June, 2014 recorded a number of the convective rain type occurrence with 21 and 17 days respectively.

Figure 4 presents the comparison of cumulative distribution of the predicted rain rate and the measured rain rate. It is observed that the cumulative rain rate statistics is underestimated by the ITU-R prediction and this is pronounced at lower time percentages (0.001 – 0.1%). Apart from the sharp contrast observed below 0.01% of the time, predictions based on other selected models show better agreement with the measured data. The measured rainfall rates for the 0.01% is 120 mm/h. Rain rate estimates for the same time percentage for the ITU-R, MOUPFOUMA, RH and KITAMI models are 55 mm/h, 103.7 mm/h, 113.2 mm/h and 120 mm/h respectively.

Table 1. *Days one-minute accumulation exceeds or equal to 0.2mm.*

Month	Days one-minute acc. exceeds 0.2mm	Days 1-minute acc. tips equals 0.2mm
March, 13	6	15
April, 13	21	4
May, 13	4	9
June, 13	11	4
July, 13	5	21
August, 13	2	13
September, 13	8	12
October, 13	8	24
November, 13	2	11
April, 14	2	18
May, 14	15	13
June, 14	17	6
July, 14	4	15
Total	105	165

This result agrees with the estimate for Osogbo based on the ITU-R rain climatic zone [29], which groups Osogbo with other locations in the South-western part of Nigeria in the P zone. However, measurement is still ongoing at this station and the result will be validated from time to time.

Figure 5 shows the cumulative distributions of the rain attenuation predicted using the surface rainfall data for all the models selected. This presents the one-to-one hypothetical behavior of the rain attenuation models at varying percentages of the time and under the same design condition. 52.50 was maintained as the antenna look angle for typical reception of digital television content at 12.245 GHz via EUTELSAT W4/W7 satellite, which is geostationary at longitude 360 East. The attenuation induced by rain at 0.01% of the time is 16.2 dB, 14.75 dB, 14.56 dB, 10.61 dB and 5.46 dB respectively for ITU, Garcia, Australian, Svjatogor and Bryant models.

At lower time percentage, 0.001% of the time, the Garcia model presents the highest estimate of 29.8 dB. The Bryant model still retains the lowest estimate of 12.24 dB, while the Australian, ITU-R and Svjatogor models predicted 29.4 dB, 25.2 dB and 22.2 dB respectively.

The models show dissimilar estimates at lower percentages of the time and it is still difficult to present the model that outperforms all others in this scenario. Their performance can be better ascertained with experimental data obtained over a practical link.

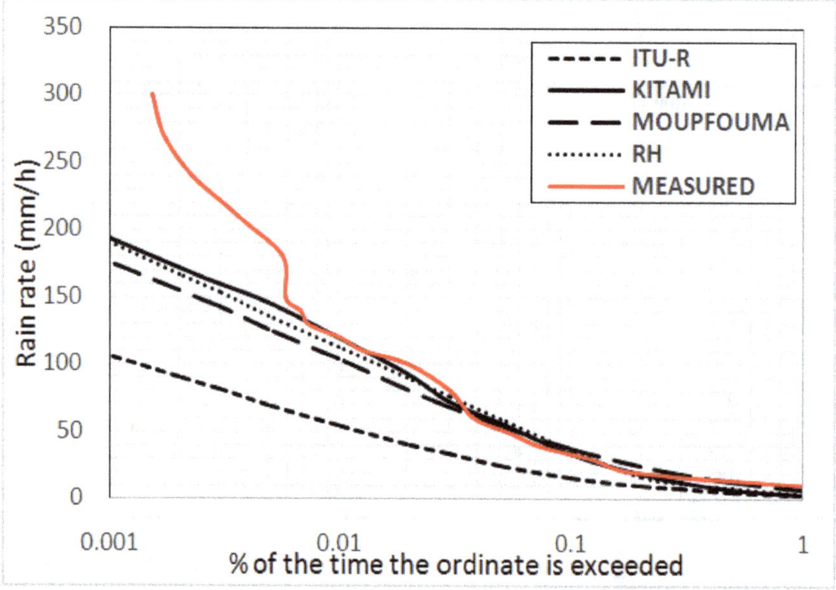

Figure 4. *Cumulative rain rate distribution over the observation period.*

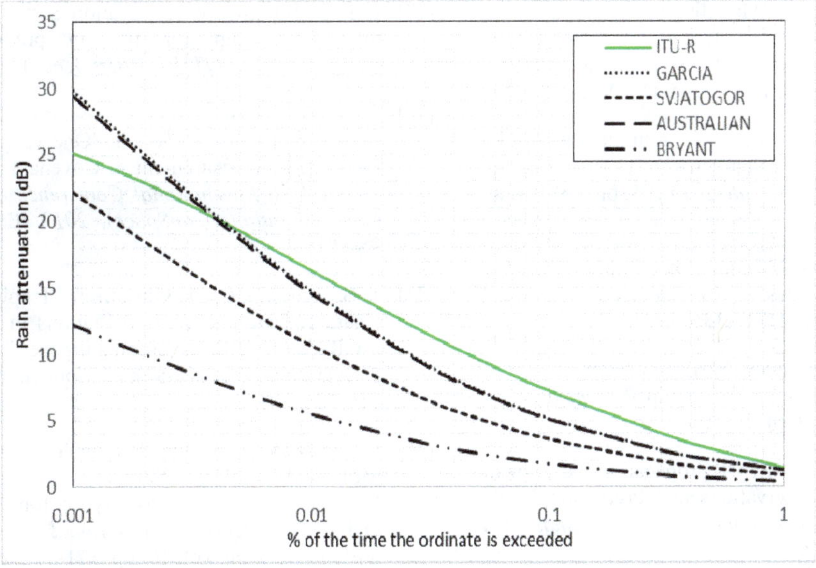

Figure 5. *Rain attenuation predicted for a typical DTH link (12.245 GHz, EUTELSAT 36B).*

5 Conclusions

In the quest to reduce the problem of rainfall attenuation and to update global precipitation database for radio-climatic and other applications, the first point rain rate estimate from surface data collected in Osogbo is presented here and will be very useful for modeling rain attenuation and for planning both terrestrial and earth space microwave links. The rainfall rate exceeded at 0.01% of the time is estimated as 120 mm/h for the 17-month observation, and this estimate was employed in predicting the attenuation induced by rain over a hypothetical DTH link for digital television reception at 12.245 GHz. Estimates presented over time percentages ranging between 0.001% and 1% are dissimilar. However, their suitability for predicting fade margins over this location could be ascertained via a performance analysis, based on experimental attenuation estimates over the link. Interestingly, precipitation measurements is ongoing at this station and the result presented is a useful tool for national and regional communication link designs.

References

[1] A. W. Kassa and B. B. Beyene, "Climate variability and malaria transmission – Fogera district, Ethiopia, 2003-2011," *Science Journal of Public Health,* vol. 2, pp. 234-237, 2014.

[2] I. S. Akhmedovic, "The unique method to prevent hail," *Science Discovery,* vol. 2, pp. 43-46, 2014.

[3] J. Chebil and A. Rahman, "Development of 1 min rain rate contour maps for microwave applications in Malaysian Peninsula," *Electronics Letters,* vol. 35, pp. 1772-1774, 1999.

[4] G. O. Ajayi and E. B. C. Ofoche, "Some tropical rainfall rate characteristics at Ile-Ife for microwave and millimeter wave applications," *Journal of Applied Meteorology,* vol. 23, pp. 562-567, 1984.

[5] R. K. Crane, "Prediction of Attenuation by Rain," *IEEE Transactions on Communications,* vol. 28, pp. 1717-1733, 1980.

[6] I. T. U. Radiowave Propagation Series, "Propagation Data and Prediction Methods Required for the Design of Earth-Space Telecommunication Systems," in *Recommendation ITU-R P.618-10*, ed: I TU-R, Geneva, 2009.

[7] J. M. Garcia-Rubia, J. M. Riera, P. Garcia-del-Pino, and A. Benarroch, "Propagation in the Ka Band: Experimental Characterization for Satellite Applications," *IEEE Antennas and Propagation Magazine,* vol. 52, pp. 65-76, 2011.

[8] R. K. Crane, *Electromagnetic Wave Propagation Through Rain*: John Wiley, New York, 1996.

[9] H. Y. Chen and D. P. Lin, "Volume Integral Equation Solution of Microwave Absorption and Scattering by Raindrops," in *Antennas and Propagation Society International Symposium, IEEE*, Orlando, FL, USA 1999, pp. 2676-2679.

[10] J. S. Ojo, M. O. Ajewole, and S. K. Sarkar, "Rain rate and rain attenuation prediction for satellite communication in Ku and Ka bands over Nigeria," *Progress In Electromagnetics Research B,* vol. 5, pp. 207-223, 2008.

[11] A. Pawlina-Bonati, "Essential Knowledge of Rain Structure for Radio Applications Based on available Data and Models," in *Proceedings of the Third Regional Workshop on Radio Communication in Africa,* Botswana Conference and Exhibition Centre, 1999, pp. 96 - 106.

[12] T. V. Omotosho and C. O. Oluwafemi, "One-minute rain rate distribution in Nigeria derived from TRMM satellite data," *Journal of Atmospheric and Solar-Terrestrial Physics,* vol. 71, pp. 625-633, 4// 2009.

[13] J. S. Ojo, M. O. Ajewole, and L. D. Emiliani, "One-minute rain-rate contour maps for microwave-communication-system planning in a tropical country: Nigeria," *Antennas and Propagation Magazine, IEEE,* vol. 51, pp. 82-89, 2009.

[14] J. S. Ojo and T. V. Omotosho, "Comparison of 1-minute rain rate derived from TRMM satellite data and raingauge data for microwave applications in Nigeria," *Journal of Atmospheric and Solar-Terrestrial Physics,* vol. 102, pp. 17-25, 2013.

[15] F. Moupfouma and L. Martin, "Modeling of the Rainfall Rate Cumulative Distribution for the Design of Satellite and Terrestrial Communication Systems," *International Journal of Satellite Communications,* vol. 13, pp. 105-115, Mar-Apr 1995.

[16] C. Ito and Y. Hosaya, "Worldwide 1-min. rain rate distribution prediction method which uses thunderstorm ratio as regional climatic parameter," *Electronic Letters,* vol. 35, pp. 1585-1587, 1999.

[17] P. Rice and N. Holmberg, "Cumulative time statistics of surface-point rainfall rates," *Communications, IEEE Transactions on,* vol. 21, pp. 1131-1136, 1973.

[18] I. T. U. Radiowave Propagation Series, "Characteristics of precipitation for propagation modelling," in *Recommendation ITU-R P.837-6*, ed: ITU-R, Geneva, 2012.

[19] J. S. Ojo and S. E. Falodun, "NECOP Propagation Experiment: Rain-Rate Distributions Observations and Prediction Model Comparisons," *International Journal of Antennas and Propagation,* 2012.

[20] O. O. Obiyemi, J. S. Ojo, and T. S. Ibiyemi, "Performance Analysis of Rain Rate Models for Microwave Propagation Designs Over Tropical Climate," *Progress In Electromagnetics Research M,* vol. 39, pp. 115-122, 2014.

[21] T. S. Ibiyemi, M. O. Ajewole, J. S. Ojo, and O. O. Obiyemi, "Rain rate and rain attenuation prediction with experimental rain attenuation efforts in south-western Nigeria," *20th Telecommunications Forum (TELFOR),* pp. 327-329, 20-22nd, November 2012.

[22] R. I. Olsen, "Radioclimatological Modeling of Propagation Effects in Clear-Air and Precipitation Conditions: Recent Advances and Future Directions," in *Third Regional Workshop on Radio Communication in Africa (Radio Africa '99),* Gborone- Botswana, 1999, pp. 81 - 87.

[23] I. T. U. Radiowave Propagation Series, "Specific attenuation model for rain for use in prediction methods," in *Recommendation ITU-R P.838-3*, ed: ITU-R, Geneva, 2005.

[24] T. V. Omotosho, A. A. Willoughby, M. L. Akinyemi, J. S. Mandeep, and M. Abdullah, "One year results of one minute rainfall rate measurement at Covenant University, Southwest Nigeria," in *International Conference on Space Science and Communication (IconSpace), 2013 IEEE* Melaka 2013, pp. 98-101.

[25] J. S. Ojo and E. O. Olurotimi, "Tropical Rainfall Structure Characterization over Two Stations in Southwestern Nigeria for Radiowave Propagation Purposes," *Journal of Emerging Trends in Engineering and Applied Sciences,* vol. 5, pp. 116-122, 2014.

[26] F. A. Semire, R. Mohd-Mokhtar, T. V. Omotosho, I. Widad, N. Mohamad, and J. Mandeep, "Analysis of Cumulative Distribution Function of 2-year Rainfall Measurements in Ogbomoso, Nigeria," *International Journal of Applied Science and Engineering,* vol. 10, pp. 171-179, 2012.

[27] O. O. Obiyemi, T. J. Afullo, and T. S. Ibiyemi, "Equivalent 1-Minute Rain Rate Statistics and Seasonal Fade Estimates in the Microwave Band for South-Western Nigeria," *Int. Journal of Scientific & Engineering Research,* vol. 5, pp. 239-244, 2014.

[28] B. Arbesser-Rastburg, E.-e. Tos-eep, and K. N.-A. Noordwijk, "Radiowave propagation modelling for new satcom services at Ku-band and above," *COST 255,* 2002.

[29] I. T. U. Radiowave Propagation Series, "Characteristics of precipitation for propagation modelling," in *Recommendation ITU-R P.837-1*, ed: ITU-R, Geneva, 1994.

Improving the OFDMA performance with common carrier frequency offset correction

Dah-Chung Chang, Yen-Heng Lai

Department of Communication Engineering, National Central University, Jhongli City, Taoyuan 320, Taiwan

Email address:
dcchang@ce.ncu.edu.tw (D.-C. Chang)

Abstract: In an OFDMA system, different carrier frequency offsets (CFOs) are possibly raised due to the mismatch of local oscillators and the Doppler effect because of multiple-antenna channels. In spite of an initial CFO synchronization scheme able to be applied at transmitters, the compensation of residual CFOs is required at the receiver in order to eliminate the inter-carrier interference (ICI) for individual subscribers, especially with an interleaved subcarrier allocation scheme. Unfortunately, conventional ICI mitigation methods cannot simultaneously remove the CFOs caused by multiple subscribers, and therefore, the multiuser interference (MUI) remains. In this case, a common CFO (CCFO) existing among the multiple CFOs is found in relation to the overall OFDMA system performance. In this paper, a CCFO estimation method is proposed at the OFDMA receiver, and the CCFO is then corrected to reach the minimum weighted mean square error (MSE) performance. Numerical results show that new OFDMA receivers after correcting the estimated CCFO significantly improve the overall bit error rate (BER) performance over the conventional receivers.

Keywords: OFDMA, Carrier Frequency Offset, Inter-Carrier Interference, Multiuser Interference, Mean Square Error

1. Introduction

Orthogonal frequency-division multiplexing (OFDM) is widely used in modern wireless communications for its good ability to reduce the multipath effect. As OFDM is used in a multiple access (MA) system, the combination of the frequency division multiple access (FDMA) method draws a lot of attention to next generations of wireless communications. The OFDM multiple access (OFDMA) technology separates groups of OFDM subcarriers allocated to different subscribers for simultaneous uplink transmission from subscriber stations (SS) to a base station (BS). WiMAX and LTE are typical OFDMA systems proposed for the application of wireless metropolitan area networks (MANs) [1]. However, in an OFDMA system, imperfect synchronization due to different carrier frequency offsets (CFOs) at individual transmitting terminals can introduce inter-carrier interference (ICI) among subcarriers and multiple access interference (MAI) among subscribers [2]-[4]. Although some methods can be exploited to initiate the synchronization at transmitters, the CFOs are hard to be completely eliminated since different local oscillators are implemented at the transmitters. Hence, a CFO tracking loop and certain MAI reduction process are usually required at the OFDMA receiver even though some initial synchronization scheme can be employed to reduce the residual CFOs.

Some approaches to dealing with CFOs in an OFDMA system can be found in the literature [5]-[10]. The CFO correction method that is conventionally used in single-user OFDM can be directly applied for different subscribers to cancel the estimated CFOs before the discrete-time Fourier transform (DFT) [5]. But the direct method requires multiple DFT blocks and causes MAI due to different offsets among subscribers. In [6], an alternative method, called the *CLJL scheme* (abbreviated for the names of the authors), was proposed to compensate for the CFOs effect after the DFT with using circular convolution. Although the CLJL scheme reduces the required number of DFT blocks, the multiuser interference (MUI) components still remain in the compensated results. Huang and Letaief [7] proposed an iterative interference cancellation scheme, called the *HL scheme*, to reduce the MUI effect. The method proposed in [7] can be regarded as a parallel interference cancellation scheme and the authors showed that only a few of iterations are required to obtain a satisfying performance. Other methods [8][9] considered a return path for control information based on maximum likelihood estimation of synchronization parameters.

In previous works [5]-[9], CFOs for different subscribers are estimated and then the correction is performed independently. However, the MUI due to different CFOs still affects the bit error rate (BER) performance even with the use of MUI cancellation schemes [7][10]. It is found that a common CFO (CCFO) existing among the multiple CFOs is related to the overall OFDMA system performance [11][12]. In this paper, we propose a feasible CCFO estimation and correction method in an OFDMA system. As we remove the estimated CCFO before the DFT at the receiver, the overall weighted mean square error (MSE) performance for multiple subscribers can be minimized such that the average BER performance is improved as well. Simulation results show that the modified CLJL or HL based OFDMA receivers together with the proposed CCFO estimation and correction method have better BER performance than those without employing CCFO correction.

The rest of this paper is organized as follows. Section II introduces the system model and the CLJL/HL methods. The proposed CCFO estimation scheme is also described in this section. Section III contains the simulation results. Section IV concludes this paper.

2. Proposed CCFO Correction Scheme

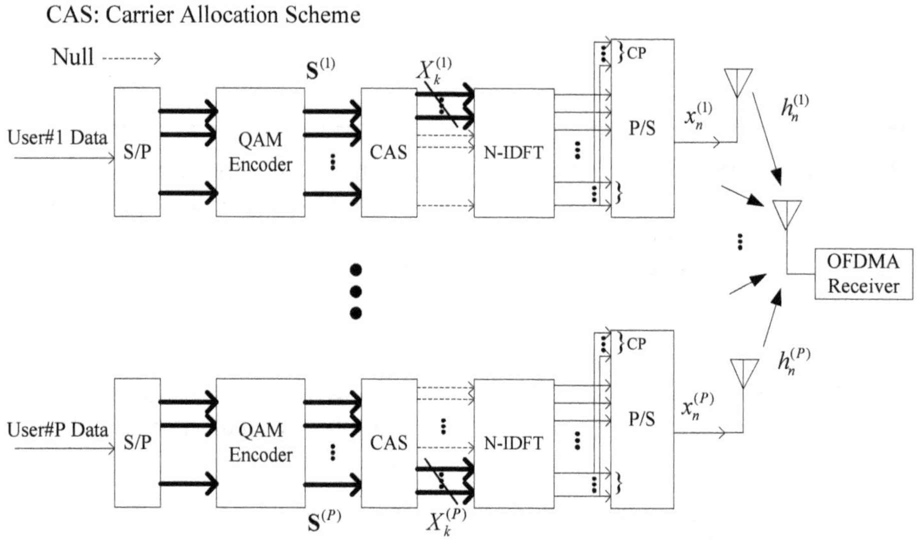

Fig. 1. Structure of the OFDMA Transmitter.

2.1. System Model and CLJL/HL Methods

Consider an N-point interleaved OFDMA system with P subscribers as depicted in Fig. 1. Each SS communicates with the BS over an independent multipath channel, which is allocated M subcarriers such that $N = M \times P$. The original data symbol for the mth subscriber is denoted by $\mathrm{S}^{(m)}$, $m = 1, 2, \cdots, P$ and $\mathrm{S}^{(m)} = [S_0^{(m)}\ S_1^{(m)} \cdots S_{M-1}^{(m)}]$. The M signals to be transmitted in the OFDMA system are first mapped into a set of N modulation samples $\{X_k^{(m)}\}$, $k = 0, 1, \cdots N-1$, by the scheme of interleaved subcarrier allocation according to

$$X_k^{(m)} = \begin{cases} S_j^{(m)}, & k = j \cdot P + m - 1 \\ 0, & \text{otherwise} \end{cases} \tag{1}$$

where $j = 0, 1, \cdots, M-1$.

Suppose $y_n^{(i)}, n = 0, 1, ..., N-1$, is the ideal ith subscriber's symbol after passing through the channel. Let ε_i, $i = 1, 2, \cdots, P$, denote the residual CFO for subscriber i with respect to the BS and ε_C the CCFO to be corrected at the BS before the DFT. The signal after correcting the CCFO at the BS consists of P subscribers' symbols accompanied by corresponding CFO effects and the additive white Gaussian noise (AWGN), which can be given as

$$r_n = \sum_{i=1}^{P} y_n^{(i)} e^{j2\pi(\varepsilon_i + \varepsilon_C)n/N} + z_n,\ n = 0, 1, ..., N-1 \tag{2}$$

where z_n is the AWGN. The vector form of the DFT output signal R_k, where k is the subcarrier index and $0 \le k \le N-1$, can be expressed as

$$\begin{aligned} \mathrm{R}(\varepsilon_C) &= \mathrm{DFT_N}(\mathrm{r}(\varepsilon_C)) \\ &= \sum_{i=1}^{P} \mathrm{Y}^{(i)} \otimes \mathrm{C}^{(i)} + \mathrm{Z} \\ &= \mathrm{Y}^{(m)} \otimes \mathrm{C}^{(m)} + \sum_{\substack{i=1 \\ i \ne m}}^{P} \mathrm{Y}^{(i)} \otimes \mathrm{C}^{(i)} + \mathrm{Z} \end{aligned} \tag{3}$$

where $\otimes$ denotes the circular convolution, $\mathrm{R}(\varepsilon_C)$

$=[R_0, R_1,..., R_{N-1}]^T$, $\mathbf{r}(\varepsilon_C)=[r_0, r_1,..., r_{N-1}]^T$, the $N\times 1$ vector $\mathrm{Y}^{(i)}$ contains the signal for the *i*th subscriber, $Y_k^{(i)}$, with $Y_k^{(i)}=\mathrm{DFT_N}(y_n^{(i)})$, $\mathrm{C}^{(i)}$ is an $N\times 1$ vector containing the value of the equivalent CFO effects, $C_k^{(i)}$, with $C_k^{(i)}=\mathrm{DFT_N}(e^{j2\pi(\varepsilon_i+\varepsilon_C)n/N})/N$, and the vector Z contains the *N*-point DFT results of the AWGN, Z_k, with $Z_k=\mathrm{DFT_N}(z_n)$. In (3), the first term is the *m*th subscriber's received signal and the second term is the MUI due to the CLJL scheme. If the MUI can be ignored and the AWGN power is small compared with the signal power, we can approximate the *m*th subscriber's received signal as

$$\begin{aligned}\hat{\mathrm{R}}^{(m)}(\varepsilon_C)&=\mathrm{Y}^{(m)}\otimes\mathrm{C}^{(m)}\\&\approx\mathrm{A}^{(m)}\mathrm{R}(\varepsilon_C)\end{aligned}\tag{4}$$

where $\mathrm{A}^{(m)}$ is a diagonal matrix with the diagonal elements defined as $\mathrm{A}^{(m)}(i+1,i+1)=1$ for $i\in\Omega_m$ and 0 for $i\notin\Omega_m$, where Ω_m is the set of subcarriers allocated to the *m*th subscriber [6][7]. Here, $\mathrm{A}^{(m)}$ acts as a filter that keeps most of the output power for the *m*th subscriber. From (4), we can restore the *m*th subscriber's signal $\mathrm{Y}^{(m)}$ from $\hat{\mathrm{R}}^{(m)}(\varepsilon_C)$ by removing the circular convolution operation in the following equation:

$$\begin{aligned}\hat{\mathrm{Y}}^{(m)}(\varepsilon_C)&=\mathrm{A}^{(m)}(\hat{\mathrm{R}}^{(m)}(\varepsilon_C)\otimes\mathrm{C}'^{(m)})\\&=\mathrm{A}^{(m)}(\mathrm{A}^{(m)}\mathrm{R}(\varepsilon_C)\otimes\mathrm{C}'^{(m)})\end{aligned}\tag{5}$$

where $\mathrm{C}'^{(m)}$ denotes the inverse of $\mathrm{C}^{(m)}$, which has components $C_k'^{(m)}$ and $C_k'^{(m)}=\mathrm{DFT_N}(e^{-j2\pi(\varepsilon_m+\varepsilon_C)n/N})/N$. The structure of the OFDMA receiver after correcting the CCFO for the CLJL scheme [6] is depicted in Fig. 2.

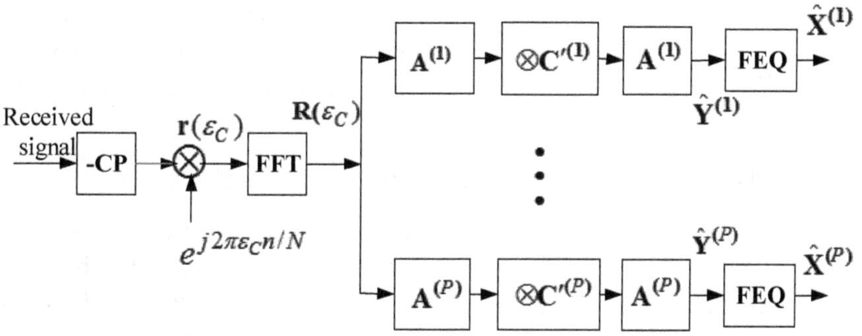

Fig. 2. *The OFDMA receiver after correcting the CCFO for the CLJL scheme.*

Let $\hat{\mathrm{Y}}^{(i),j}(\varepsilon_C)$ denote the estimate of $\hat{\mathrm{Y}}^{(i)}(\varepsilon_C)$ after performing the *j*th step of the iterative interference cancellation algorithm proposed in [7]. By ignoring the noise effect, the MUI term can be calculated by

$$\hat{\mathrm{M}}^{(m),j}(\varepsilon_C)=\sum_{\substack{i=1\\i\neq m}}^{P}\hat{\mathrm{Y}}^{(i),j}(\varepsilon_C)\otimes\mathrm{C}^{(i)}\tag{6}$$

The MUI cancellation algorithm employing the HL scheme can be summarized as follows:

$$\begin{aligned}&\textit{Initialization}: \textit{Set } j=0 \textit{ and}\\&\quad\hat{\mathrm{Y}}^{(m),j}(\varepsilon_C)=\mathrm{A}^{(m)}\left(\left(\mathrm{A}^{(m)}\mathrm{R}(\varepsilon_C)\right)\otimes\mathrm{C}'^{(m)}\right),\\&\quad\textit{for } m=1,\ldots,P\end{aligned}\tag{7}$$

$$\begin{aligned}&\textit{Loop}: j=j+1\\&\quad\textit{Set } \bar{\mathrm{Y}}^{(m),j}(\varepsilon_C)=\mathrm{R}(\varepsilon_C)-\hat{\mathrm{M}}^{(m),j}(\varepsilon_C),\\&\quad\hat{\mathrm{Y}}^{(m),j}(\varepsilon_C)=\mathrm{A}^{(m)}\left(\left(\mathrm{A}^{(m)}\bar{\mathrm{Y}}^{(m),j}(\varepsilon_C)\right)\otimes\mathrm{C}'^{(m)}\right),\\&\quad\textit{for } m=1,\ldots,P\\&\textit{Go back to Loop}\end{aligned}\tag{8}$$

2.2. Effect of CCFO

Suppose $H_k^{(m)}$ is the channel frequency response on the *k*th subcarrier of the *m*th subscriber. From (3), after some mathematical manipulation we have

$$R_k(\varepsilon_C)=X_k^{(m)}H_k^{(m)}+ICI_k^{(m)}(\varepsilon_C)+MUI_k^{(m)}(\varepsilon_C)+Z_k\tag{9}$$

and

$$ICI_k^{(m)}(\varepsilon_C) = \sum_{j\in\Omega_m} X_j^{(m)} H_j^{(m)} D\left(\frac{2\pi(\varepsilon_m+\varepsilon_C-j+k)}{N}\right)\times e^{\frac{j\pi(\varepsilon_m+\varepsilon_C-j+k)(N-1)}{N}} - X_k^{(m)} H_k^{(m)} \quad (10)$$

$$MUI_k^{(m)}(\varepsilon_C) = \sum_{\substack{i=1\\ i\neq m}}^{P} \sum_{j\in\Omega_i} X_j^{(i)} H_j^{(i)} D\left(\frac{2\pi(\varepsilon_i+\varepsilon_C-j+k)}{N}\right)\times e^{\frac{j\pi(\varepsilon_i+\varepsilon_C-j+k)(N-1)}{N}} \quad (11)$$

where $D(x) = \sin(Nx/2)/N\sin(x/2)$. From (10) and (11), we can notice that ICI and MUI are influenced by the CCFO.

2.3. Estimation of CCFO with Minimum Weighted MSE

From (5) or (8), we can obtain the estimate of the transmitted symbol on the kth subcarrier of the mth subscriber

$$\hat{X}_k^{(m)}(\varepsilon_C) = \hat{Y}_k^{(m)}(\varepsilon_C)/H_k^{(m)} \quad (12)$$

We note that $H_k^{(m)}$ can be estimated from the preamble or pilots in practical applications. Given the transmitted signal $X_k^{(m)}$, we define the weighted MSE as

$$\xi(\varepsilon_C) = \sum_{m=1}^{P} \sum_{k\in\Omega_m} \rho_m |X_k^{(m)} - \hat{X}_k^{(m)}(\varepsilon_C)|^2 \quad (13)$$

where ρ_m is the pre-determined weighting coefficient for the mth subscribers, and

$$\sum_{m=1}^{P} \rho_m = 1,\ 0 \le \rho_m \le 1. \quad (14)$$

The purpose of the weighting calculation in (13) is to emphasize the performance for specified subscribers. If all subscribers are equally weighted for the overall performance, we can set $\rho_m = 1/P$. In some situations, it may be occurred that in an OFDMA system, not all of allocated subscribers constantly occupy the channel. Hence, the base station can adaptively specify subscribers' performance by changing the weighting coefficients. By the steepest descent approach, the minimization of the weighted MSE can be obtained by calculating the following recursion at time instant n:

$$\hat{\varepsilon}_C(n+1) = \hat{\varepsilon}_C(n) - \mu\cdot\nabla_{\varepsilon_C}\xi(\varepsilon_C(n)) \quad (15)$$

where μ is a step size which controls the convergence rate and the steady-state estimation accuracy. However, it is difficult to obtain the exact formulation for a convergent innovation $\nabla_{\varepsilon_C}\xi(\varepsilon_C(n))$. We approach the derivative by the following numerical method:

$$\nabla_{\varepsilon_C}\xi(\varepsilon_C(n)) = \lim_{\Delta\varepsilon\to 0}\frac{\xi(\varepsilon_C(n)+\Delta\varepsilon)-\xi(\varepsilon_C(n))}{\Delta\varepsilon} \quad (16)$$

where $\Delta\varepsilon$ is chosen as a small value approaching zero, for example, $\Delta\varepsilon = 10^{-7}$ is used in our numerical simulation.

It should be noted that $\xi(\varepsilon_C(n))$ is not a quadratic function of $\varepsilon_C(n)$, and hence, $\xi(\varepsilon_C(n))$ may have some local minimums with respect to $\varepsilon_C(n)$. Some routine is required to search for the initial value in the region of global minimum in order to guarantee optimal convergence. A simple method is to divide the whole searching region into several sub-regions for possible initial CCFO values $\hat{\varepsilon}_C(0)$ at first. Then, we can determine the optimal CCFO initial value by finding the minimum value of their MSE metrics from (13). Since this method is proposed for improving performance when coarse synchronization and channel estimation have been achieved by the preamble, pilots or decision feedback data can be used to avoid the requirement of pre-knowing $X_k^{(m)}$ shown in (13).

3. Simulation Results

We consider an OFDMA system with the DFT size N=512, the length of guard-interval is 64, the number of subscribers is 4 with 128 block subcarriers allocated to each subscriber, eight pilots are used for each subscriber, and the Gray-coded 16-QAM signals are transmitted. The power profile of the impulse response of the multipath channel is assumed to be exponentially decaying with the characteristics $E\{|h(n)|^2\}$ $=\exp(-n/5)$, n=0, 1,…, 11. For the purpose of estimating the coarse CFOs and channel responses of the four subscribers, two pilot symbols as the preamble lead in advance of the transmitted data. To avoid the interference among subscribers in using the preambles for initial estimation of CFOs and channel responses [7], the preambles for different subscribers are assigned at non-overlapped symbol slots for simplicity.

We randomly choose two sets of CFO values to see the influence of CCFO correction in this OFDMA system: One is with large CFO values for the four subscribers and denoted by CFO1=[-0.1 0.3 0.25 -0.15], the other is with smaller values and denoted by CFO2=[0.1 -0.1 -0.05 0.05]. After the coarse CFOs and channel responses are obtained from preambles, we search for the optimal initial value for CCFO estimation based on the first 30 decision feedback data symbols. Note that during these 30 data symbols, the CCFO effect is temporarily ignored such that the system performance is equivalent to the conventional methods that do not apply CCFO correction. Thereafter, the estimated initial CCFO is used in (15) for recursive estimation.

In Fig. 3, we show the MSE performance with respect to the CCFO for the four subscribers in the CFO1 case. That is, for a specified subscriber we assume that its weighting coefficient is unity while the coefficients for others are zero. We compare the performance for the CLJL and HL schemes. It is obvious from the result of the HL scheme that the optimal CCFO value is about to cancelling the CFO effect before the DFT. By adopting the proposed approach for the initial value, we get the initials [0.05 -0.25 -0.15 0.15] for the four subscribers. As shown in Fig. 4, the proposed CCFO estimation algorithm for the HL scheme finally converges close to the CCFO values with MMSE as shown in Fig. 3. Note that the number of iterations for convergence depends on the choice of the step size. It is a tradeoff between the convergence rate and the steady state performance. A globally optimum CCFO value can be also found under the consideration of an overall minimum weighted MSE performance, i.e. setting $\rho_i = 0.25$ for i=1, 2, 3, and 4. For example, the optimum CCFO value is found about -0.1 for CFO1.

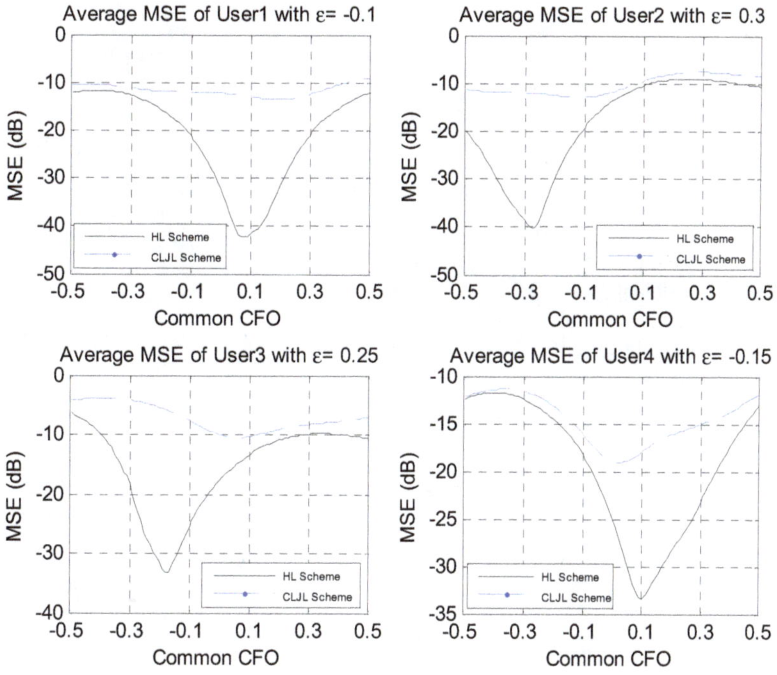

Fig. 3. MSE performance with respect to the CCFO for different subscribers in the CFO1 case.

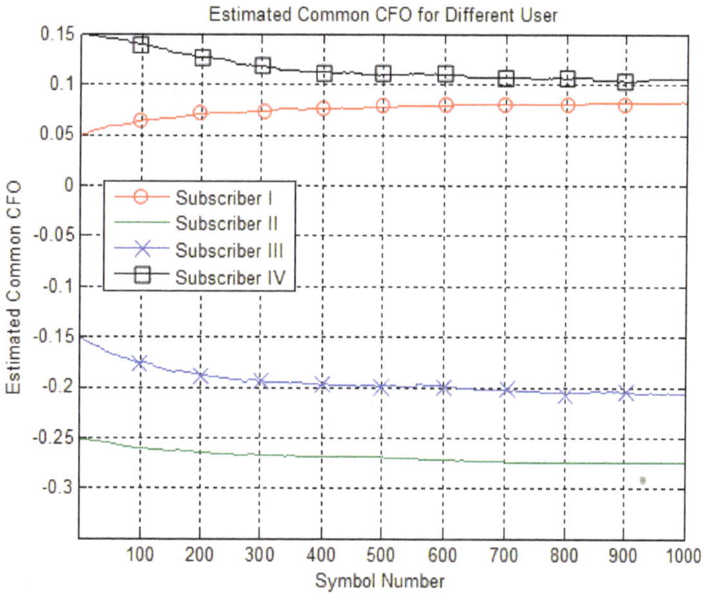

Fig. 4. The learning curves of CCFO estimation for different subscribers.

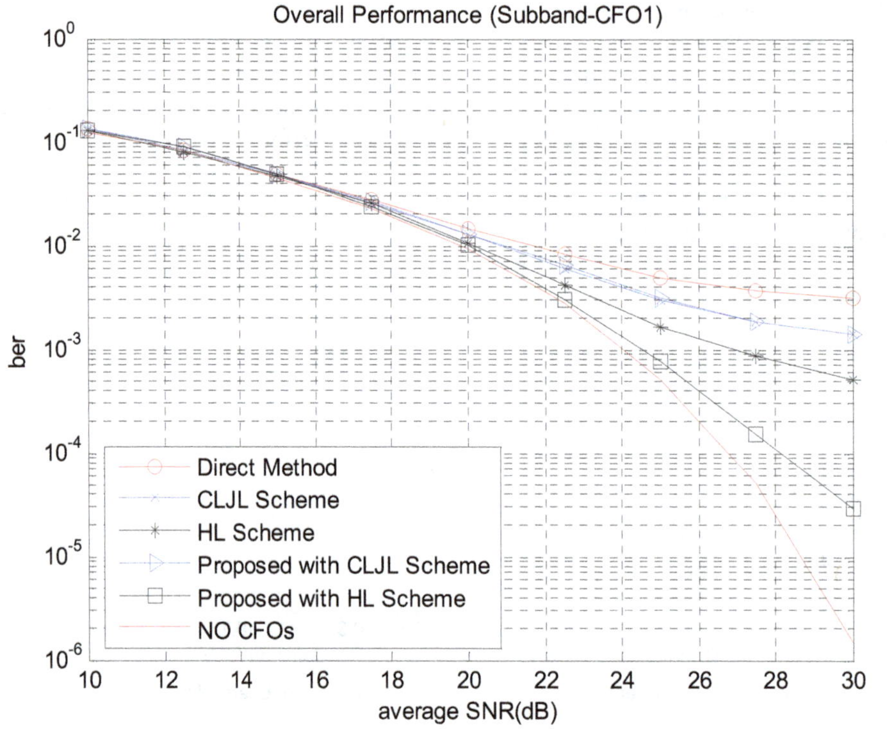

Fig. 5. *BER comparison of the overall performance for different CFO compensation methods in OFDMA with CFO1.*

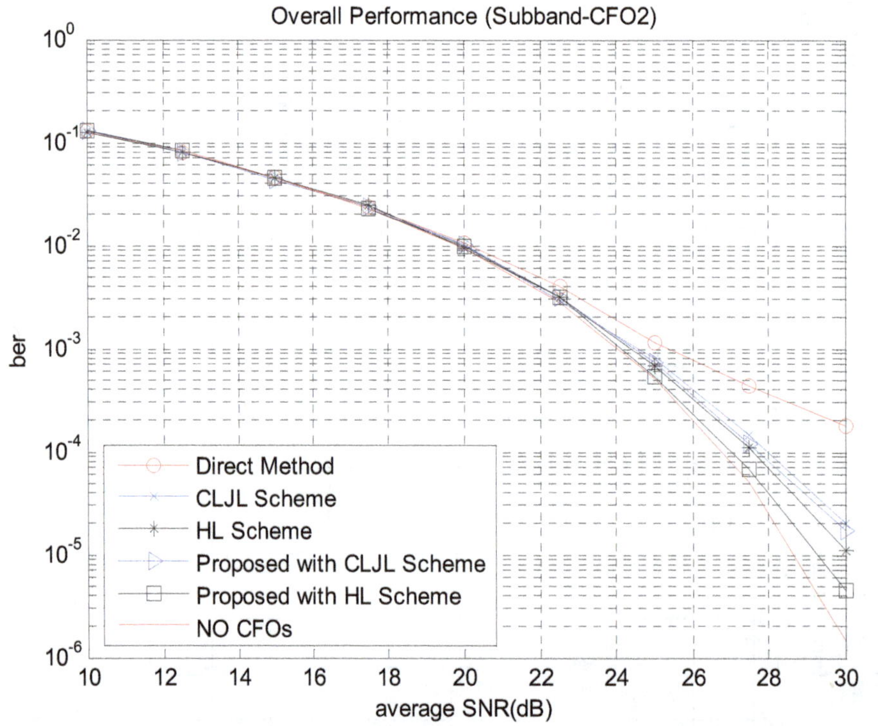

Fig. 6. *BER comparison of the overall performance for different CFO compensation methods in OFDMA with CFO2.*

We compare the BER performance for different CFOs mitigation schemes in Figs. 5 and 6. The direct method [5] and the CLJL method only compensate for the effect of CFOs, while the MUI effect is not solved. Thus, the HL method that can reduce MUI has better performance than other two methods. However, the proposed CCFO estimation scheme can search for a proper CCFO to minimize the weighted MSE performance. Hence, the proposed method along with the HL's MUI cancellation scheme outperforms others. In Fig. 5, the CFO1 case is simulated to show that a significant performance difference can be noticed. While in Fig. 6 the CFO2 case is simulated, we can observe that the BER performance is also improved with applying the proposed method together with the HL scheme.

4. Conclusion

In conventional methods, the CFO effect can be compensated for by the CLJL scheme at the BS in the OFDMA system. The MUI effect can be reduced by the HL scheme. In this paper, we show that the MSE performance of the demodulated signals is related to a CCFO which should be corrected in advance of the DFT to reach the optimum weighted MSE. Therefore, we propose a new method for CCFO estimation and correction. From simulation results, a better BER performance can be obtained by applying the CCFO adjustment method along with the CLJL or the HL scheme. Moreover, the performance of the proposed method together with the HL scheme is superior to that of others from our simulation results.

Acknowledgement

This work was supported in part by the Research Center for Advanced Science and Technology, National Central University as well as the National Science Council of Taiwan under contract NSC 103-2221-E-008-034.

References

[1] S. Srikanth, P. A. Murugesa Pandian, and X. Femando, "Orthogonal frequency division multiple access in WiMAX and LTE: a comparison," *IEEE Commun. Mag.*, vol. 50, no. 9, pp. 153-161, Set. 2012.

[2] S.-W. Hou and C. C. Ko, "Intercarrier interference suppression for OFDMA uplink in time- and frequency-selective fading channels," *IEEE Trans. Veh. Technol.*, vol. 58, no. 6, pp. 2741-2754, July 2009.

[3] Z. Zhang and C. Tellambura, "The effect of imperfect carrier frequency offset estimation on an OFDMA uplink," *IEEE Trans. Commun.*, vol. 57, no. 4, pp. 1025-1030, Apr. 2009.

[4] T. Yucek and H. Arsian, "Carrier frequency offset compensation with successive cancellation in uplink OFDMA systems," *IEEE Trans. Wireless Commun.*, vol. 6, no. 10, pp. 3546-3551, Oct. 2007.

[5] J.-J. van de Beek et. al, "A time and frequency synchronization scheme for multiuser OFDM," *IEEE J. Sel. Areas Commun.*, vol. 17, no. 11, pp.1900-1913, Nov. 1999.

[6] J. Choi, C. Lee, H. W. Jung, and Y. H. Lee, "Carrier frequency offset compensation for uplink of OFDM-FDMA systems," *IEEE Commun. Lett.*, vol. 4, no. 12, pp. 414-416, Dec. 2000.

[7] D. Huang and K. B. Letaief, "An interference-cancellation scheme for carrier frequency offsets correction in OFDMA systems," *IEEE Trans. Commun.*, vol. 53, no. 7, pp. 1155-1165, July 2005.

[8] M. Morelli, "Timing and frequency synchronization for the uplink of an OFDMA system," *IEEE Trans. Commun.*, vol. 52, no. 2, pp. 296-306, Feb. 2004.

[9] M.-O. Pun, M. Morelli, and C.-C. Jay Kuo, "Maximum-likelihood synchronization and channel estimation for OFDMA uplink transmissions," *IEEE Trans. Commun.*, vol. 54, no. 4, pp. 726-736, Apr. 2006.

[10] S. Manohar, D. Screedhar, V. Tikiya, and A. Chockalingam, "Cancellation of multiuser interference due to carrier frequency offsets in uplink OFDMA," *IEEE Trans. Wireless Commun.*, vol. 6, no. 7, pp.2560-2571, July 2007.

[11] D.-C. Chang, Y.-H. Lai, and Y.-C. Hsu, "Effect of common carrier frequency offset at the OFDMA receiver," in *Proceeding of IEEE International symposium on Circuits and Systems* (ISCAS), May 2009, pp.201-204.

[12] S. Gaum and R. Kumar, "Performance analysis of OFDMA receiver with common carrier frequency offset (CCFO)," in Proceeding of 2012 World Congress on Information and Communication Technologies (WICT), Oct. 2012, pp. 266-271.

Presenting Solutions to Increase Simultaneous Call in VOIP System by SIP Protocol - Based Media Server

Sajad Gharaguozloo[1], Abdolhamid Zahedi[2], Mohammad Norouzi[2], Hamid Chegini[2, *]

[1]Telecommunication, of Non-profit Institution of Higher Education, ABA, Abyek, Qazvin, Iran
[2]Non-profit Institution of Higher Education, ABA, Abyek, Qazvin, Iran

Email address:
sajadgharegozloo@yahoo.com (S. Ghareguozloo), H.zahedi.62@gmail.com (A. Zahedi), noroozi.62@gmail.com (M. Norouzi), fmirzaei_91@yahoo.com (H. Chegini)
[*]Corresponding author

Abstract: By development of multi-media in networks, the borders among networks are changed and all networks are approaching to be united. Unity of data, video and voice networks in one network has many advantages and disadvantages for users and servers. One of the disadvantages is unwanted events in network including load increase, jitter, information packet loss, delay and etc. and all these lead into low quality of voice and disconnection during simultaneous call. The use of multimedia server is one of the efficient ways to improve VOIP. We can enable the video conferences to transit information packets by media servers, so we can say: media servers can be used as core component for VOIP. In this research work, assessing of media servers is done by simulators that they produce RTP's connections, in additions as an experimental components SEMS that it's a source of media server, is used for asserting the quality by doing packets information with SIP. We can observe that the more increase of connections and pass of the certain threshold, the less of quality. In addition, the other performance metrics such as error rate And packet lost are asserted. The identification of load saturation points and the efforts to eliminate disturbing factors during the increase of simultaneous call in this telephone system, can present quality-based approach to servers of these networks.

Keywords: VOIP, SEMS, SIP, Increase of Simultaneous Call

1. Introduction

Today, with the technology progress and creating data networks (e.g. internet) and using these networks for voice transmission, a new method is crated in telecommunication connection and it is called VOIP technology or "Voice. Over Internet Protocol" and the networks based on this technology are called VOIP networks.

Due to various facilities providing for users, these networks are welcomed as they are considered as the alternative of current telecommunication networks.

Any new technology has some limitations. One of the problems of data-based networks is the problem of increase of simultaneous call in them. This study attempts to evaluate that in a SIP [1] -based VOIP telephone system, by increasing participants in a voice conference simultaneously, media server behavior is changed and then by the results of experiment and experience, proposed solutions to increase calls in this telephone system as network can be defined. In the test, open source media service as called SEMS[2] is applied. The results of test can be used to increase call in each network-based telephone system.

Regarding the evaluation of media service performance, there are various researches as follows:

In reference [1], a hash table is used to test the trend of SIP on virtual media server. In this study by a SIP proxy, about 5 thousands calls were established with different times and after

[1]Session Initiation Protocol
[2]SIP Express Media Server

the evaluation of outputs, it was shown that by the increase of calls about 300 calls per seconds, due to losing signal, unsuccessful call rates were increased.

Based on hash table system, this trend is improved and is initiated from 300 to 500 calls per seconds.

In reference [2], two parts of SIP software as one server and another one customer is used to send signaling traffic and the rate of calls is increased from 5 calls to 60 calls simultaneously per seconds and a call control system is considered to be used for terminating the remaining calls exceeding 3.5m, about 37% of calls are in time out and are finished before being terminated by controller [3].

The paper is structured as follows. The second section is dedicated to the introduction of VOIP telephone system and its advantages to old communication lines. Third section is about the evaluation of SIP protocol as the protocol playing the main role in creating calls. Fourth section introduces media server and voice conferences and constituent elements of a conference. Finally, fifth section is regarding the evaluation of media server and its test in terms of RTP[3] load. Final section is about the results of discussions and solutions to increase simultaneous call in VOIP system by SIP protocol-based media.

2. What Is VOIP

VOIP is the technology providing data network for telephone conversations. By VOIP, human voice is sent via IP[4] information packets and via data-based network as internet [4]. In other words, VOIP technology is a set of hardware and software enabling us to use data network as transition mediator for telephone calls [5].

2.1. The Advantages of VOIP over PSTN

In case of using PSTN[5], users pay the time cost by the company responsible for PSTN line (telecommunication), more than one person is not communicated simultaneously [6]. In VOIP technology, we can talk simultaneously with more than one without any extra cost. Also data is exchanged during talking (e.g. image, chart and video images). VOIP service provides all services presented by ordinary phone [7] [8].

2.2. VOIP Defects in Contrast with PSTN

Because of connecting VOIP to the internet, invaders may be able to do something such as cut the connection, over hearing or interrupting in this service [9] [10].

Unlike old phone lines, when the electricity goes out, we cannot use VOIP. This causes some problems in security system in the houses and in that time, there is lack of access to emergency calls.

Also, as the internet's issues, perhaps we can't receive internet packets that they were sent. This problem causes some disadvantages in VOIP system.

3Real Time Protocol
4Internet Protocol
5Public Switched Telephone Network

3. SIP Protocol

Protocol means a set of rules determining the information exchange. VOIP system works on network and we need a communication language or protocol for data transmission.

SIP is an application-layer control (signaling) protocol for creating, modifying and terminating sessions with one or more participants. [11] [12] The invitations sent by SIP are used for sessions on IP network [13].

3.1. Important Elements of SIP

- User agent (UA) creates SIP transactions or responds alone.
- User agent client (UAC) creates SIP requests and accepts responses and reactions of SIP.
- User agent server (UAS) gets SIP requests and sends the responses.

3.2. SIP Requests

SIP requests are the messages being sent from customer to server to call for SIP operation and most important examples are as follows:

- INVATE or invitation: It is a method showing that invitation client is invited to participate in a network.
- ACK[6]: ACK request shows that a client agent of the last request has received an invitation. ACK is used continuously for 200 OKs.
- BYE: An UA uses BYE request to terminate old session.

Connection by SIP is shown in Figure 1.

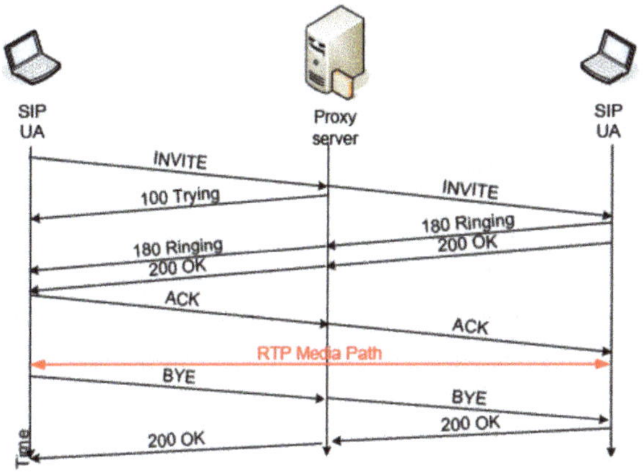

Figure 1. *Connection by SIP.*

For a two-party session, if a message is sent by user agent, the proxy sends two messages of Trying and Ringing to the user agent and if the invitation is accepted, a 200OK message is sent to user agent and user agent acknowledges the message of OK. RTP[7] or voice and video stream can be exchanged among them. [14] If the user agent attempts to terminate the session, BYE message is sent to server proxy and to respond this request, by sending message acknowledges the termination of session.

6Acknowledge
7Real Time Protocol

4. The Role of Media Server in Voice Conference

The history of media server was at the same time with the history of automatic clock and automatic teller in 1960. In 1990, a great revolution was occurred in independent message transmission from fax and voice conference servers and in 1998-1999, media servers were similar to new forms.

The example of the performance of media server is conference services. These services by combining voice streams send it for all participants in conference.

4.1. Constituents of Voice Conferences

Generally, connection sessions with multiple participants are recognized as conference [15]. A voice conference is composed of conference server and participants. Conference server is composed of a focus and a mixer. Focus is in control of conference and mixer attempts to transfer multi-media RTP stream among participants [16].

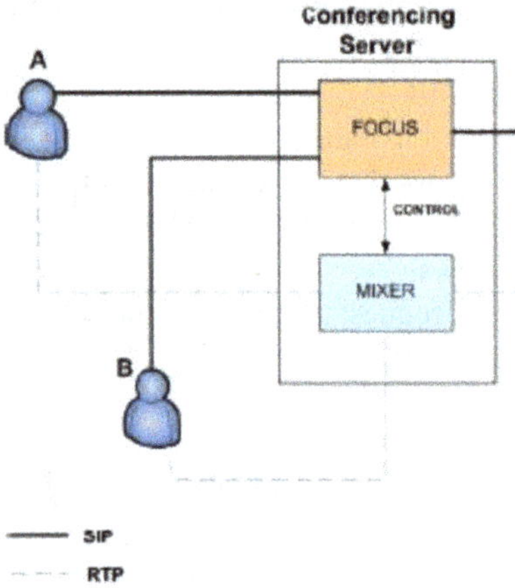

Figure 2. Diagram of a voice conference.

Mixer receives control orders from focus unit. Voice streams are received by mixer from participants and streams are distributed among them again by mixer. Each participant receives the sum of media streams of other participants minus RTP stream.

4.2. SEMS Media Server

The main duty of a media server is processing mass media stream.

The following Figure shows the architecture of a simple media server.

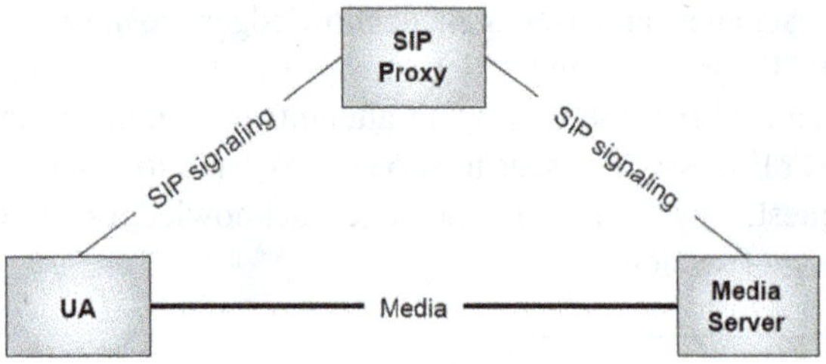

Figure 3. Architecture of a media server.

Some of the most important duties of a media server are as follows:

- Combining media stream in conference and its re-sending to participants
- Conversion of text to speech
- Automatic detection of voice

SEMS media server [17] in VOIP services based on SIP can be used. SEMS has an internal focus acting as its control sector.

5. A Method to Evaluate Media Server and Presenting Solution to Increase Simultaneous Calls in It

In this paper, we evaluate SEMS media server and use the conference service of this media server.

The following Table expresses the test conditions. To test this test, encoder-decoder G711 as free and standard is used.

***Table 1.** Test conditions Table.*

Explanation	Type
Pentium IV- 1.86GHZ / 3096MB	System
Pentium IV- 1.86GHZ / 3096MB	Directing system
Pentium IV- 1.86GHZ / 3096MB	Traffic transmitter
Pentium IV- 1.86GHZ / 3096MB	Media server

Figure 4 shows the location of media server and SIP in test.

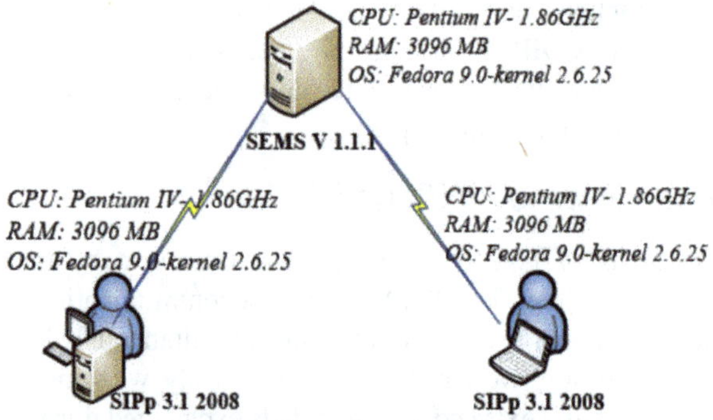

***Figure 4.** The location of media server and participants in test.*

All tests as explained later are performed under equal conditions and after sending background load or warming media server. It is worth to mention that in these tests, rate of 10 calls per seconds is considered and 40 seconds RTP on server is sent. In the first test in this study, the aim of study is to test load, we want to evaluate what happens to CPU[8] load by increasing the number of participants in conference?

As shown, when the number of participants in conference is more than 190, the processer load reaches 100% and the number of participants in this conference shouldn't be more than 75 and the load is not above 80%. CPU load can be dependent upon applied hardware and type of directing system.

In this study, we try to investigate the cause of losing

[8]Central Personal Unit

packets and jitter in this system.

Generally, the reasons of packets loss, signal reduction in network, saturation of communication links, defect hardware and the method of routing in routers. As the media server has limited memory and limited processing resources, by increasing the number of participants in server conference, it reaches saturation level and the percent of packets loss is increased. As shown in the following Figure, if the number of participants exceeds 75, packets loss is increased as sudden but the number of participants achieves 75, the percent of packets loss is about 3% as acceptable for an encoder-decoder G 711 acceptable. In some papers [18], quality of maximum value service is by one percent.

By packets loss, we can expect jitter phenomenon.

One of the most important indices in evaluation of VOIP service quality is jitter. Jitter in network can be due to traffic, change of routing and even bad configuration of device. In this test, as server capacity is limited, by increasing the number of callers, disturbance is made between media streams combination with their re-distribution among participants and jitter is occurred.

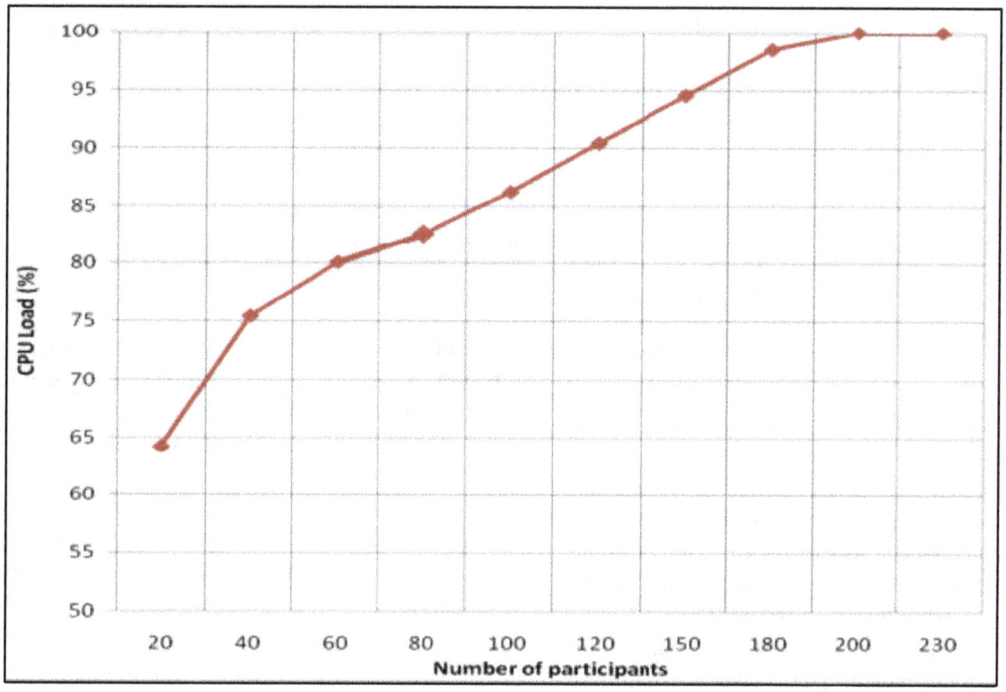

Figure 5. *CPU load chart by increasing the number of participants.*

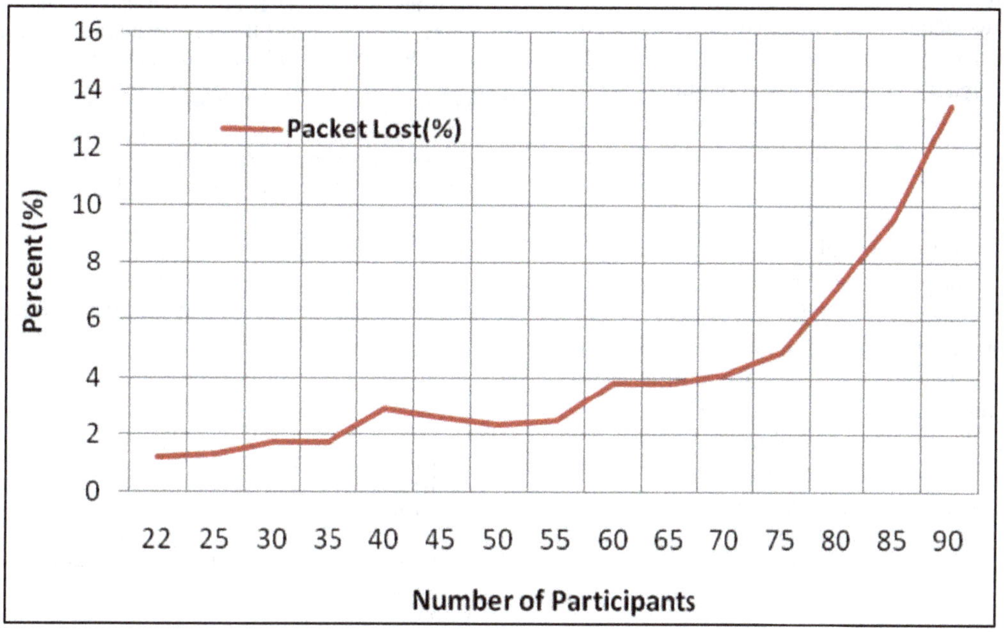

Figure 6. *The chart of number of lost information packets by increasing the number of participants.*

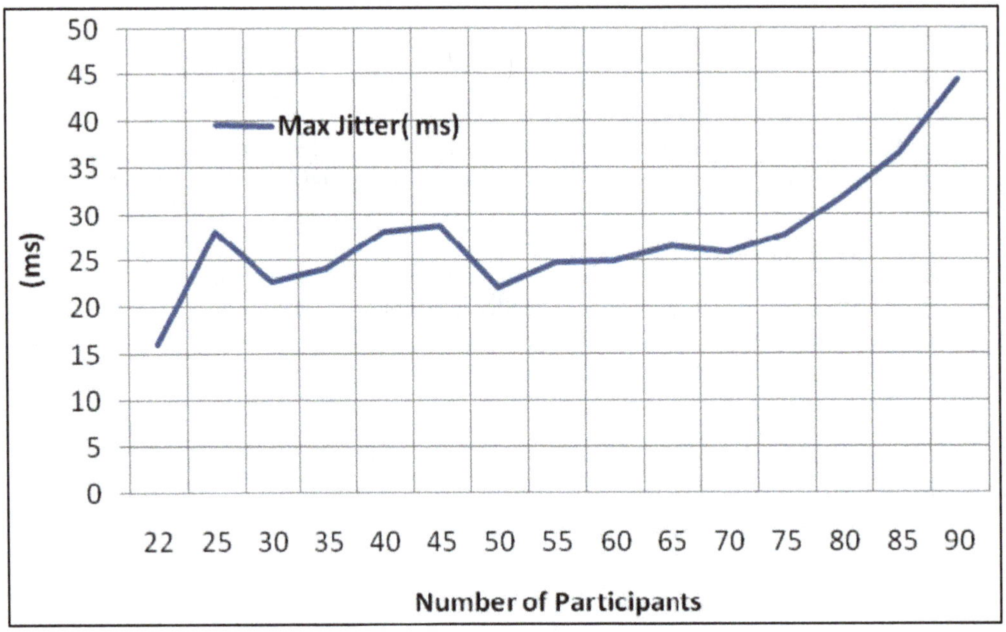

Figure 7. *The chart of jitter by increasing the number of calls.*

6. Conclusion

In this study, it was shown that by increasing the number of participants in a conference, media server is saturated after a definite point and no new participant is accepted. The main goal of simultaneous calls is quality. Thus, evaluation of RTP loan on media server can present service-quality based attitude but at signaling level, it is only zero and one and we can be aware of making or not making calls (number of calls).

It is observed that after passing saturation point, connection is established in media but lack of quality avoids continuing work.

Thus, based on the test and experiences, we can use the following methods to achieve the increase of number of simultaneous calls in a VOIP system:

- Increasing the capacity of servers by combining two or some servers as the number of callers is increased and no problem is made in combination and distribution of media streams.
- In writing programming for VOIP telephone system, we should try to have an optimal application. The main purpose of efficient plan is only support your needs of one call or conference call in VOIP.
- During telephone conference, if the number of callers is high, recording conversation is avoided to prevent server to be involved other affairs.
- For the entire system, only a protocol is used, for example applied protocols in receiver and transmitter are both SIP. Use of different protocols in transmitter and receiver causes spending more time to exchange their language to each other and the function of system is decreasing.
- Codecs of voice transmission are similar to make the application as optimal. As codecs of voice transmission aren't same, it causes waste of time and waste of some calls when at the same time the calls increase.
- We shouldn't use compressing Codecs of voice transmission as these Codecs reduce bandwidth, they reduce quality in information packets and number of callers are similar at a time.

Knowing the methods of increasing the number of simultaneous calls in VOIP systems and reaching saturation points and acquiring them can be of great importance for VOIP servers.

References

[1] H. Wook, S. Kang, D. Kim, "Performance Enhancement of SIP proxy server by using Ihash for matching transaction", IEEE, ISBN 978-89-5519-131-8 93560, Feb 2007

[2] Mauro Femminella, Roberto Francescangeli, Francesco Giacinti, Emanuele Maccherani, "Design, Implementation, and performance of an advanced SIP-based call control for VoIP services", IEEE, ISBN 978-1-4244-3435, 2009

[3] http://sipp.sourceforge.net/doc/reference.html

[4] Montoro, P, Casilari, E, 2009, *A* Comparative Study of VoIP Standards with Asterisk, Fourth International Conference on Digital Telecommunications

[5] Iseki, F, Sato, Y, Kim, M. 2011, *VoIP* System based on Asterisk forEnterprise Network, ICACT2011

[6] Pantelis A. Frangoudisa, George C. Polyzosb, On the performance of secure user-centric VoIP communication, 2014, Computer Networks Volume 70, 9 September Pages 330–344

[7] Abdul Qadeer, M, Shah, K, Goel, U, 2012, Voice - Video Communication on Mobile Phonesand PCs' using Asterisk *EPBX*, International Conference on Communication Systems and Network Technologies

[8] WWW.VOIP-IRAN.COM

[9] Nikos Vrakasa, Dimitris Geneiatakisb, Costas Lambrinoudakisa, Obscuring users' identity in VoIP/IMS environme, 2014, Computers & Security, Volume 43, June, Pages 145–158

[10] Ryan Farley, Xinyuan Wang, Exploiting VoIP softphone vulnerabilities to disable host computers: Attacks and mitigation, 2014, International Journal of Critical Infrastructure Protection, Volume 7, Issue 3, September, Pages 141–154

[11] Liping Zhang, Shanyu Tang, Shaohui Zhu, An energy efficient authenticated key agreement protocol for SIP-based green VoIP networks, 2016, Journal of Network and Computer Applications, Volume 59, January, Pages 126–133

[12] Jinzhu Wanga, et al, Probe-based end-to-end overload control for networks of SIP servers, 2014, Journal of Network and Computer Applications, Volume 41, May, Pages 114–125

[13] Basicevic, M. Popovic, D. Kukolj, 2008, Comparison of SIP and H. 323 Protocols, Proc. of The Third International Conference on Digital Telecommunications (ICDT'08), Bucharest (Romania), Jul. pp.162-167

[14] Regis J. (Bud) Bates, Chapter 6 – Other protocols SRTP, ZRTP, and SIPS, 2015, Securing VOIP Keeping your VOIP Network Safe, Pages 123–150

[15] J. Rosenberg, H. Schulzrinne, G. Camarillo, A. Johnston, J. Peterson, R. Sparks, M. Handley, E. Schooler, "SIP: Session Initiation Protocol", RFC 3261, June 2002

[16] J. Rosenberg, "A Framework for Conferencing with the Session Initiation Protocol (SIP)", RFC 4353, February 2006

[17] R. Even, N. Ismail, "Conferencing Scenarios", RFC 4597, August 2006

[18] http://www.iptel.org/sems.

[19] C. Partridge, “Isochronous Applications Do Not Require Jitter-Controlled Networks”, RFC 1257, September 1991

[20] H. Schulzrinne, S. Casner, R. Frederick, V. Jacobson, "RTP: A Transport Protocol for Real-Time Applications", RFC 3550, July 2003

[21] A. H. Ashouri, F. Samsami, A. Akbari, “E-Learning Media Server Evaluation and its architecture modeling with signaling load tests,” ICeLT, IUST, Tehran, Iran, Dec 2009

[22] A. H. Ashouri. “Media Server Evaluation and Real-Time Tests” Iran University of Science and Technology, B. Sc Thesis, p46-61, Sep 2009

A BLE Communication Design of Glucose Monitor based on x73-PHD Standards and Continua Design Guidelines

Yuan-Fa Lee

Biomedical Technology and Device Research Laboratories, Industrial Technology Research Institute, Taiwan, R.O.C.

Email address:
YuanFaLee@itri.org.tw

Abstract: The ISO/IEEE 11073 (a.k.a. x73) specifications adopted by the Continua Health Alliance constitute the international personal healthcare device (PHD) communication standards. These standards allow medical devices to intercommunicate and exchange measurement data within a single system. The Bluetooth special interest group (SIG) introduced a Bluetooth Low Energy (LE) technology with reduced power consumption. The Bluetooth LE can be used in medical devices with low data transmission rates. The research aims to identify a practical solution for the Continua BLE glucose monitor. Based on the Bluetooth LE standards as well as the Continua design guidelines and x73-PHD standards, the proposed Continua BLE glucose monitor is developed. The Continua BLE glucose monitor takes glucose measurements from the user and transmits these data to the gateway via the Bluetooth LE interface, in compliance with the Bluetooth GATT-based data format and protocol standards. Our results show that glucose measurements can be successfully transmitted to an Android-based gateway and Continua-compliant gateway through the Bluetooth LE interface. The practical solution is also feasible for other medical devices, for instance, blood pressure monitor or thermometer.

Keywords: Bluetooth LE, Continua, Glucose Monitor

1. Introduction

Personal healthcare systems are being aggressively marketed, and medical devices with a communication function are being widely discussed. Currently, healthcare devices have six types of transmission interface: RS-232, USB personal health device class (PHDC), Bluetooth serial port profile (SPP), Bluetooth health device profile (HDP), the ZigBee health care profile (HCP), and Bluetooth LE. In which, the transport interfaces supported by x73-PHD standards are USB PHDC, Bluetooth HDP, ZigBee HCP, and Bluetooth LE. Wireless transmission interfaces currently appear to be the most popular.

In 2010, we [1] proposed a novel personal healthcare system comprising a service gateway, an adapter, and a legacy device. The adapter, when combined with the legacy device, converts measured data from the legacy device, and then transmits these data to the service gateway via a Bluetooth interface. The adapter communicates with the legacy device via a universal asynchronous receiver/transmitter (UART) interface. In 2012, Park and his colleagues [2] presented an implementation model of standardization for legacy healthcare devices. The proposed system generates standard PHD protocol message blocks using legacy device information via a UART interface.

In 2013, we [3] introduced a personal medical monitoring system based on the international x73-PHD standards [4]. The proposed solution, a two-in-one blood pressure plus blood glucose monitoring system, enables wireless transmission functionality by adopting the IEEE 802.15.1 Bluetooth standard and a new HDP [5]. In 2015, Jiho Kim and colleagues [6] proposed an ISO/IEEE 11073 standardization system, including a ZigBee adaptation module (ZAM), and a PHD manager. The ZAM communicates with a healthcare device through a UART interface and transmits measurement data to the PHD manager via the ZigBee interface. These studies proposed x73-PHD solutions but focused on the Bluetooth HDP or ZigBee HCP interface.

In 2013, Park and colleagues [7] proposed a patch-type ECG sensor system (ECG node) and an android application. The ECG sensor node makes a packet which is transmitted to the smart device after sensing ECG data.

In 2014, Lin and colleagues [8] proposed a Bluetooth low energy based blood pressure monitoring system, which integrated a homemade blood pressure measuring device and a smartphone. The device uses single BLE System-on-Chip to process the algorithm of blood pressure measurement and data transmission. However, these solutions didn't adopt x73-PHD standardizations and Continua design guidelines. They didn't consider the interoperability of medical devices. Thus, different types of medical devices can't all communicate with the same gateway or application host device (AHD).

In this study, we propose a Continua BLE glucose monitor that follows the Bluetooth LE standards and Continua design guidelines v4.0 [9-12]. This system converts the communication protocol of legacy devices to the Bluetooth LE protocols [10-12]. In addition, an interoperability design of the x73-PHD standardizations is developed. It is further integrated into the Bluetooth smart ready gateway, which is also an IEEE 11073 PHD compatible manager.

Our research aims to identify a practical solution for the Continua BLE glucose monitor. The proposed method including *System Architecture, Use Case Analysis, Message Flow Sequence, and GATT Profile Design* is introduced in the *Methods* section of this paper. Our test results from the lab work are presented in the *Results and Discussion* section, and our concluding remarks are given in the *Conclusions* section.

2. Methods

2.1. System Architecture

The Continua BLE glucose monitor consists of a Continua BLE system and a glucose sensor system. The Continua BLE system connects with a gateway via the Bluetooth GATT-based interface and connects with the glucose sensor system via a wired UART interface. The measurement data are exchanged between the Continua BLE glucose monitor and the gateway. Figure 1a gives an operational overview of the proposed Continua BLE glucose monitor. It offers an interoperability platform for measurement data exchange that is compliant with the international Bluetooth LE and x73-PHD standards via a standard low power (LP) wireless PAN interface.

Figure 1b illustrates the software system architecture of the Continua BLE glucose monitor based on a Texas Instruments BLE CC2540/41 SDK (Software Development Kit) [13]. The Continua BLE glucose monitor consists of several modules including a link layer, the logical link control and adaptation protocol (L2CAP), the generic access profile (GAP), the generic attribute profile (GATT), a security manager, the GAP role, the GAP bond manager, a glucose profile/service, and an uart_if. The uart_if module is designed to communicate with the glucose sensor system.

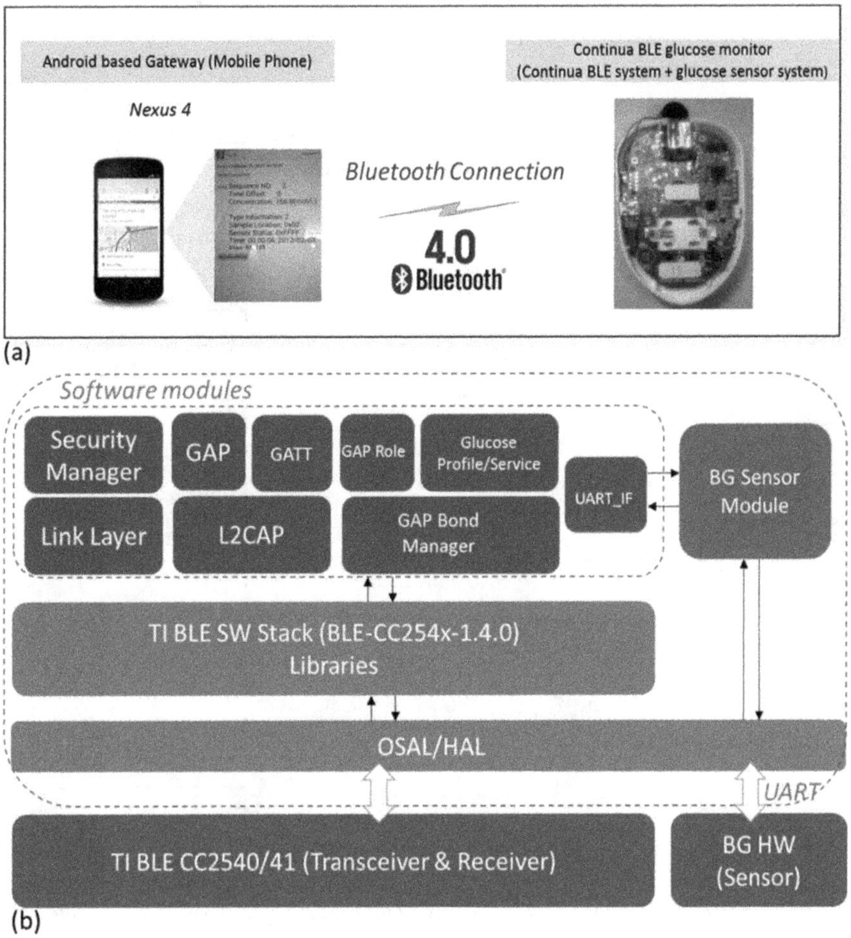

Fig. 1. *(a) A operational overview of the Continua BLE glucose monitor connected to an Android-based mobile phone via a Bluetooth LE interface. (b) A system architecture overview of the Continua Bluetooth LE glucose monitor.*

2.2. Use Case Analysis

The proposed Continua BLE glucose monitor consists of a Continua BLE system and a glucose sensor system in which the Continua BLE system adopts the iMCC2541 module (TI BLE CC2540/41 based) as the Bluetooth LE solution. The glucose sensor is supported by Wisdom IOT-Biomedical Technology Inc. The Continua BLE glucose monitor uses the wireless transmission of Bluetooth LE and meets the x73-PHD standards. Figure 2a illustrates the use case analysis.

Taking the Continua BLE system as the kernel, there are six main functions: start system, stop system, send advertisement, get vital sign information, establish BLE data channel, and send out BLE data. The glucose sensor system provides power to the Continua BLE system at system start and cuts off the power at system stop. An advertisement message is periodically sent out by the Continua BLE system when the RF button is pressed. At the same time, the Continua BLE glucose system becomes discoverable to the GATT client (i.e., gateway). A Continua-compliant gateway connects with the Continua BLE system. The Continua BLE system acquires measurement data from the glucose sensor system and sends them to the gateway after parsing and encapsulating the data in Bluetooth LE message format.

2.3. Message Flow Sequence

Figure 2b illustrates the overall message flow sequence of the system components: gateway, Continua BLE system, and glucose sensor system. After the Continua BLE system wakes up from sleep mode, it starts to send out advertising messages if system initialization is successful. A gateway attempts to connect with the Continua BLE system by sending a connect request message. If the Continua BLE system accepts the connection request, the devices start to exchange information. The Continua BLE system notifies the glucose sensor system that a Bluetooth LE data channel has been established by sending an *online test* message. The glucose sensor system responds to a successful message with *online ready*. The Continua BLE system starts to read measured data from the glucose sensor system and sends the data, with an indication or notification procedure to the gateway. The Continua BLE system notifies the glucose sensor system that the measurement data have been transmitted successfully by sending an *upload complete* message, and the glucose sensor system labels the measurement data as transmitted. The Continua BLE system reads the measurement data sequentially and sends them out to the gateway until all measurement data have been transmitted successfully. The Continua BLE system disconnects the Bluetooth LE connection automatically and enters the sleep mode again.

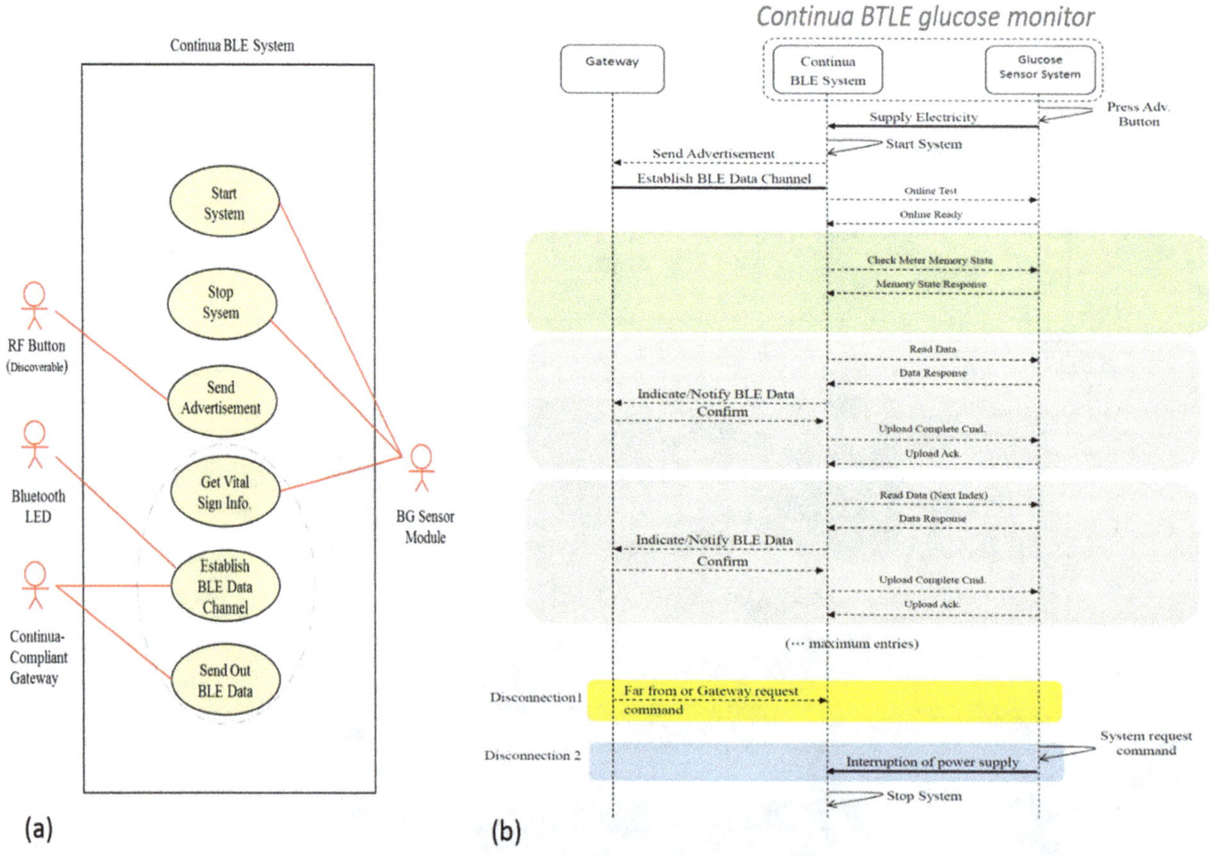

Fig. 2. (a) Analysis of the Continua BLE system use case. (b) A work flow overview of the Continua BLE system with a glucose sensor system conjoined for connecting to the gateway.

2.4. GATT Profile Design

Figure 3 illustrates the GATT profile design of the Continua BLE glucose monitor. The GATT profile is designed to be used by an application, allowing the client to communicate with a server. The server has a number of attributes, and the GATT profile defines how to use the attribute protocol to discover, read, write, and obtain other details of these attributes, as well as configuring the broadcast of the attributes.

Based on the Bluetooth glucose protocol and service specifications [11, 12], the GATT profile of the Continua BLE glucose monitor has three primary services: the GAP service (whose universally unique identifier (UUID) is 0x1800), the glucose service (whose UUID is 0x1808), and the device information service (whose UUID is 0x180A). The GAP service has two characteristics: device name and appearance. The glucose service has four characteristics: glucose measurement, glucose feature, record access control point, and date time. The device information service has six characteristics: system identity, model name, manufacturer's name, firmware revision, serial number, and IEEE 11073-20601 regulatory certification data list [4]. The Continua BLE glucose monitor supports a GATT client to access these attributes using ATT commands, including discovery service, read a characteristic value, write a characteristic value, indication of a characteristic value and notification of a characteristic value [10].

In addition, the date time characteristic of the glucose service and the IEEE 11073-20601 regulatory certification data list of the device information service are added to the GATT profile to meet the requirements of the Continua design guidelines v4.0.

ConHnd	Handle	Uuid	Uuid Description	Value
0x0000	0x0001	0x2800	GATT Primary Service Declaration	00:18
0x0000	0x0002	0x2803	GATT Characteristic Declaration	02:03:00:00:2A
0x0000	0x0003	0x2A00	Device Name	42:47:20:4D:65:74:65:72
0x0000	0x0004	0x2803	GATT Characteristic Declaration	02:05:00:01:2A
0x0000	0x0005	0x2A01	Appearance	00:00
0x0000	0x0006	0x2800	GATT Primary Service Declaration	08:18
0x0000	0x0007	0x2803	GATT Characteristic Declaration	10:08:00:18:2A
0x0000	0x0008	0x2A18	Glucose Measurement	
0x0000	0x0009	0x2902	Client Characteristic Configuration	
0x0000	0x000A	0x2803	GATT Characteristic Declaration	02:0B:00:51:2A
0x0000	0x000B	0x2A51	Glucose Feature	00:00
0x0000	0x000C	0x2803	GATT Characteristic Declaration	28:0D:00:52:2A
0x0000	0x000D	0x2A52	Record Access Control Point	
0x0000	0x000E	0x2902	Client Characteristic Configuration	
0x0000	0x000F	0x2803	GATT Characteristic Declaration	02:10:00:08:2A
0x0000	0x0010	0x2A08	Date Time	DD:07:0A:1F:0E:0C:00
0x0000	0x0011	0x2800	GATT Primary Service Declaration	0A:18
0x0000	0x0012	0x2803	GATT Characteristic Declaration	02:13:00:23:2A
0x0000	0x0013	0x2A23	System ID	00:00:00:00:00:00:00:00
0x0000	0x0014	0x2803	GATT Characteristic Declaration	02:15:00:24:2A
0x0000	0x0015	0x2A24	Model Number String	54:44:33:32:36:31:46
0x0000	0x0016	0x2803	GATT Characteristic Declaration	02:17:00:29:2A
0x0000	0x0017	0x2A29	Manufacturer Name String	57:2D:69:4F:54:20:43:6F:2E
0x0000	0x0018	0x2803	GATT Characteristic Declaration	02:19:00:26:2A
0x0000	0x0019	0x2A26	Firmware Revision String	31:2E:30:30:2D:42:30:31
0x0000	0x001A	0x2803	GATT Characteristic Declaration	02:1B:00:25:2A
0x0000	0x001B	0x2A25	Serial Number String	30:30:30:30:30:30:30:30:30:30
0x0000	0x001C	0x2803	GATT Characteristic Declaration	02:1D:00:2A:2A
0x0000	0x001D	0x2A2A	IEEE 11073-20601 Regulatory Certificati...	00:02:00:12:02:01:00:08:02:00:00:01:00:02:11:80:02:02:00:02:00:00

Fig. 3. *The GATT profile design for the Continua BLE glucose monitor.*

3. Results and Discussion

The system proposed in this study was developed based on a previous implementation [1, 3]. The Android based gateway was developed on a Google Nexus 4 mobile phone, while the Continua AHD was developed and integrated with CESL [14] using a low power wireless PAN-IF solution on a Windows-based platform. The Continua BLE system was developed in an embedded system with RTOS support and used protocol and device specifications defined in the Bluetooth GATT based standards [10-12]. The Continua BLE system utilized MCU (TI BLE CC2541) as its micro-controller unit. It communicated with the glucose sensor system through the UART interface. The Continua BLE glucose monitor was tested using Continua Test Management Lite, version 4.0.0.0, and Continua CESL Manager, version 4.0 [15]. For the duration of the test process the air traffic between the Continua BLE glucose monitor and the gateway/Continua AHD was sniffed by a Frontline Bluetooth protocol analyzer (BPA). The captured traffic was used for diagnosis and troubleshooting of problems. All test results were positive. Table 1 shows the laboratory equipment.

Table 1. Laboratory equipment.

Device Name	Personal Healthcare System	
	Item	Used in the test
Android-based Gateway	Hardware	Google Nexus 4 (Phone) *Support Bluetooth LE Feature (as Master)
Continua AHD	Hardware/	Intel® Core 2 Duo CPU 2.40GHz (Windows XP)
		CSR 8510 USB dongle (as Master)
	Software	CESL Manager & Test Management Lite
Continua BLE Glucose Monitor	Hardware	iMCC2541 module (as Slave)
TI CC2540/41 BLE Evaluation Board	Hardware	TI SmartRF05 Evaluation Board CC2540 Evaluation Module (with BLE-CC254x-1.4.0)
Frontline BPA	Hardware	Bluetooth Protocol Analyzer

When the RF button was pressed the Continua BLE glucose monitor started periodically to send an advertising message. The gateway connected with the Continua BLE glucose monitor by sending a connect request message, and a physical Bluetooth channel was successfully created. The gateway started to interrogate the GATT profile of the Continua BLE glucose monitor with ATT request commands, to which the monitor responded with corresponding ATT response commands. The transaction continued until the gateway had captured the GATT profile of the Continua BLE glucose monitor. A GATT based data channel was concurrently created. The Continua BLE glucose monitor then started to transmit its measured data to the gateway.

Because of the low power consumption and limited message transmission of Bluetooth LE, the system operation of the Continua BLE glucose monitor was divided into three phases: advertisement, GATT profile interview, and measurement data transmission.

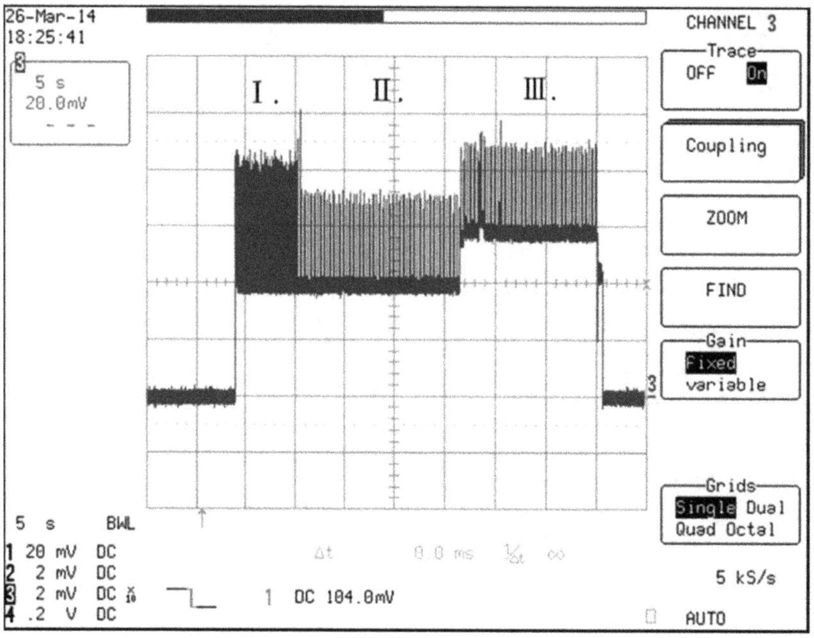

Fig. 4. The three phases of current consumption for the Continua Bluetooth LE glucose monitor: advertising, interrogating GATT profile, and transmitting measurement data.

3.1. Advertisement

The Bluetooth specifications [11, 12] recommend that the GAP peripheral role advertising follows two time durations: fast connections (first 30 seconds) and slow connections (after 30 seconds). For fast connections, the advertising interval is 20 to 30 ms. For slow connections, the advertising interval is 1 to 2.5 s. The interval values in the first connection are designed to attempt fast connection during the first 30 s; however, if a connection is not established within that time, the interval values in the second connection are designed to the reduce power consumption of devices that continue to advertise. It is also recommended that the GAP central role use the recommended scan interval and scan window values. For the first 30 s, the GAP central should use the first scan window (30 ms) and scan interval (30 to 60 ms) pair to attempt fast connection. However, if a connection is not established within that time, the GAP central should switch to one of the other scan window/scan interval options to reduce power consumption. The Continua BLE glucose monitor only advertises itself in the first 30s. After that, the monitor will stop advertising and enter the *stop system* phase. It will enter the advertisement phase and start sending advertisement message only if the user presses the RF button again. The current design offers low power consumption in the advertisement phase.

3.2. GATT Profile Interview

GATT profiles vary according to the type of medical device used and the product features. Some BLE medical devices need to include more services in their GATT profiles, and some need to add or subtract service characteristics. The interview time and the current consumption vary according to the size of the GATT profile. To reduce the interview time of a GATT profile, the GATT client currently supports a cache mechanism to remember the profile of the last BLE medical device that was connected. The GATT profile is interrogated only at the first time of connection. The interview time of the profile at the next connection is ignored, and the Continua BLE glucose monitor enters the measurement data transmission phase directly. The total connection time is shortened and total power consumption is reduced. Our Android-based gateway supported the cache mechanism, allowing the interview time of the GATT profile to be ignored. However, the test tools (Continua CESL manager and Continue test management lit) did need to interview the GATT profile at every connection time because they need to verify the interoperability and functionalities of the BLE medical device being tested.

3.3. Measurement Data Transmission

Depending on the Bluetooth specification [10], the BLE measurement data has a fixed message format and size. The criterion for power consumption in this phase is that the same measured data is transmitted repeatedly. For example, the use case of some BLE medical devices is that measurement data is not be erased and can be transmitted repeatedly. The total transmission time is then higher, and the power consumption is also higher. The Continua BLE glucose monitor has a limited maximum volume of measurement data. The measurement data is labeled as transmitted if it is transmitted successfully. The transmitted measurement data is not transmitted again and can be erased by the user, or automatically by the system. Therefore, the total transmission time is lowest and the power consumption is lowest for the Continua BLE glucose monitor.

Based on our implementation and experience, we found the Bluetooth related specifications [10-12] cover the design requirement of GATT profile almost completely. In addition, based on Continua design guidelines v4.0, only two characteristics (date time and IEEE 11073-20601 regulatory certification data list) need to be added to the GATT profile to meet Continua product requirements. These are that the date time characteristic should be added to the glucose service of the GATT profile, and that the IEEE 11073-20601 regulatory certification data list characteristic should be added to the device information service of the GATT profile. The Continua BLE glucose monitor meets both the requirements of the Bluetooth LE specifications and the Continua low power wireless PAN specifications.

4. Conclusions

This study introduced a Bluetooth LE based Continua glucose monitor. We have demonstrated that the Continua BLE glucose monitor is capable of connecting to and exchanging measurement data with an Android-based mobile phone with Bluetooth smart ready capability. The Continua BLE glucose monitor offers an interoperable platform for personal healthcare ecosystems based on the ISO/IEEE 11073 transport-independent personal-health data and protocol standards with low power wireless PAN-IF features.

In the experiment, we created a Continua BLE system based on TI BLE CC2540/41 chipset. This standardization system can be adapted to many legacy healthcare devices. We demonstrated that the system offers a practical approach to constructing an e-health service environment based on the international x73-PHD standards and Bluetooth LE technology. Future research will focus on supporting other types of healthcare devices and other kinds of device specifications, for example ECG and cardiovascular, and fitness devices.

References

[1] Y.-F., Lee and Y.-S. Huang, "Novel Personal Healthcare System," 4th Intl. Symposium on Medical Info. and Communication Technology, 2010.

[2] C.-Y. Park, J.-H. Lim, and S.-J. Park, "ISO/IEEE 11073 PHD Adapter Board for Standardization of Legacy Healthcare Device," IEEE Intl. Conf. on Consumer Electronics, 2012.

[3] Y.-F. Lee, "Personal Medical Monitoring System Based on x73-PHD Standards," IEEE IT Professional, vol. PP, Issue 99, September 2012.

[4] IEEE, Health Informatics - Personal Health Device Communication. Part 20601: Application Profile - Optimized Exchange Protocol, 2010.

[5] Bluetooth SIG, Health Device Profile, 2008.

[6] J. Kim and O. Song, "ISO/IEEE 11073 interoperability for person health device based on ZigBee healthcare service," IEEE Intl. Conf. on Consumer Electronics (ICCE), pp. 263-264, 2015.

[7] Y.-J. Park, and H.-S. Cho, "Transmission of ECG Data with the Patch-Type ECG Sensor System using Bluetooth Low Energy," Intl. Conf. on ICT Convergence (ICTC), pp. 289-294, 2013.

[8] Z.-M. Lin, C.-H. Chang, N.-K. Chou, and Y.-H. Lin, "Bluetooth Low Energy (BLE) Based Blood Pressure Monitoring System," Intl. Conf. on Intelligent Green Building and Smart Grid (IGBSG), pp. 1-4, 2014.

[9] Continua Health Alliance, Continua Design Guidelines Version 4.0, 2013.

[10] Bluetooth SIG, Bluetooth Specification Version 4.0, 2010.

[11] Bluetooth SIG, Glucose Service, 2012.

[12] Bluetooth SIG, Glucose Profile, 2012.

[13] (2015, July.). Bluetooth low energy software stack and tools. [Online]. Available: http://www.ti.com/product/CC2540/toolssoftware

[14] (2015, July.). Continua Enabling Software Library (CESL). [Online]. Available: https://cw.continuaalliance.org/wg/members/home/cesl-download.

[15] (2015, July.). Continua Test Tool. [Online]. Available: https://cw.continuaalliance.org/wg/members/home/test-tool-download.

23

Hands-on Analysis of 802.11ac Modulation and Coding Scheme

Doru Gabriel Balan, Alin Dan Potorac, Radu Cezar Tărăbuţă

Ştefan cel Mare University of Suceava / Computers, Electronics and Automation Department, Suceava, Romania

Email address:
dorub@usv.ro (D. G. Balan), alinp@eed.usv.ro (A. D. Potorac), radut@stud.usv.ro (R. C. Tărăbuţă)

Abstract: Current communication networks are full of mobile devices that are capable of performing data transfers at high data rates or access high resolution video streams. For such requirements, in the data communications there is a permanent concern of wireless communications development to keep up with the wired communication networks. Because the 802.11n technology is not fast enough for increasingly more users willing games and online broadcasts, the 802.11ac technology was developed. This new IEEE standard, along with the future technology 802.11ad, is aiming to achieve a new level of performance, called VHT (Very High Throughput). The goal is to reach transfer rates (over 1Gbps for now) comparable to those establish in wired networks. This article is proposing a study over 802.11ac technology by exploring the performances of the specific MCS (Modulation and Coding Scheme) with a handy digitally modulated signals investigation method, CCDF (Complementary Cumulative Distribution Function).

Keywords: Wireless, 802.11ac, Modulation, Coding Scheme

---

1. Introduction

IEEE 802.11ac [1] is one of the ongoing WLAN standards aiming to support very high throughput (VHT) with data rate of up to 6 Gbps below the 6GHz band. [2].

The increasing demands on modern Wi-Fi networks are based on several trends that are requiring ever greater levels of performance, scalability, and availability [3]:

- Proliferation of mobile devices (devices as smartphones and tablets rely exclusively on Wi-Fi for network connectivity);
- Multiple devices per user (most users are using concurrently multiple devices with networking capabilities.);
- Always-on connectivity (it is common that today users have permanent connectivity to the Internet);
- Wi-Fi as the primary network access method;
- IoT (Internet of Things), IoE (Internet of Everything) and BYOD (Bring Your Own Device) expansion;
- Cellular network offload (3G, 4G and future 5G network protocols are avoiding network congestion);
- Hotspots are becoming increasingly more important.
- The theoretical maximum data rate and actual user throughput, that will be lower due to the shared medium, and transmission, management and control overhead, depend on a number of factors, such as:
- Physical-layer (PHY) connection rates;
- Number of spatial streams;
- Allocated channel width;
- Modulation and coding scheme (MCS) used;
- Guard interval (time between transmitted characters).

2. Wireless Technology Elements

2.1. 802.11ac Characteristics

From this perspective, several characteristics of 802.11ac technology can be easily identified [4, 5], such as:

- Introduces the new VHT operating mode, that is a mixt mode, supporting 802.11a and 802.11n clients in the 5Ghz band;
- Supports from 1 to 8 (up to 4 per client) spatial streams between infrastructure access points and wireless clients, using several principles, like:
    - Space multiplexing (signal is splinted in multiple signal streams, each transmitted in a separate spatial stream),
    - Space Time Block Coding - STBC (uses more antennas to redundantly transmit a single traffic

stream over multiple RF paths),

- Multi-User Multiple Input Output (MU-MIMO) use a multiple antenna array to direct beams and transmit simultaneously to multiple clients.
- Wider RF transmission channels, it adds 80MHz and 160MHz channels to the 5GHz band;
- Higher Modulation and Coding Schemes (MCS), introducing the 256 QAM signal modulation;
- Specifies 10 MCS indices (0-9, as in Figure 1);

When all specified enhancements are enabled, 802.11ac delivers a maximum PHY data rate of 6933 Mbps, as can be seen in Figure 1, in the last filled cell.

2.2. Specific MCS - Modulation and Coding Scheme

As seen before, network communications via 802.11ac can be realized using one of the 10 MCS provided with maximum 8 spatial streams. Figure 1 is presenting the 802.11ac characteristics, regarding PHY data rates and MCS, for only two situations of user number of spatial streams, namely for 1 spatial stream and for 8 spatial streams. Obviously there are also data for the other cases, (the cases with 2-7 spatial streams), which were excluded from organizing considerations.

MCS	Modulation	Bits per Symbol	Coding Ratio	20-MHz		40-MHz		80-MHz		160-MHz	
				800ns	400ns	800ns	400ns	800ns	400ns	800ns	400ns
1 Spatial Stream				Data Rate (Mbps)							
MCS 0	BPSK	1	1/2	6.5	7.2	13.5	15.0	29.3	32.5	58.5	65.0
MCS 1	QPSK	2	1/2	13.0	14.4	27.0	30.0	58.5	65.0	117.0	130.0
MCS 2	QPSK	2	3/4	19.5	21.7	40.5	45.0	87.8	97.5	175.5	195.0
MCS 3	16-QAM	4	1/2	26.0	28.9	54.0	60.0	117.0	130.0	234.0	260.0
MCS 4	16-QAM	4	3/4	39.0	43.3	81.0	90.0	175.5	195.0	351.0	390.0
MCS 5	64-QAM	6	2/3	52.0	57.8	108.0	120.0	234.0	260.0	468.0	520.0
MCS 6	64-QAM	6	3/4	58.5	65.0	121.5	135.0	263.3	292.5	526.5	585.0
MCS 7	64-QAM	6	5/6	65.0	72.2	135.0	150.0	292.5	325.0	585.0	650.0
MCS 8	256-QAM	8	3/4	78.0	86.7	162.0	180.0	351.0	390.0	702.0	780.0
MCS 9	256-QAM	8	5/6	N/A	N/A	180.0	200.0	390.0	433.3	780.0	866.7
8 Spatial Streams				Data Rate (Mbps)							
MCS 0	BPSK	1	1/2	52.0	57.8	108.0	120.0	234.0	260.0	468.0	520.0
MCS 1	QPSK	2	1/2	104.0	115.6	216.0	240.0	468.0	520.0	936.0	1040.0
MCS 2	QPSK	2	3/4	156.0	173.3	324.0	360.0	702.0	780.0	1404.0	1560.0
MCS 3	16-QAM	4	1/2	208.0	231.1	432.0	480.0	936.0	1040.0	1872.0	2080.0
MCS 4	16-QAM	4	3/4	312.0	346.7	648.0	720.0	1404.0	1560.0	2808.0	3120.0
MCS 5	64-QAM	6	2/3	416.0	462.2	864.0	960.0	1872.0	2080.0	3744.0	4160.0
MCS 6	64-QAM	6	3/4	468.0	520.0	972.0	1080.0	2106.0	2340.0	4212.0	4680.0
MCS 7	64-QAM	6	5/6	520.0	577.8	1080.0	1200.0	2340.0	2600.0	4680.0	5200.0
MCS 8	256-QAM	8	3/4	624.0	693.3	1296.0	1440.0	2808.0	3120.0	5616.0	6240.0
MCS 9	256-QAM	8	5/6	N/A	N/A	1440.0	1600.0	3120.0	3466.7	6240.0	6933.3

Figure 1. *802.11ac PHY data rates for 1 and 8 spatial streams [3].*

Figure 2 shows the current state of technology development 802.11ac identified by the initially developed set of 802.11ac equipment (802.11ac 1st wave). As can be seen, the actual data rate is 1300 Mbps.

In the results included in this article we have concentrated over 802.11ac specific Modulation and Coding Schemes (MCS). Our determinations are based on the CCDF (Complementary Cumulative Distribution Function) investigation method.

Channel Width (MHz)	Spatial Streams	802.11ac MCS Index	Guard Interval	PHY Data Rate (Mbps)	
20	1	5	LGI	52	
40	4	7	SGI	600	802.11n Max
40	4	9	SGI	800	
80	1	9	SGI	433	
80	2	9	SGI	867	
80	3	9	SGI	1300	802.11ac 1st wave
80	4	9	SGI	1733	
80	8	9	SGI	3467	
160	1	9	SGI	867	
160	2	9	SGI	1733	
160	3	8	SGI	2340	
160	4	9	SGI	3467	
160	8	9	SGI	6933	

Figure 2. *802.11 PHY data rates in real use [6].*

3. Investigation Method

Measurement of the CCDF is often used for evaluating nonlinearities of amplifiers or transmitter output stages, for instance. This measurement indicates how often the observed signal reaches or exceeds a specific level. From a physical point of view, the CCDF measurement is the integral of the distribution function versus the level (integration of the observed level to infinity).

Comparison of evaluated or measured values and theoretical reference values (as CCDF Gaussian), as it can be seen in Figures 5-14, quickly yields information on the nonlinear response of all types of active elements. However, the great advantage of the CCDF measurement is that the useful signal itself is analyzed; as a result, it is not necessary to transmit complex test sequences. CCDF determination is a very important measurement in RF transmission systems. Measuring the CCDF is a simple and effective method of determining the nonlinear characteristics of active elements. [6].

4. Testbed Elements

To achieve the aimed performance analysis on 802.11ac modulation and coding scheme, it was used a testing space consisting of a Keysight (formerly Agilent) set of tools.

There are two key elements of the test bench that provided the results presented in this paper:

The first element (software) is the N7617B Signal Studio for WLAN 802.11a/b/g/n/ac [7], used to design the 802.11ac signals;

The second element (hardware) is a vector signal generator, N5182A MXG [8], with N7617B option, for testing loaded signals from software.

The interconnection of these components, as in Figure 3, allows controlling the theoretical elements for 802.11ac technology-specific signals description and the signal characteristics validation, achieved through vector signal generator.

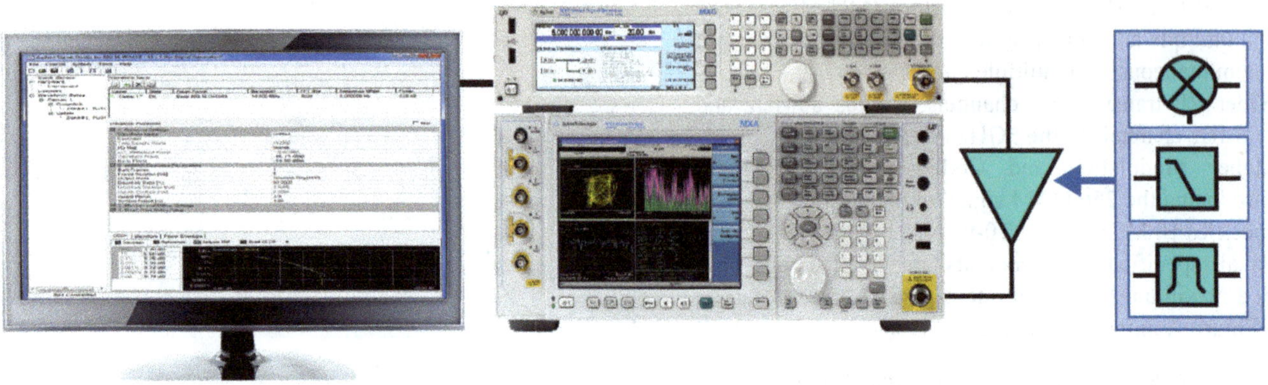

Figure 3. *Component test configuration [9].*

5. Mathematics Behind

As mentioned in mathematical theories, expressed by [10] and [11], in probability theory and statistics, the cumulative distribution function (CDF), or just distribution function, describes the probability that a real-valued random variable X with a given probability distribution will be found to have a value less than or equal to x. In the case of a continuous distribution, it gives the area under the probability density function from minus infinity to x. Cumulative distribution functions are also used to specify the distribution of multivariate random variables.

The cumulative distribution function of a real-valued random variable X is the function given by (1):

$$F_X(x) = P(X \le x) \quad (1)$$

where the right-hand side represents the probability that the random variable X takes on a value less than or equal to x. The probability that X lies in the semi-closed interval $(a, b]$, where $a < b$, is expressed by (2):

$$P(a < X \le b) = F_X(b) - F_X(a) \quad (2)$$

The CDF of a continuous random variable X can be expressed as the integral of its probability density function f_X as in (3):

$$F_X(x) = \int_{-\infty}^{x} f_X(t)dt \quad (3)$$

In the case of a random variable X which has distribution having a discrete component at a value b, stated by (4),

$$P(X = b) = F_X(b) - \lim_{x \to b^-} F_X(x) \quad (4)$$

If F_X is continuous at b, this equals zero and there is no discrete component at b.

Sometimes, it is useful to study the opposite question and ask how often the random variable is above a particular level. This is called the *Complementary Cumulative Distribution Function* (CCDF) or simply the *tail distribution* or *exceedance*, and is defined as in (5):

$$\overline{F}(x) = P(X > b) = 1 - F(x) \quad (5)$$

This has applications in statistical hypothesis testing, for example, because the one-sided p-value is the probability of observing a test statistic at least as extreme as the one observed. Thus, provided that the test statistic, T has a continuous distribution, the one-sided p-value is simply given by the CCDF, for an observed value t of the test statistic will be expressed by (6):

$$p = P(T \ge t) = P(T > t) = 1 - F_T(t) \quad (6)$$

In survival analysis, $\overline{F}(x)$ is called the survival function and denoted $S(x)$, while the term reliability function is common in engineering.

6. Practical Determinations

The objective of practical determinations involves passing the generated signal through all 802.11ac available MCS. The analysis result is the digitally modulated signals characteristics identification, based on CCDF curves observation for each generated signal. Figure 4 shows the spectrum analysis diagram for a sample 802.11ac signal.

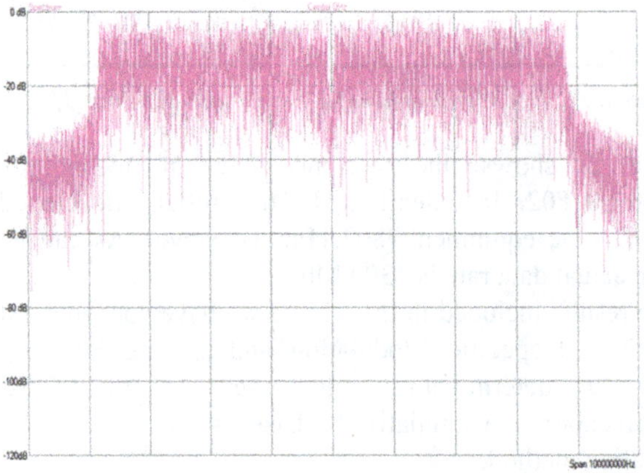

Figure 4. *Signal spectrum MCS 0-9.*

7. Results

Diagrams from Figures 5-14 are providing an intuitive analysis of 802.11ac signals, for each specific MCS, emerged from the use of CCDF analysis functions over entire signal or only the burst part of the signal. The analysis is made by comparing the signal power with the Gaussian CCDF reference line (red line on Figs. 5-14).

To select the portion of the waveform to calculate the CCDF data the analysis charts are containing the both components available on signal design operations (Waveform CCDF and Burst CCDF, green and blue lines). Waveform CCDF will include all components of the configured frame including gaps and non-transmitted portions (both RF burst on and off portions). Burst CCDF will include the configured bursts only (this does not includes gaps or times when the RF burst is off).

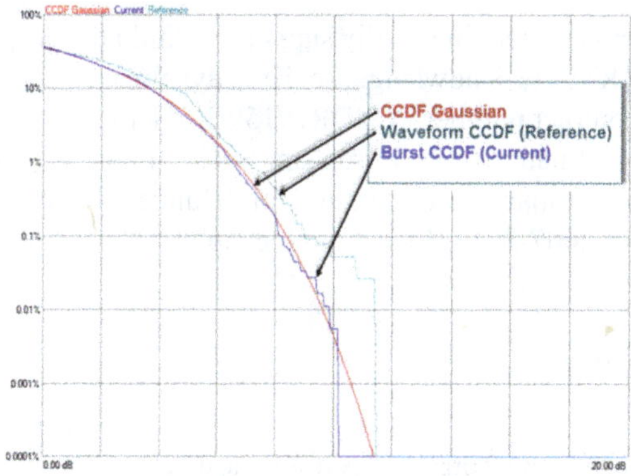

Figure 5. *CCDF analysis for MCS 0.*

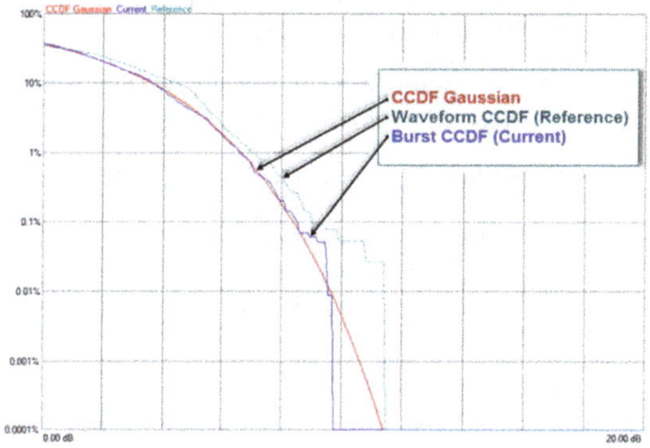

Figure 6. *CCDF analysis for MCS 1.*

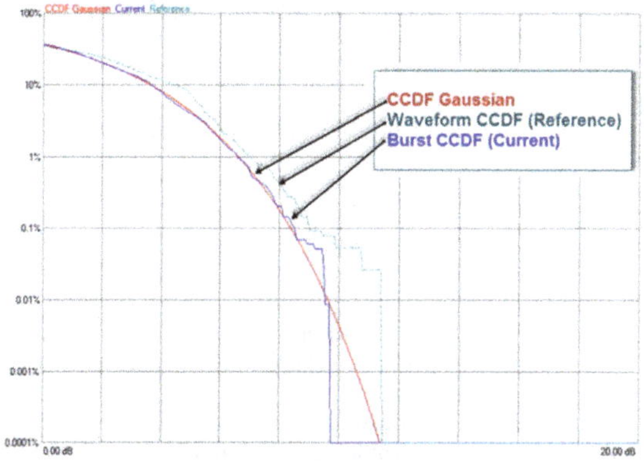

Figure 7. *CCDF analysis for MCS 2.*

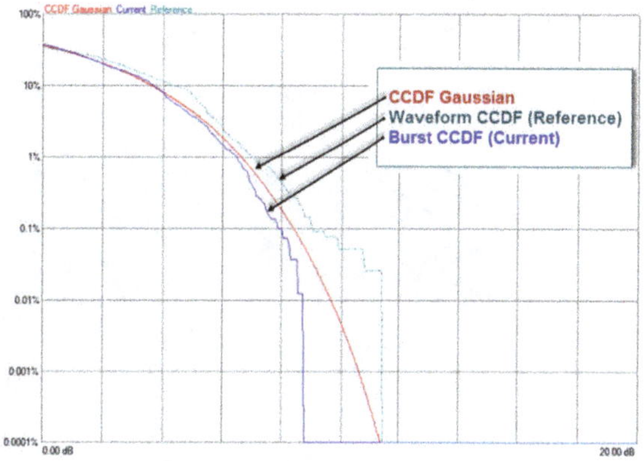

Figure 8. *CCDF analysis for MCS 3.*

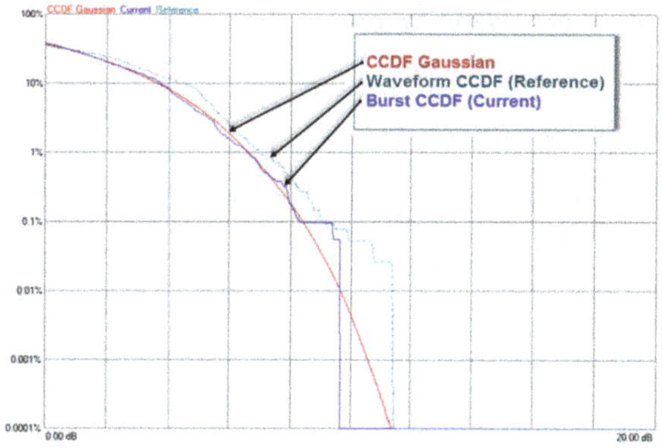

Figure 9. *CCDF analysis for MCS 4.*

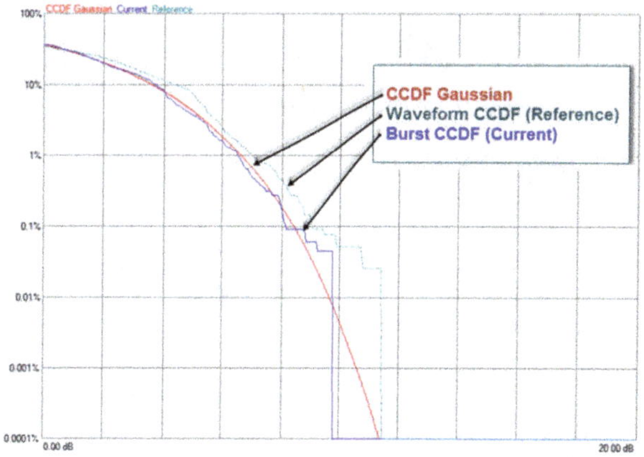

Figure 10. *CCDF analysis for MCS 5.*

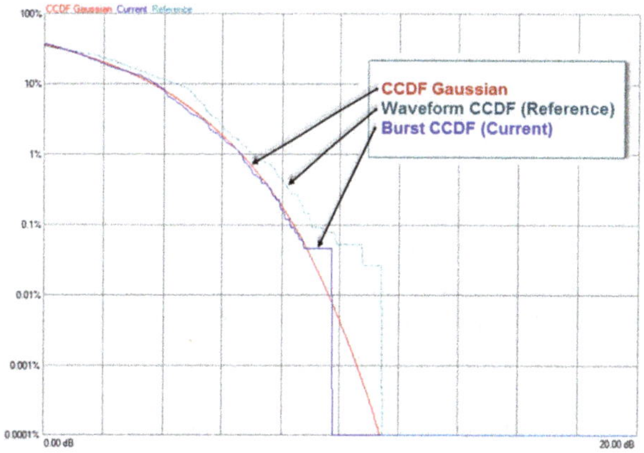

Figure 11. *CCDF analysis for MCS 6.*

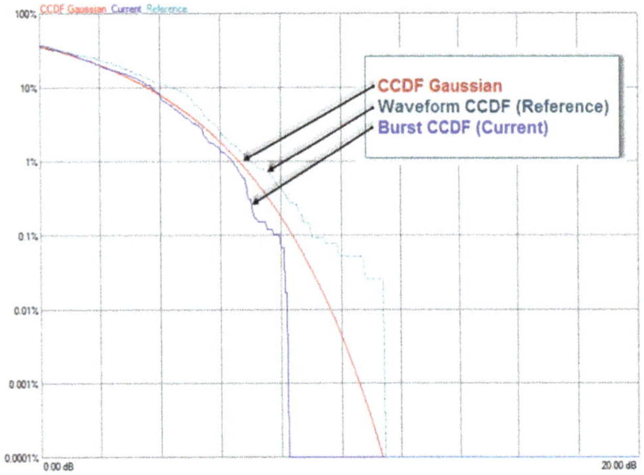

Figure 12. *CCDF analysis for MCS 7.*

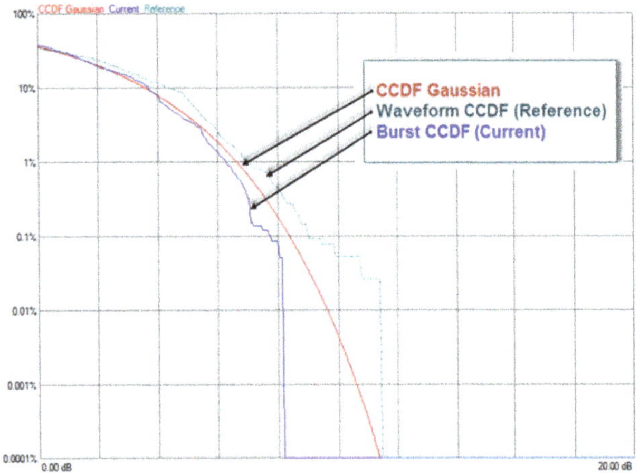

Figure 13. *CCDF analysis for MCS 8.*

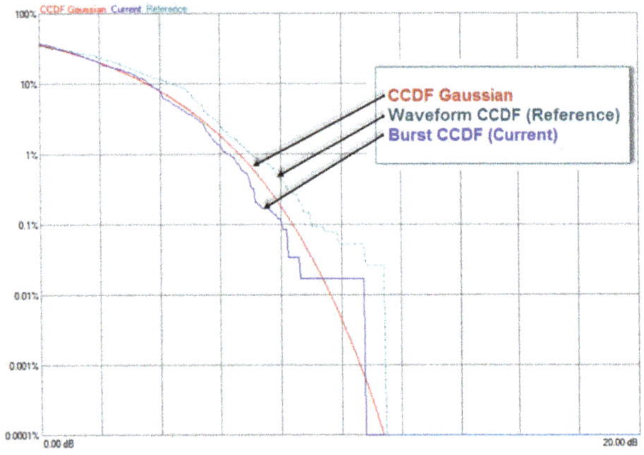

Figure 14. *CCDF analysis for MCS 9.*

8. Conclusions

It is well known that the modulation format of a signal affects its power characteristics. Using CCDF curves, we can fully characterize the power statistics of different modulation formats, and compare the results of choosing one modulation format over another.

The determinations contained in this article are guided on the behavior of radio signals in the technology 802.11ac. The outcomes obtained and presented in this paper can be used as a good base for identifying 802.11ac characteristics. The analysis of MCS through CCDF gives valuable information about signal levels on each available MCS on 802.11ac, taking as comparative reference the band-limited Gaussian noise reference curve.

The results presented in this study are presenting the exploration step that is a part of a started and more ample research initiative regarding wireless interference detection and prevention. Future operations are aiming the analysis of the crowded signal spectrum, faced by the wireless communications and the opportunities offered by 802.11ac technology to solve such problems, regarding the benefits of using multiple antenna techniques.

Acknowledgements

This paper has been financially supported within the project entitled „SOCERT. Knowledge society, dynamism through research", contract number POSDRU/159/1.5/S/132406. This project is co-financed by European Social Fund through Sectoral Operational Programme for Human Resources Development 2007-2013. Investing in people!

References

[1] IEEE 802.11ac, http://www.ieee802.org/11/Reports/tgac/update.htm.

[2] Verma, L.; Fakharzadeh, M.; Sunghyun Choi, "Wifi on steroids: 802.11AC and 802.11AD," Wireless Communications, IEEE, vol.20, no.6, pp.30, 35, December 2013.

[3] Andrew von Nagy, "Aerohive High-Density Wi-Fi Design & Configuration Guide v2 ", Aerohive Networks, 2013.

[4] Eng Hwee Ong; Kneckt, J.; Alanen, O.; Zheng Chang; Huovinen, T.; Nihtila, T., "IEEE 802.11ac: Enhancements for very high throughput WLANs," Personal Indoor and Mobile Radio Communications (PIMRC), 2011 IEEE 22nd International Symposium on, vol., no., pp.849,853, 11-14 Sept. 2011.

[5] Fluke Networks White Paper, Implementing 802.11ac. 2014.

[6] Agilent Application Note, Characterizing Digitally Modulated Signals with CCDF Curves, 2000.

[7] N7617B Signal Studio for WLAN 802.11a/b/g/n/ac, http://www.keysight.com/en/pd-803256-pn-N7617B/signal-studio-for-wlan-80211a-b-g-n-ac?cc=US&lc=eng.

[8] N5182A MXG, http://www.keysight.com/en/pd-797248-pn-N5182A/mxg-rf-vector-signal-generator.

[9] Agilent Technologies,, Signal Studio for WLAN 802.11a/b/g/n/ac - N7617B, Technical Overview 5990-9008EN, January 16, 2014.

[10] Cumulative distribution function, http://en.wikipedia.org/wiki/Cumulative_distribution_function

[11] Zwillinger, Daniel; Kokoska, Stephen, CRC Standard Probability and Statistics Tables and Formulae, CRC Press. p. 49. ISBN 978-1-58488-059-2, 2010.

PAPR Reduction in LFDMA Using Hadamard Transform Technique

Mohamed M. El-Nabawy[1], Mohamed A. Aboul-Dahab[2], Khairy El-Barbary[3]

[1]Modern Academy for Eng. & Tech in Maadi (M.A.M)/ Electronic and Communication Dept., Cairo, Egypt
[2]Arab Academy for Science and Technology and Maritime Transport (AAST)/ Electronic and Communication Dept., Cairo, Egypt
[3]Canal University, Electronic and Communication Dept., Cairo, Egypt

Email address:
eng.mohamed.elnabawy@gmail.com (M. M. El-Nabawy), mdahab@aast.edu (M. A. Aboul-Dahab),
khbar2000@yahoo.com (K. El-Barbary)

Abstract: The power consumption is an essential issue for designers of mobile devices. Orthogonal Frequency Division Multiplexing (OFDM) suffers from its Peak – to – Average Power Ratio (PAPR) problem. The Discrete Fourier Transmission Spread OFDM (DFTS-OFDM) based on Single Carrier Frequency Division Multiple Access (SC-FDMA) has been widely adopted due to its lower peak-to-average power ratio (PAPR) of transmits signals compared with OFDM. There are three types of mapping in SCFDMA. These namely Localized FDMA (LFDMA), Distributed FDMA (DFDMA), and Interleaved FDMA (IFDMA). In this paper, we propose a modified SCFDMA. This is based on using Hadamard transform instead of Fast Fourier Transform (FFT). Utilizing LFDMA Mapping techniques this is proposed for uplink scenario of LTE. Simulation results show that the proposed scheme reduces PAPR compared to the conventional SCFDMA while the BER does not degraded.

Keywords: Orthogonal Frequency Division Multiplexing (OFDM), Single Carrier Frequency Division Multiple Access (SC-FDMA), Localized FDMA (LFDMA), Distributed FDMA (DFMA), Interleaved FDMA (IFDMA), Cumulative Complementary Distribution Function (CCDF), Peak to Average Power Ratio (PAPR), Bit Error Rate (BER)

1. Introduction

Broadband wireless communication systems are designed to provide high-data-rate services to satisfy the increasing demands of the future wireless networks. As the bit rate increases, the problem of inter-symbol interference becomes more serious. Orthogonal frequency division multiple access (OFDMA) is an attractive technology to deal with the detrimental effects of multipath fading, but it has several inherent drawbacks such as the large peak-to-average-power ratio (PAPR) and the sensitivity to carrier frequency offsets [1, 2]. Recently, much attention has been focused on another broadband wireless communication system, which is the single carrier frequency division multiple access (SC-FDMA) system [3, 4].

The SC-FDMA has two main advantages over orthogonal frequency division multiplexing (OFDM), namely, a lower PAPR and a lower sensitivity to carrier frequency errors [3]. There are two methods to choose the subcarriers in SC-FDMA systems: interleaved subcarrier mapping and localized subcarrier mapping. We will refer to the localized subcarrier mapping mode of SC-FDMA as localized frequency division multiple access (LFDMA) and the interleaved subcarrier mapping mode as interleaved frequency division multiple access (IFDMA). The LFDMA system incurs a higher PAPR compared to the IFDMA system but, compared to OFDM; it is lower, though not significantly [3]. Thus, we will be concerned with the LFDMA system only.

For OFDM systems, PAPR reduction techniques have been introduced to reduce problems at the high-power amplifier and/or alleviate its back-off specifications by companding or clipping the amplitude of the OFDM signals [5–9].

In this paper, we will discuss and evaluate the PAPR of a modified transceiver scheme of SCFDMA. The proposed system is similar to SCFDMA except that it uses Hadamard Transform and Inverse Hadamard Transform instead of DFT,

FFT, IDFT, and IFFT. This paper is organized as follows: In section II presents the SCFDMA system model and the concept PAPR problem. In section III the proposed system is discussed. Simulation results are reported in section IV and conclusions are presented in V.

2. SC-FDMA System Model

Fig. 1 shows the block diagram of a basic LTE SC-FDMA system. SC-FDMA is a modified structure of OFDMA system [8]. SC-FDMA can be viewed as Discrete Fourier Transform (DFT)-spread OFDMA or frequency-spread OFDMA, where time domain data symbols are transformed to frequency domain by DFT before going through OFDMA system [9].

The input to the SC-FDMA transmitter is a stream of x bits that is converted to multilevel sequences of complex number x(s), where x(s) is the data symbol, and s is the sample index. Then, the transmitter concatenates the modulation symbols into blocks through serial to parallel converter (S/P) block, each containing M symbols. These modulated symbols perform M-point DFT to produce a frequency domain representation as follows:

$$X(k) = \sum_{m=0}^{M-1} x(n) e^{\frac{-j2\Pi mk}{M}} \quad (1)$$

The output of the DFT is then applied to a subcarrier mapping block which maps each of the M-DFT outputs to one of the N orthogonal subcarriers (N >M) that can be transmitted. The ratio between N and M is called the bandwidth expansion factor of the symbol sequence or the spreading factor Q (Q=N/M) [4]. If all terminals transmit M symbols per block, the system can handle Q simultaneous transmissions without co-channel interference. The outcome of the subcarrier mapping is N of complex subcarrier amplitudes $\tilde{X}(l)$ (l = 0, 1, 2…, N-1),The outputs of subcarrier mapping block are used for N-point Inverse Fast Fourier Transform (N-IFFT) which transforms the N subcarriers to a signal $\tilde{x}(m)$ in time domain, then each $\tilde{x}(m)$ is transmitted sequentially.

In DFT spreading technique, the allocation of subcarriers to users is done using one of the most schemes, Localized FDMA (LFDMA), Distributed FDMA (DFDMA) and Interleaved FDMA (IFDMA). In LFDMA scheme, M consecutive subcarriers are allocated in a total of N subcarriers and the consecutive unused subcarriers are filled with zeroes. DFDMA scheme involves allocation of M outputs over the entire band of N subcarriers and the remaining unused subcarriers are filled with zeroes. IFDMA scheme involves allocation of M DFT outputs over total of N subcarriers with equidistance.

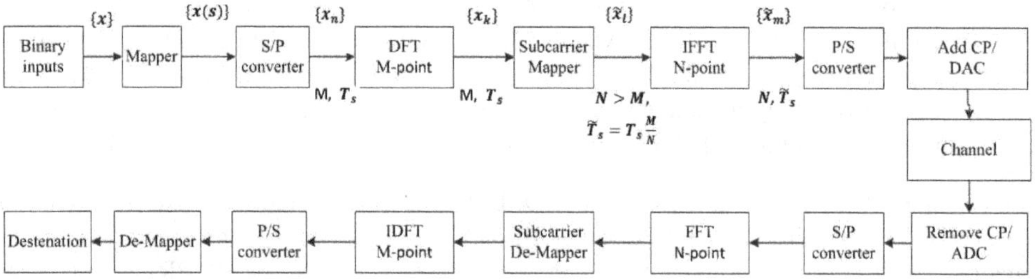

M number of assigned subcarriers, N number of available subcarriers, and T_s , $\tilde{T}_s$ symbol duration

Figure 1. *Basic block diagram of SC-FDMA*

The allocation of subcarriers using the three schemes is shown in Fig (2). The total number of subcarriers is N=12. Here we assume that 4 DFT outputs are distributed (M=4). The spreading factor is Q=3. X(0),X(1),X(2),X(3) represent the DFT outputs that are distributed using LFDMA, DFDMA and IFDMA schemes. In the following sections, we derive the IFFT outputs for LFDMA scheme. The IFFT output in each scheme is obtained after the DFT and subcarrier mapping operations. The significance of IFFT outputs is that it gives the information about the time domain signal depending on certain factors like scaling, phase rotation, complex weighting factor, etc.

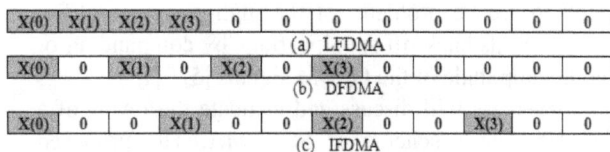

X(0)	X(1)	X(2)	X(3)	0	0	0	0	0	0	0	0

(a) LFDMA

X(0)	0	X(1)	0	X(2)	0	X(3)	0	0	0	0	0

(b) DFDMA

X(0)	0	0	X(1)	0	0	X(2)	0	0	X(3)	0	0

(c) IFDMA

Figure 2. *DFT Spreading Technique showing distribution of DFT outputs (N=12,M=4,Q=3).*

2.1. IFFT Output for LFDMA

Let us assume a total number of N subcarriers and M be the consecutive DFT outputs that are allocated with the unused subcarriers filled with zeroes.

Let us define $n = Qm + l$ where Q is the Spreading factor where $0 \leq l \leq L - 1$ and $0 \leq m \leq M - 1$.

The output if SC-LFDMA after the subcarrier mapping is

$$X_{LFD}(k) = \begin{cases} X(k), & k = 0,1,.....,(M-1) \\ 0, & \text{otherwise} \end{cases} \quad (2)$$

so that the IFFT output sequence is given by

$$x_{LFD}(n) = x_{LFD}(Qm + l), \quad 0 \leq n \leq N\text{-}1 \quad (3)$$

Now

$$x_{LFD}(n) = \frac{1}{N}\sum_{k=0}^{N-1} X_{LFD}(k)e^{\frac{j2\pi kn}{N}} \quad (4)$$

$$\therefore x_{LFD}(n) = \frac{1}{QM}\sum_{k=0}^{M-1} X(k)e^{\frac{j2\pi k(Qm+l)}{QM}} \quad (5)$$

If l = 0, then from (5)

$$\therefore x_{LFD}(n) = \frac{1}{QM}\sum_{k=0}^{M-1} X(k)e^{\frac{j2\pi km}{M}}$$

$$\therefore x_{LFD}(n) = \frac{1}{Q}x(m) \quad (6)$$

From (6) It can be seen that the IFFT output for LFDMA depends on the scaled version of the input signal by 1/Q.

2.2. Concept of PAPR

Peak to Average Power Ratio (PAPR) occurs due to the summing of carriers together. The maximum peak power increases proportionally to the number of carriers in the system. After linear region, the scalar relationship is lost and the amplifier moves into saturation region. The use of amplifiers in the saturation region leads to distortion which is a major drawback.

PAPR is expressed as

$$PAPR = \frac{\sum_{n=0}^{N-1} \max\{x^2(n)\}}{P_{av}\{x(n)\}} \quad (7)$$

Where $x(n)$ is the original signal, $\max\{x^2(n)\}$ Indicates the peak signal power, N is the total number of data symbols and

$$P_{av}\{x(n)\} = \frac{1}{N}\sum_{n=0}^{N-1} E\{x^2(n)\} \quad (8)$$

indicates the average signal power $E\{x^2(n)\}$ in (8) is the mean square value of x(n) and 'E' stands for expectation.

3. The Proposed Scheme

A baseband model of proposed system is shown in Figure 3. The proposed system is similar to SCFDMA except that it uses Hadamard Transform and Inverse Hadamard Transform instead of DFT, FFT, IDFT, and IFFT.

3.1. Hadamard Transform

The proposed scheme is based mainly on the use of hadamard transform replaced of FFT module. Park et.al; have proposed in [10] a scheme for PAPR reduction in OFDM transmission using hadamard transform. The proposed hadamard transform scheme may reduce the occurrence of the high peaks comparing the original OFDM system. The idea to use the hadamard transform is to reduce the autocorrelation of the input sequence to reduce the peak to average power problem and it requires no side information to be transmitted to the receiver. In the section, we briefly review hadamard transform. We assume H is the hadamard transform matrix of N orders, and hadamard matrix is standard orthogonal matrix. Every element of hadamard matrix only is 1 or -1. The hadamard matrix of 2 orders is stated by

$$H_2 = \frac{1}{\sqrt{2}}\begin{pmatrix} 1 & 1 \\ 1 & -1 \end{pmatrix} \quad (9)$$

hadamard matrix of 2N order may be constructed by

$$H_{2N} = \frac{1}{\sqrt{2N}}\begin{pmatrix} H_N & H_N \\ H_N & -H_N \end{pmatrix} \quad (10)$$

Where $-H_N$ is the complementary of H_N. Hadamard matrix satisfy the relation

$$H_{2N}H_{2N}^T = H_{2N}^T H_{2N} = I_{2N} \quad (11)$$

Where H_{2N}^T is the transport matrix, I_{2N} is the unity matrix of 2N order. Note that Hadamard transform is an orthogonal linear transform and can be implemented by a butterfly structure as in FFT. This means that applying Hadamard transform does not require the extensive increase of system complexity.

After the sequence $x = [x_1 x_2 \ldots\ldots x_N]^T$ is transformed by hadamard matrix of N order, the new sequence is

$$Y = HX \quad (12)$$

3.2. Description of Proposed Scheme

M is the number of subcarriers assigned to each user. N is the total number of available subcarriers. N = Q.M, where Q is the spreading factor and N > M. Hadamard transform performs an orthogonal, symmetric, involution, linear operation on 2m real numbers (or complex numbers) which can be generated by, H0 = 1 and then define for m> 0. The Hadamard transform is 2^mX 2^mmatrix.

$$H_m = \frac{1}{\sqrt{2}}\begin{pmatrix} H_{m-1} & H_{m-1} \\ H_{m-1} & -H_{m-1} \end{pmatrix} \quad (13)$$

Compared to the Fast Fourier Transform (FFT), the Hadamard transform requires less storage space and is faster to calculate because it uses only real additions and subtractions, while the FFT requires complex values.

We choose Localized Frequency Division Multiple Access (LFDMA) subcarrier mapping mode. An example of Hadamard SCFDMA transmit symbols for M = 4, Q = 3, N =

12 is illustrated in Figure 4. x_n are the transmit symbols in time domain passed through Hadamard transform to get x_k which are time domain samples. Therefore, subcarrier mapping is done in time domain. After the subcarrier mapping is done, the data is passed through inverse Hadamard transform.

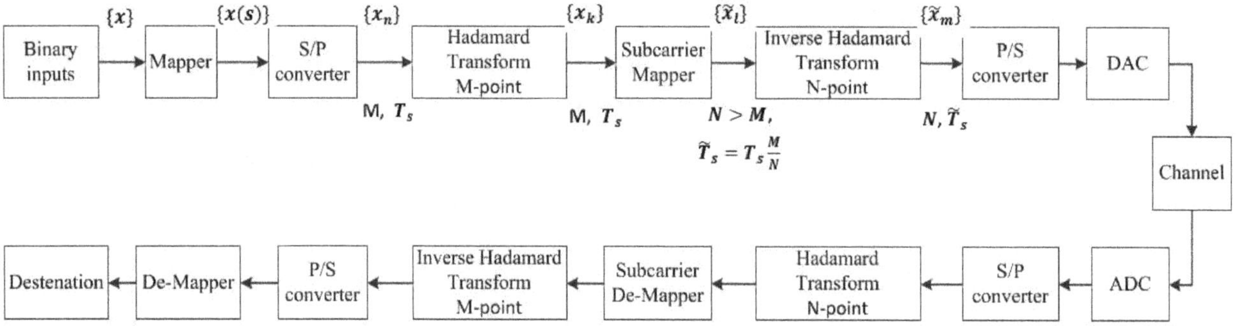

M number of assigned subcarriers , N number of available subcarriers , and T_s , $\tilde{T}_s$ symbol duration

Figure 3. *Block diagram of proposed SC-FDMA system*

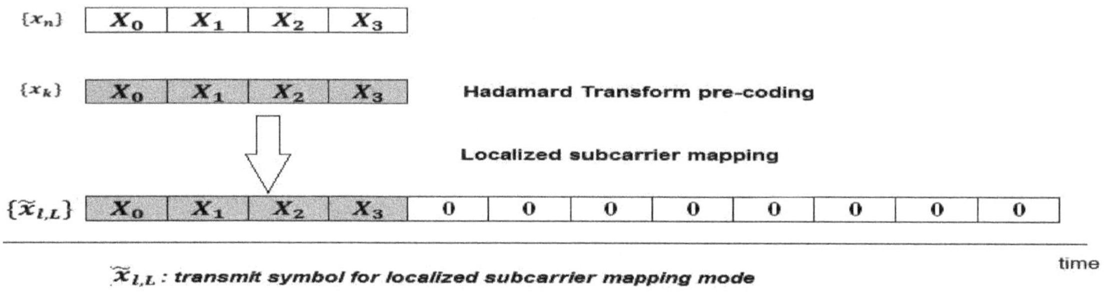

Figure 4. *An example of Hadamard SCFDMA transmit symbol for N=4, Q=3 and M=12.*

We analyze the PAPR of Hadamard SCFDMA signal with companding technique for LFDMA subcarrier mapping mode. In the subsequent derivations, we will assume N = Q.M and follow the notations in Figure 3. Let the data block of M complex valued symbols to be modulated is $\{x_n : \text{n=0,...., M-1}\}$.

Then, by applying M-point Hadamard transform to, x_n, $\{x_k : \text{k=0,...., M-1}\}$ the time domain symbols can be described as

$$x_k = H_N x_n^T ; \quad \text{k=0, ..., N-1} \tag{14}$$

H_N is the Hadamard matrix and can be obtained from (13). $\{\tilde{x}_l : l = 0, \dots\dots, N-1\}$ are time domain samples after subcarrier mapping. For IFDMA (subcarrier mapping technique) the time domain samples $\tilde{x}_l$ after mapping can be represented as

$$\tilde{x}_l = \begin{cases} x_{l/Q} & l = Q.k \quad k = 0, \dots\dots, M-1 \\ 0 \quad \text{otherwise} & l = 0, \dots\dots, N-1 \end{cases} \tag{15}$$

$\{\tilde{x}_m\}$ are time domain symbols after inverse Hadamard transform of $\{\tilde{x}_l\}$ given by

$$\tilde{x}_m = \text{inv}(H_N).\tilde{x}_l \quad ; m, l = 0, \dots\dots, N-1 \tag{16}$$

where H_N is the Hadamard matrix and can be obtained from (13).

The PAPR is the measure of the peak-to-average power ratio. Battery operated terminal must have low PAPR when transmitting a signal so as the power amplifier operates in its linear region.

The PAPR of the transmitted signal can be expressed as:

$$CCDF(N, PAPR_0) = \Pr\{PAPR > PAPR_0\} = 1 - (1 - PAPR_0)^N \tag{17}$$

Where N is the number of subcarriers and $PAPR_0$ is a certain value of PAPR.

4. Simulation Results

To show the overall effect of the proposed scheme on reducing PAPR, we consider 500000 Number of iteration blocks to calculate the CCDF (Complementary Cumulative Distribution Function) of the PAPR. CCDF of PAPR for SCFDMA, Hadamard SCFDMA and OFDMA are evaluated and compared. The subcarrier mapping is LFDMA. In the simulations, the total number of subcarriers N is set to 256; input data block size M to 64, the spreading factor Q to 4, and compered CCDF of PAPR at Q-PSK, 16-QAM, and 64-QAM mapper.

In figure 5, The SCFDMA system incurs a lower PAPR compared to OFDM, and the performance of the proposed scheme provides better performance than classical SCFDMA.

We can see in figure 5(a) the PAPR of proposed scheme is almost 0 dB, which mean the probability to occur the PAPR problem when used Q-PSK QAM is equal zero, from the same figure we can ensure that the PAPR of SC-FDMA is less than OFDM system.

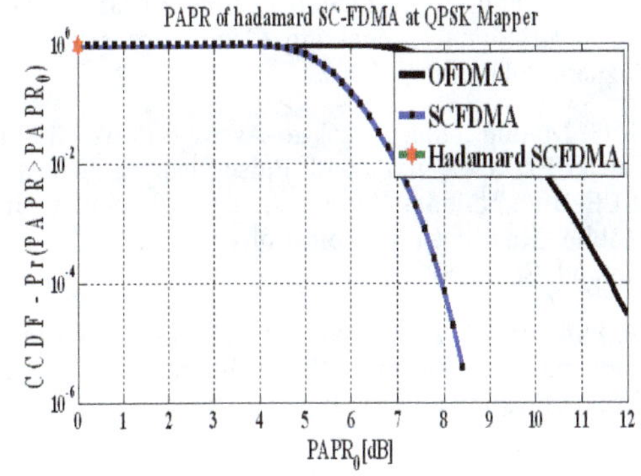

(a) With Q-PSK

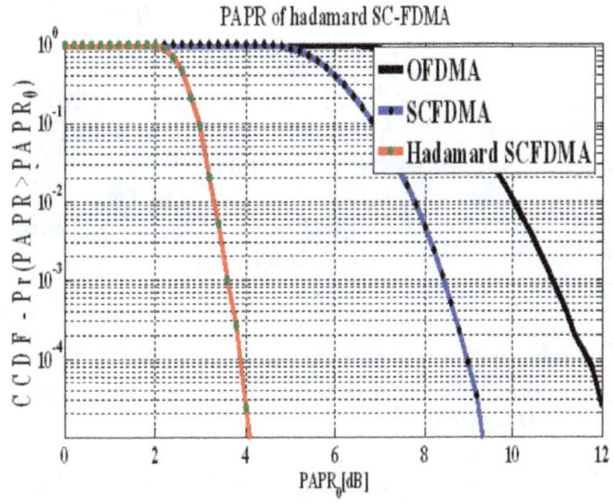

(b) With 16-QAM

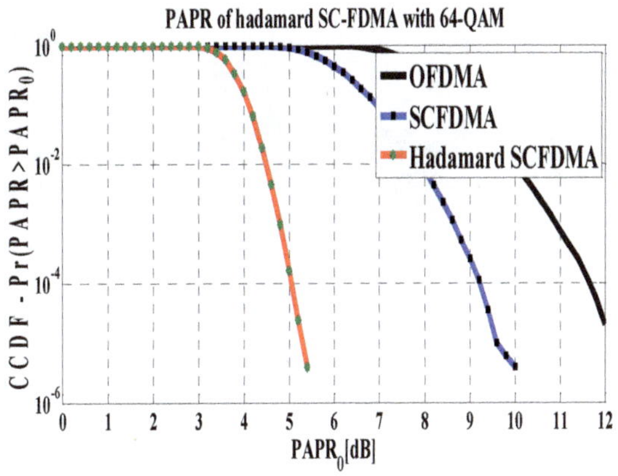

(c) With 64-QAM

Figure 5. *Comparison of PAPR for OFDMA, SCFDMA and proposed scheme of hadamard SCFDMA*

We can see in figure 5(b) at CCDF= 10^{-4}, the PAPR of the proposed scheme is almost 5.1 dB smaller than that of classical SCFDMA scheme. And in figure 5(c) at CCDF= 10^{-4}, the PAPR of the proposed scheme is almost 4.1 dB smaller than that of classical SCFDMA scheme. From the result show in figures and discuss of these results we can see the PAPR is varieties with the type of modulation order techniques, at the modulation order increase the reduction value of PAPR decrease.

In the other hand the BER rate is the most important parameter to evaluate this proposed scheme compared to the conventional SCFDMA. From Figure 6 we can see the BER is not degraded.

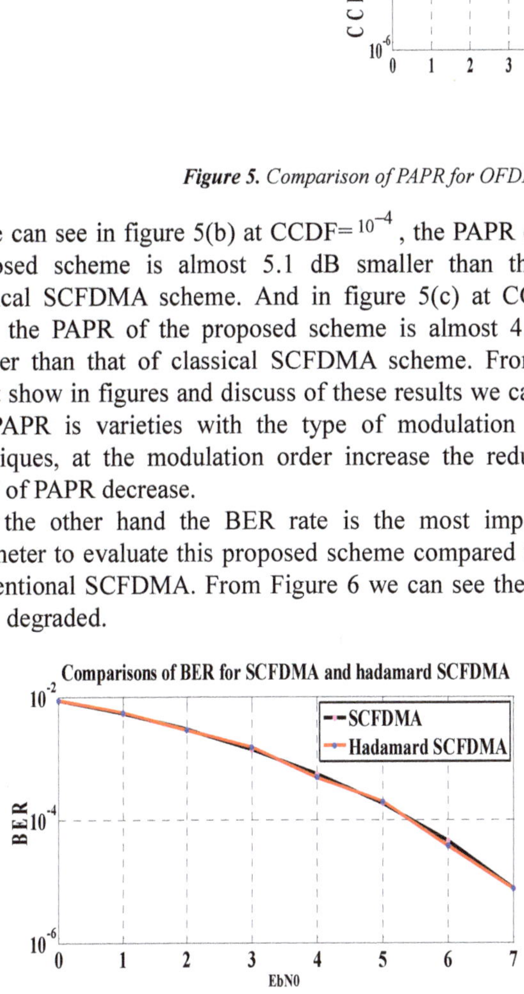

Figure 6. *Comparisons of BER for SCFDMA and proposed scheme of Hadamard SCFDMA.*

5. Conclusion

In this paper, a PAPR reduction scheme based on the Hadamard transform instead of FFT is proposed. Note that Hadamard transform is an orthogonal linear transform and can be implemented by a butterfly structure as in FFT. This means that applying Hadamard transform does not require the extensive increase of system complexity. The DFT spreading techniques of SCFDMA is based on Localized FDMA technique. The PAPR reduction performance and BER performance are evaluated by computer simulation. Simulation results state that the PAPR reduction performance is improved in the proposed techniques compared with conventional SCFDMA and is varied with the modulation order used as mapper techniques. On the other hand, the BER of system using proposed PAPR reduction scheme is not degraded and the complexity of the system is reduced.

References

[1] D. Falconer, et.al. ,"Frequency domain equalization for single-carrier broadband wireless systems", IEEE Commun. Mag., 2002, 40, pp. 58–66.

[2] F. Adachi, et.al. ,"Broadband CDMA techniques", IEEE Wirel. Commun., 2005, 12, (2), pp. 8–18.

[3] H.G. Myung, et.al. ,"Single carrier FDMA for uplink wireless transmission", IEEE Veh. Technol. Mag., 2006, 1, (3), pp. 30–38.

[4] M. Vittal., et.al., "PAPR analysis of single carrier FDMA signals with pulse shaping", International conference on Communication and Signal Processing, April 3-5, 2013, India.

[5] Li X., Cimini L.J., "Effects of clipping and filtering on the performance of OFDM". Proc. IEEE Vehicular Technology Conf., May 1997, vol. 3, pp. 1634–1638.

[6] J. Armstrong., "Peak-to-average power reduction for OFDM by repeated clipping and frequency domain filtering", Electron. Lett., 2002, 38, pp. 246–247

[7] X. Li., Cimini L.J, "Effects of clipping and filtering on the performance of OFDM", IEEE Commun. Lett., 1998, 2, (5), pp. 131–133

[8] H. G. Myung, "Introduction to Single Carrier FDMA," 15th European Signal Processing Conference (EUSIPCO) 2007, Poznan, Poland, Sep. 2007.

[9] H. G. Myung, et.al., "Peak-to-Average Power Ratio of Single Carrier FDMA Signals with Pulse Shaping", The 17th Annual IEEE International Symposium on Personal, Indoor and Mobile Radio Communications (PIMRC " 06), Helsinki, Finland, Sep. 2006.

[10] M. Park, et.al., "PAPR reduction in OFDM transmission using Hadamard transform", IEEE International Conference o Communications, Vol.1, Jun 2000, pp.430-433.

Identification and Classification of Processing Unit Eligibility for Ubiquitous Computing Using Feature Selection Mechanism and Artificial Neural Network

Patience Spencer[1], Enoch O. Nwachukwu[1, 2]

[1]Department of Computer Science, Ignatius Ajuru University of Education, Rumuolumeni, Rivers State, Nigeria

[2]Department of Computer Science, University of Port Harcourt, Rivers State, Nigeria

Email address:

patsyspency2013@hotmail.co.uk (P. Spencer), enoch.nwachukwu@uniport.edu.ng (E. O. Nwachukwu)

Abstract: Ubiquitous Computing is a trending innovation that allows a user to have access to many computers in a transparent manner anytime anywhere thereby enhancing computing confidence. However, the full potential of ubiquitous computing is not yet realised due to challenges including changing location of mobile users, poor network infrastructure, limited system resources, and poor transaction processing model. This work is concerned with the development of a proactive support for active transaction coordination in ubiquitous computing environment. The specific objectives are to identify relevant values of predefined key features of processing units that greatly impact on ubiquitous computing and to predict the processing capability of processing units using relevant values of the predefined features. An object-oriented analysis and system design methodology is employed and the proposed processing unit eligibility identification mechanism and neural network-based classifier is shown to effectively support ubiquitous computing.

Keywords: Ubiquitous Computing, Transaction, Multi-Layer Perceptrons, Neural Network, Feature Selection

1. Introduction

The shift from Mainframe-based Computing paradigm to PC-based Computing paradigm and then to Ubiquitous Computing is indeed a welcome development. Ubiquitous computing allows a user to have access to many computers in a transparent manner anytime anywhere [1] thereby enhancing computing confidence. The full potential of ubiquitous computing has not been realised [2] due to challenges including changing location of mobile users, poor network infrastructure, limited system resources, and poor transaction processing model.

This work presents a proactive platform for the coordination of active transactions in such a way that transaction processing is not affected by any form of system or network challenge. For example, the negative effect of transaction processing systems having frequent disconnections from distributed databases while executing a transaction results in low throughput and response time in ubiquitous computing environment. This is not acceptable as it creates unreliable access to data sources and unnecessary process delays.

The specific objectives of this work are to identify relevant values of predefined key features of processing units that greatly impact on ubiquitous computing and to predict the processing capability of processing units using the relevant values of the selected system attributes. To achieve these objectives, an object-oriented analysis and design methodology have been employed. The objects [3] of the proposed system and their behaviours are modelled using a combination of feature selection mechanism and Artificial Neural Networks. This approach is shown to enable a proactive support platform for transaction management where the scheduling component of the transaction server is given accurate prediction of the most eligible ` processing unit to be scheduled for transaction execution. This in turn impacts on users whose operations depend on information systems.

1.1. Transaction Processing and Ubiquitous Computing

Basically, ubiquitous computing paradigm is concerned with the ability of a user with a mobile computing device (wearable and handheld) to be able to access information residing in different computers as though the information is in the user's computer [1]. Advancements made in wireless networking technology, portable computing devices, and sophisticated mobile software applications greatly contributed to the paradigm shift from PC-based computing to ubiquitous computing. Software applications developed to support ubiquitous computing do not only take advantage of the advancement made in the field to do complex task but also face the challenges that come with the technology [4]. The challenges are mainly related to mobility, interconnectivity, and context-awareness. To achieve effective ubiquitous computing, the physical environment must be influenced digitally through collaborative signal and information processing sensors and control devices installed everywhere in the environment [5]. These devices perceive our environment and automatically adapt to the environment resulting in easy access to data and services. It also takes advantage of Positioning Systems to determine current location of mobile users and also link to other information services. Contrary to traditional mobile computing, context-based (physical, Human, and IT state properties) transaction management components are influenced by active environmental factors dynamically.

A database user (That is, the physical user) sees a transaction as a single operation whereas a database management system sees a transaction as a collection of several operations that form a single logical unit of work" [6]. Each logical unit of work is referred to as a sub-transaction. Sub-transactions can be distributed to different processing units (database hosts). In a distributed database system, different processing units are connected to one another through a communication network infrastructure [5]. Distributed transaction processing and wireless network infrastructure form a backbone for ubiquitous computing. Transaction management for ubiquitous computing aims at providing mobile users with reliable services in a transparent way anytime anywhere [7]. However, the processing of a single logical unit of work (That is, a transaction) in Ubiquitous Computing Environment as earlier mentioned is faced with unbearable challenges resulting in unreliable access to data sources (that is distributed databases) by transaction processing components [8]. This simply implies that the series of events that are carried out on a single unit of work (such as switching on a fan, registering courses to be taken by a student, retrieving a student's CGPA, confirming stock level of an item in a refrigerator, and the likes) from anywhere and at anytime need a favourable environment to perform optimally.

1.2. Changing Location of a Mobile User and Ubiquitous Computing

In ubiquitous computing environment, the activities of a mobile user cause the user to move about and the user may want to access variety of remote homogenous or heterogeneous databases. To avoid unreliable or denied access to data sources, a transaction management model that supports learning, collaboration, and autonomy is required. Existing database management systems may not have shown these characteristics sufficiently enough to effectively support ubiquitous computing especially as it affects mobile and distributed transaction processing.

2. Materials and Methods

In the proposed work, feature identification is achieved through Perception Reasoning technique [9] whereas transaction protocols execution is achieved through intelligent control [9]. These are key Artificial Intelligence behaviours [9] and are of great concern to innovative researchers in the field of Machine Learning. The perception reasoning mechanism eliminates noisy, irrelevant and redundant data [10]. In so doing, only meaningful features or feature values are fed into the neural network thereby accelerating input data capturing and classification processes.

For example, should a mobile user request for Location Dependent Information, the proposed transaction processing model ensures that the request is processed to a logical conclusion without loss of connectivity to data source. A question that is addressed in this work is "How will transaction processing components predict the most eligible processing unit to be allocated a new transaction or a migrated transaction?" In answering this question, varying relevant values of selected attributes of processing units are identified and used as the classification parameters. Tables 1a, 1b, and 1c show the processing unit parameters used in defining processing units eligibility ratings. Only participating processing units in the ubiquitous computing environment are considered in the classification process.

Table 1a. *Predefined Network Signal Strength Rating.*

Processing Unit Rating	1	2	3	4	5
Network Signal Strength (%)	0 - 20	20 - 40	40 - 60	60 - 80	80 - 100

Table 1b. *Predefined Processor Speed Rating.*

Processing Unit Rating	1	2	3	4	5
Processor speed in GHz	0 - 0.6	0.6-1.2	1.2 - 1.8	1.8-2.4	2.4 - 3.0

Table 1c. *Predefined Available RAM Size Rating.*

Processing Unit Rating	1	2	3	4	5
Available Ram Size in GB	0 - 1.6	1.6 - 3.2	3.2 - 4.8	4.8-6.4	6.4 - 8.0

Information in tables 1a, 1b, and 1c describe the rating of possible input values of Network Signal Strength (NSS), Processor Speed (PS) and Available RAM Size (ARS) of a participating processing unit. The values of these parameters vary at run-time thereby changing the behaviour of the classifier dynamically. However, the classifier is expected to predict the set of features that qualifies a participating processing unit to be the most eligible for online transaction execution within a specified time.

To ensure that the right processing unit is selected without human intervention in a transparent manner, Artificial Neural Network (ANN) approach is employed. ANN is also known as Neural Network and can be described as a computational network that is implemented on a digital computer resulting in Digital Signal Processing (DSP) with interconnected processing entities that mimic biological structures (nervous system) in processing information. It is an innovative and Powerful modelling tool that can be applied to develop predictive models [11]. One of the most attractive properties of ANNs that instigated our choice of approach is the possibility of ANNs to adapt their behaviour to the changing characteristics of the modelled system [12]. Another useful characteristic is the ability of ANNs to effectively manage inter-related outputs in a better way. For example, the eligibility of processing units in ubiquitous computing environment is a function of measurable state information about the processing units (NSS, ARS and PS). In this context, NSS, ARS and PS are interrelated. Different combinations of the values of these system features results in different classes that define the eligibility of a processing unit.

The use of neural network technology to optimize classification processes is not novel but overcoming the challenge of getting the right neural network model to optimally solve the classification problem is what really matters. In this work, a Muiti-Level Perceptron (MLP) Artificial Neural Network is used to systematically solve the classification problem. The optimization of the performance of the neural network is done by applying Cross Entropy-based Back Propagation Algorithm.

The processing capabilities are then used to define the eligibility levels of participating processing units. If the eligibility of a participating processing unit falls under the category with the highest processing capability, that processing unit is presented to the transaction processing model as the most eligible processing unit within a predefined period of time. With this, online migration of active transaction from a processing unit with compromised processing capability to a processing unit with better processing capability is made possible to ensure continues access to data source.

2.1. The Proposed Model

Using a combination of feature selection mechanism made up of collaborative agent and MLP neural network, relevant values of processing units attributes (NSS, ARS and PS) are collected and processed (classified) in order to be used as the basis for prediction of the eligibility level of processing units in the ubiquitous computing environment. Fig. 1 shows the feature selection and input data normalization stages of the best processing unit prediction neural network model whereas fig. 2 is used to represent the neural network-based intelligence model. In fig 1, dynamic values of selected features are collected and separated into five different classes based on the rating of the sensed values of selected system features from the feature space *N* (shown as context info 1, 2, 3…n in fig 1). In this work, the system features are also referred to as context or state information.

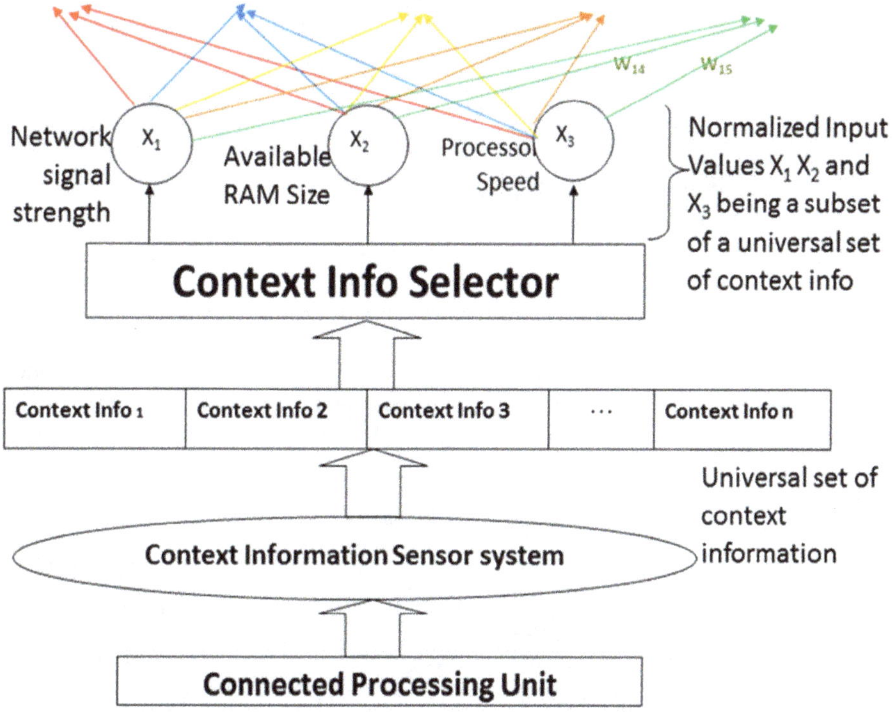

Figure 1. *Feature selection and input data normalization model.*

The "Context Info Selector" is designed to capture values of the predefined system features (That is, NSS, ARS and PS). The dynamic values of these features are then processed for the identification of relevant and significant values that are then fed into the neural network via the input layer of the neural network. The input nodes manage the non-classed vector dataset. Outputs of the input layer are then fed into the hidden layer of the network.

The formal presentation of the non-class vector (that is, a multi-dimensional input) is given by (1).

$$X = \{x_1 x_2 ... x_n\} \quad (1)$$

where: x represents the sensed input value and n represents the dimension of the input space. Applying it to the proposed model, $n = 3$. This means that the proposed model deals with a vector X of three input values of NSS, ARS and PS.

Every input value is first normalised to have an equivalent value that falls within the range of [0, 1]. Equation 2 is an adopted mathematical model used to determine the normalized input values:

$$v' = \frac{v - minA}{maxA - minA}(new\_maxA - new\_mind) + new\_minA \quad (2)$$

where:

- Attribute A corresponds to "Network Signal Strength" or Available RAM Size" or "Processor Speed".
- v is an element of all possible elements of A
- MinA is the minimum value of attribute A
- MaxA is the highest value of attribute A
- New_maxA is 1 and
- New_minA is 0.

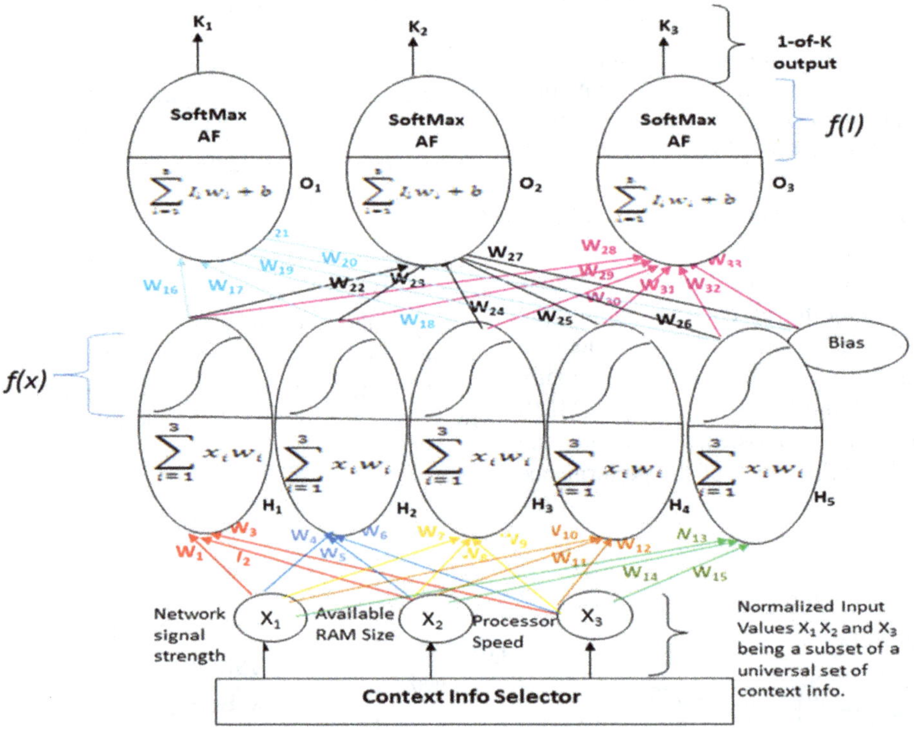

Figure 2. *A multi perceptron feed forward neural network influenced by Log-sigmoid and Softmax activation function used to build back-end mobile agent's intelligence in classifying processing units' eligibility.*

Each input value is sent to all the nodes in the hidden layer. Each node in the hidden layer is limited to a particular task using the logistic sigmoid function. The logistic sigmoid function decides whether an input (represented by a non-class vector of numbers I_i) belong to one class or another. For example, the first processing node in the hidden layer tagged H_1 is designed to identify input values (x_1, x_2, x_3) of the predefined features of participating processing unit under investigation. Expected output Vectors I_i of the nodes in the hidden layer representing five separate categories of perceived input values with rating from 1 to 5 where rating 5 represents the most significant value allowed and rating 1 representing the least significant value allowed. Nodes in the hidden layer are influenced by logistic Sigmoid Activation Function. The log-Sig activation function is a mathematical function used to identify the state of nodes in the hidden layer. Table 2 represents the expected states of nodes in the hidden layer H_1, H_2, H_3, H_4 and H_5 (presented in their un-normalized format).

Table 2. *Expected output vector data set at nodes H_1 to H_5 of the hidden layer (in their un-normalized format).*

Hidden layer Node Name	Dataset of expected values of predefined attributes		
	NSS	ARS	PS
H_1	[0, 20]	[0, 1.6]	[0, 0.6]
H_2	[20, 40]	[1.6, 3.2]	[0.6, 1.2]
H_3	[40, 60]	[3.2, 4.8]	[1.2, 1.8]
H_4	[60, 80]	[4.8, 6.4]	[1.8, 2.4]
H_5	[80, 100]	[6.4, 8.0]	[2.4, 30]

The connection strengths (that is, the value of weight that solves the given task optimally) of the inputs are determined arbitrarily and are normally small in value (lies between -∞ to +∞). The sum product of the input values and their associated connection strengths given by (3) are fed into the hidden layer.

$$I_i = \sum_{i=1}^{n} w_i x_i \quad (3)$$

where:

i is the node number, I is the net input of the ith node, w_i is the ith weight associated with the ith node and x_i is the ith input associated with the ith node.

The Net Inputs from the hidden layer fed into the nodes of the output layer are given by (4), (5), and (6):

$$Z_1 = w_{16}I_1 + w_{17}I_2 + w_{18}I_3 + w_{19}I_4 + w_{20}I_5 + w_{21}b \quad (4)$$

$$Z_2 = w_{22}I_1 + w_{23}I_2 + w_{24}I_3 + w_{25}I_4 + w_{26}I_5 + w_{27}b \quad (5)$$

$$Z_3 = w_{28}I_1 + w_{29}I_2 + w_{30}I_3 + w_{31}I_4 + w_{32}I_5 + w_{33}b \quad (6)$$

It means that the net input from the input layer to the hidden layer is a function of x $(f(x))$ and the net input from the hidden layer to the output layer is a function I $(f(I))$.

The net inputs I_i plus their corresponding connection weights are passed through the logistic-sigmoid activation function in the processing nodes for category validation as each node represents a different category.

The hidden layer output plus their respective weighted bias b (usually within the range of 0 to 1) are in like manner fed into the output layer where only significant categories of the predefined feature values are classified using the Softmax activation function as the category validation tool. Note that the bias b is added to normalize the net input.

The three processing nodes in the output layer represent the three most significant sets of predefined feature values. These set of values are chosen because they give significant support to ubiquitous computing.

Table 3 shows the un-normalized expected states of the three most significant categories of processing unit K_1, K_2, and K_3 (that is, output). The normalised output of each output node (O_1, O_2, and O_3) rages from 0 to 1. Zero (0) represents the lowest (or no) probability of correct match and 1 represents the highest probability of correct match.

Table 3. *Expected classed vector datasets at node O_1 O_2, and O_3 output layer (in their un-normalized format).*

significant category name	Dataset of expected values of predefined attributes		
	NSS	ARS	PS
K_1	[40, 60]	[3.2, 4.8]	[1.2, 1.8]
K_2	[60, 80]	[4.8, 6.4]	[1.8, 2.4]
K_3	[80, 100]	[6.4, 8.0]	[2.4, 30]

Only one out of the three possible output nodes (see fig. 2) is expected to give a probability value of 1 at the end of each iteration. However, there is a high possibility that, more than one output node will fire (a situation where the value of calculated output is 1). When the state of more than one output node is 1, it means that the eligibility status of the processing unit under investigation is not certain (indicating erroneous prediction). When this happens, the neural network is subjected to take error correction training in order to give a non ambiguous prediction.

The state of each output node is determined by passing their corresponding net inputs through Softmax Activation function which is given by (7):

$$K_{i=}\frac{1}{1+exp^{(-Z)}} \quad (7)$$

where: $Z = \sum w_{ji} I_{ji} + b_i$

w_{ji} *is* the connection weight from the hidden *layer to the output layer*

I_{ji} is the vector of input values from the hidden layer to the output layer.

b_i is the added bias.

K_i represents a possible class of processing unit eligibility such that:

K_1 represents a dataset of values of NSS, ARS, and PS of the processing unit that qualifies it to be the least eligible processing unit at a particular point in time

K_2 represents a dataset of values of NSS, ARS, and PS of the processing unit that qualifies it to be the second to the most eligible processing unit at a particular point in time and

K_3 represents a dataset of values of NSS, ARS, and PS of the processing unit that qualifies it to be the most eligible processing unit at a particular point in time

Table 3 shows the accepted relevant and significant values of context information used to determine the eligibility level of processing units in the ubiquitous computing environment. The output of each output node K_i is validated using (8):

$$K_i = \begin{cases} 1 \; if \; Z \geq T \\ 0 \; if \; Z < T \end{cases} \quad (8)$$

where:

$Z = \sum_{i=1}^{n} w_i I_i + b_i$ (weighted sum of inputs or observations including the bias).

T is the predefined rule also known as threshold (used to aid the supervision of the node state validation process).

2.2. Training the Network

A training sample consists of the sensed input vector X and the expected output vector of a node denoted by Y [13]. A multi-input vector, and multi-output vector is represented as $\{X_i, Y_i\}$. If the calculated output vector K_i do not match the expected output vector Y_i, the network is then turned using Cross Entropy (CE) error correction function (where the values of the weights and biases are modified guided by a threshold value). The error correction algorithm starts from the output layer through the input layer (That is, Back propagating the error) for re-computation of the neural network with adjusted connection weights and possibly biases. This process continues until the weights or biases that solve the problem optimally is found.

In Implementing a back propagation with CE error function, the sum of the log to the base e of each calculated network output K_i times their corresponding expected outputs Y_i is determine first, and then, the negative of the sum (that is, the result) is taken. For example, if the calculated output at node O_1 is a classed vector K_1 consisting of the elements NSS (%), ARS (GB) and PS (GHz) with values that fall within the expected ranges of the values of the elements at node O_1, then K_1 is said to match the expected output vector Y_1 at node O_1. The input and output vectors of the three output nodes of the proposed classifier neural network are formally represented in (9), (10), and (11). Now if the values of Ki are far from the values of their corresponding Yi, then, the state of the node is described as "0". However as the values of the elements gets closer to the expected values, the probability of getting the expected match gets higher. A probability of "1" indicates that K_i correctly matches Y_i. The CEs associated with K_1, K_2 and K_3 outputs are determined using the adopted formulas (15), (16) and (17). The expected outputs correspond to 0 (For no, there is no match) or 1 (For yes. there is a match).

$$K_1 = \{k_1, k_2, k_3\} \mid \{[40, 60], [3.2, 4.8], [1.2, 1.8]\} \quad (9)$$

Where, k_1 is the calculated value of NSS, k_2 is the calculated value of ARS and k_3 is the calculated value of PS.

$$Y_1 = \{y_1, y_2, y_3\} \mid \{[40, 60], [3.2, 4.8], [1.2, 1.8]\} \quad (10)$$

Where, y_1 is the expected value of NSS, y_2 is the expected value of ARS and y_3 is the expected value of PS.

In a similar manner, using the same state information, the outputs of node O_2 and O_3 are represented as:

Outputs at O_2

$$K_2 = \{k_4, k_5, k_6\} \mid \{[60, 80], [4.8, 6.4], [1.8, 2.4]\} \quad (11)$$

$$Y_2 = \{y_4, y_5, y_6\} \mid \{[60, 80], [4.8, 6.4], [1.8, 2.4]\} \quad (12)$$

Outputs at O_3

$$K_3 = \{k_7, k_8, k_9\} \mid \{[80, 100], [6.4, 8.0], [2.4, 3.0]\} \quad (13)$$

$$Y_3 = \{y_7, y_8, y_9\} \mid \{[40, 60], [3.2, 4.8], [1.2, 1.8]\} \quad (14)$$

The subscripts of the various vector elements are serially numbered for easy identification.

CE error at output node O_1 is given by equation 15:

$$E_{K1} = -\left((Y_1(\ln(k_1))) + (Y_2(\ln(k_2))) + (Y_3(\ln(k_3)))\right) \quad (15)$$

CEE at output node O_2 is given by equation 16:

$$E_{K2} = -\left((Y_4(\ln(k_4))) + (Y_5(\ln(k_5))) + (Y_6(\ln(k_6)))\right) \quad (16)$$

CEE at output node O_3 is given by equation 17:

$$E_{K3} = -\left((Y_7(\ln(k_7))) + (Y_8(\ln(k_8))) + (Y_9(\ln(k_9)))\right) \quad (17)$$

Mean Cross Entropy error

To determine the Mean Cross Entropy Error E_m for the three-element dataset described in section 2..2, the negative values of the node wise errors (That is, (15), (16), and (17)) are summed up and divided by 3 (since there are three classed vectors). Mathematically this is represented by (18).

$$E_m = (E_{K1} + E_{K2} + E_{K3}) \div 3 \quad (18)$$

The mean cross entropy error E_m is to be reduced to zero (0) or very close to zero for the performance of the neural network to be accepted. It is important to note that CE error function essentially ignores all computed outputs which do not correspond to an expected output of "1" [10]. It is used as error correction function in this work because it also serves as a stop condition for the neural network learning without exceeding the number of required training iterations [10]. This improves the net performance of the classifier. Now if a node's computed state is close to its expected state for all training inputs x, the error will be close to zero. For example, if the expected network state Y_1 is 0 and the calculated sate K_1 is 0 for input *I*. this means that the network is performing well on the input. In a similar way, if Y is 1 and K is very close to 1, the network is performing well. Hence, the processing cost (error) will be low in as much as K is close to Y. This implies that the error update that is propagated backwards from each output node is directly proportional to the difference between expected values of NSS, ARS and PS and the calculated values of NSS, ARS and PS.

Figure 3 is a representation of a transaction processing model showing the Transaction Server where the proposed (agent-based) feature selector and neural network-based processing unit classifier are implemented.

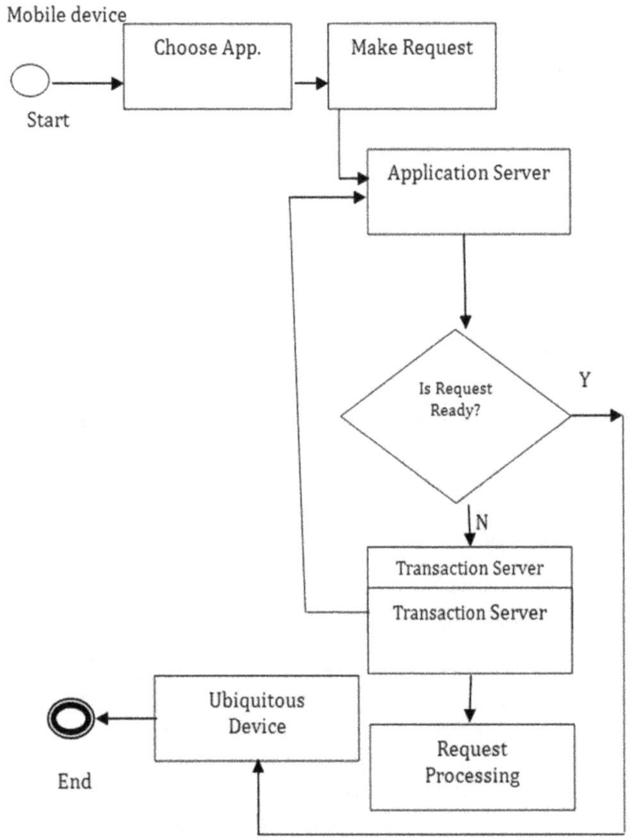

Figure 3. *A System Flow diagram showing the Transaction Server where the proposed Feature Identifier and Neural Network-based Processing Unit Classifier are implemented.*

3. Result and Discussion

For a timely and reliable performance, the proposed feature selector and neural network-based most eligible processing unit predictor is shown to intelligently create an effective platform for a scheduler in transaction processing systems for ubiquitous computing environment. The feature selection component of the proposed system transmits only predefined values of network signal strength, available RAM size, and processor speed to the multi-perceptron neural network where classification is based on processing units' processing capacity within a given period of time. It is very essential to allocate new or migrated transaction to processing units with the required processing capacity to avoid frequent disconnection from data sources and unnecessary delays resulting from network and system resource related issues. Allocating transaction to the most eligible processing unit is shown to provide uninterrupted access to data sources. This in turn enhances Response Time and Transaction Throughput.

4. Conclusion

Allocating transaction to processing units with sufficient RAM size, high Processor Speed and good Network Signal Strength affects transaction processing in ubiquitous computing environment positively. These are key transaction processing system context that must not be over looked when designing a any transaction processing model for ubiquitous computing environment.

The implementation of transaction processing models with no context-awareness [14] or poor context management skills and possible failure prediction mechanisms is unacceptable as it creates room for undesirable challenges during the execution of distributed transactions. To solve this problem, this work presents a transaction processing model with a proactive transaction server.

Using an object oriented analysis and design methodology [9], this work showed how a combination of feature selection mechanism and Artificial Neural Network are used to create an effective platform for a proactive transaction management at the backend of a transaction processing model. Implementing a back propagation with cross-entropy error function when training the proposed MLP neural network classifier gives a transaction server the context awareness strength it needs to make accurate prediction of the most eligible processing unit to be allocated a transaction at any point in time.

References

[1] Nwachukwu, E. O. (2010). *Information Technology: The Albatross of Our Time: an Inaugural Lecture*. University of Port Harcourt.

[2] Byeong-Ho, K. A. N. G. (2007). Ubiquitous computing environment threats and defensive measures. *Int. J. Multimedia Ubiquit. Eng*, *2*(1), 47-60.

[3] Filip, M. J., Karunungan, K. L., Kramer, J. C., Lee, L. C., Moore, D. L., Shih, C. C., and Sydir, J. J. (1995). *U.S. Patent No. 5,414,812*. Washington, DC: U.S. Patent and Trademark Office.

[4] Puder, A., Römer, K., and Pilhofer, F. (2006). *Distributed systems architecture: a middleware approach*. Elsevier, UK.

[5] Chong, C. Y., and Kumar, S. P. (2003). Sensor networks: evolution, opportunities, and Challenges. *Proceedings of the IEEE*, *91*(8), 1247-1256.

[6] Silberschatz, A., Korth, H. F., Sudarshan, S. (2002) *Database System Concepts* McGraw Hill, New York.

[7] Tang, F., and Li, M. (2012). Context- adaptive and energy-efficient mobile transaction management in pervasive environments. *The Journal of Supercomputing*, *60*(1), 62- 86.

[8] Silberschatz, A., Korth, H. F., and Sudarshan, S. (2006) *Database System Concepts* McGraw Hill, New York.

[9] Beetz, M., Buss, M., and Wollherr, D. (2007). Cognitive technical systems—what is the role of artificial intelligence? In *KI 2007: Advances in Artificial Intelligence* (pp. 19-42). Springer Berlin Heidelberg.

[10] Bengio, Y., and LeCun, Y. (2007). Scaling learning algorithms towards AI. *Large-scale kernel machines*, *34*(5).

[11] Rughani, A. I., Dumont, T. M., Lu, Z., Bongard, J., Horgan, M. A., Penar, P. L., and Tranmer, B. I. (2010). Use of an artificial neural network to predict head injury outcome: clinical article. *Journal of neurosurgery*, *113*(3), 585-590.

[12] Karlik, B., & Olgac, A. V. (2011). Performance analysis of various activation functions in generalized MLP architectures of neural networks. *International Journal of Artificial Intelligence and Expert Systems*, *1*(4), 111-122.

[13] Huang, G. B., Wang, D. H., and Lan, Y. (2011). Extreme learning machines: a survey. *International Journal of Machine Learning and Cybernetics*, *2*(2), 107-122.

[14] Schilit, B., Adams, N., & Want, R. (1994, December). Context-aware computing applications. In *Mobile Computing Systems and plications, 1994. WMCSA 1994. First Workshop on* (pp. 85-90). IEEE.

Performance Evaluation of Variant Error Correction Schemes in Terms of Extended Coding Rates & BER for OFDM Based Wireless Systems

Md. Nurul Mustafa, Mohammad Jahangir Alam, Nur Sakibul Huda

Department of Computer Science and Engineering (CSE), Southern University Bangladesh, Chittagong, Bangladesh

Email address:
nurul.mustafa@southern.edu.bd (Md. N. Mustafa), jahangir@southern.edu.bd (M. J. Alam), jetsakib@yahoo.com (N. S. Huda)

Abstract: Modern OFDM systems provide effective spectral usage by allowing overlapping in the frequency domain. Moreover, it is highly resistant to multipath delay spread. The suppression of inter-symbol interference (ISI) is one of the top features of OFDM. It also facilitates mobile bandwidth allocation and may increase the capacity in terms of number of users. Even though the presence of OFDM's built in error preventing mechanism; error tends to occur that averts delivery of proper signal. In this work we evaluated and adapted ways of fine-tuning for different error correction methods so that they may bring positive impact on communication systems who are developing and using OFDM. Here the performance of different known error correcting techniques for OFDM systems have been analyzed after extending and puncturing feature is applied on them. Simulations are performed in well-known simulator MATLAB to evaluate the modified techniques for different channel conditions. As the advantage of OFDM based systems are mainly the robustness to channel impairments and narrow-band interference; thus added error reduction models will improve significant efficiency of systems based on OFDM.

Keywords: Extended RS Code, AWGN, ISI, BER, Extended Convolution Code, Extended LB Code, FFT, IFFT

1. Introduction

Wireless communication, as the name suggests [1], is wireless way of transmitting information from one place to another, is replacing most of the wired transmission of today's world. Research in the field of wireless communication is still a hot topic to discover new possibilities. The goal of every research in this topic is to find more effective communication methods. Wireless communication helped the user to move freely without worrying about transfer of data. It dramatically changed the concept of information transfer in homes and in offices.

The basic building blocks of a typical wireless communication system [2], are encoder, channel coder, modulator, demodulator, and channel decoder. The signal is first converted to digital data and then source encoded. Source encoding reduces the amount of the data present in the signal to reduce the bandwidth required to transmit the associated data. Then the data proceeds to channel encoder block, which is responsible for adding extra bits in the data to help correcting errors inflicted to the data due to fading and noise. Our work in this paper contains three techniques for channel encoding that will be explained in later chapters. The modulator modulates the message signal on the transmission frequency so that the signal is ready for transmission.

Signal when transmitted, through wireless channel, faces multiple problems [3]. One major problem faced by the signal is fading. Fading can be caused by natural weather disturbances, such as rainfall, snow, fog, hail and extremely cold air over a warm earth. Fading can also be created by manmade disturbances, such as irrigation, or from multiple paths. All these factors introduce errors into the transmitted data.

When received at the receiver the signal is added with noise, since receiver antenna is designed to receive any signal present within a certain frequency range and noise is also present in that range. Now that data is passed to the demodulator whose job is convert the signal back from the carrier frequency to its normal form. After that the channel decoder helps to recover original signal from the degraded signal due to channel fading and noise. This is done by using the redundant bits that were added by the channel encoder.

This signal after recovery is passed to the source decoder, which converts the signal back to its original form.

2. The Problem in Details

OFDMis a technique, for transmission of data stream over a number of sub-carriers. In OFDM, a high rate bit stream is divided into bit streams of lower rate and each of them are modulated over one of the orthogonal subcarriers [1]. In a single carrier system a single fade can cause the entire link to fail while in a multi carrier system only a few bits will be disturbed and they can be corrected by applying error correction codes.

OFDM overcomes the problem of inter-symbol interferenceby transmitting a number of narrowband subcarriers together with a guard interval [4]. But this gives rise to another problem that all subcarriers will arrive at the receiver with different amplitudes. Some carriers may be detected without error but the errors will be distributed among the few subcarriers with small amplitude.

Channel coding can be used across the subcarriers to correct the errors of weak subcarriers. In OFDM systems error correction has a significant role since OFDM along with error correcting techniques help to deal with fading channels. Error correction helps in recovery of faded information by providing a relation between information and transmitted code such that errors occurring within the channel can be removed at the receiver. A lot of such techniques for error correction are given till date. In this work some of these techniques are considered, modified to an extent & implemented for OFDM system and then a comparison between them are made for best outcome.

3. Channel Coding

Channel codingis basically from the class of signal transformations designed to improve the communication performance by enabling the transmitted signal to better resist the effects of various channel impairments such as noise fading and jamming [5]. The goal of channel coding is to improve the bit error rate (BER) performance of power limited and/or band limited channels by adding redundancy to the transmitted data. The three channel coding schemes [13], that are modified to achieve objective of this work are described in the following section:

3.1. Linear Block Coding

A block code is defined as a code in which k symbols are input and n symbols are output and is denoted as a (n, k) code[5]. If the input is k symbols, then there are 2k distinct messages. For each k input symbols output is n symbols known as a codeword where n is greater than k. Figure 1 shows input and output of a block code and the size of codeword formed.

A block code of length n with 2 kcodewords is called a linear (n, k) code if and only if its 2 kcodewords form a k-dimensional subspace of the vector space of all n-tuples over the Galois field GF. Since there are n output bits so there are 2n combinations possible for codewords but all of them are not codewords rather 2k are codewords. Rest of the combinations usually comes forward when there is an erroneous transmission and codewords are corrupted by change in bits. There are some rules which codewords have to fulfil. One of them is that the sum of any two codewords is also a codeword and being a linear vector space, there is some basis, and all codewords can be obtained as linear combinations of the basis [6]. Anexample of [5] code is generated by generator matrix given below:

$$G = \begin{bmatrix} 1 & 0 & 0 & 0 & 0 & 1 & 1 \\ 0 & 1 & 0 & 0 & 1 & 0 & 1 \\ 0 & 0 & 1 & 0 & 1 & 1 & 0 \\ 0 & 0 & 0 & 1 & 1 & 1 & 1 \end{bmatrix}$$

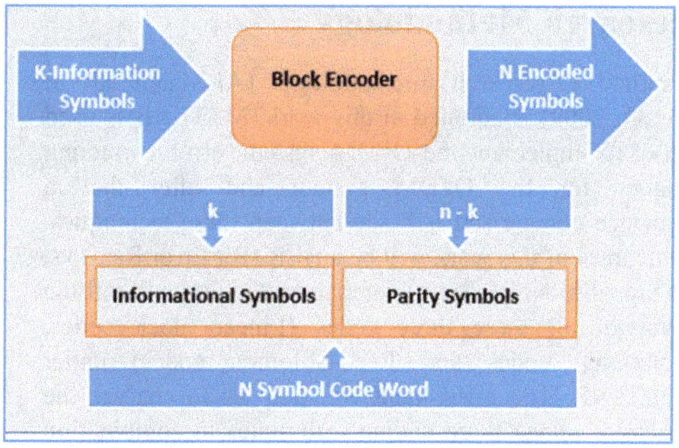

Figure 1. *A block encoder and a LB coding codeword.*

3.2. Convolution Code

Convolutional codes are different from the block codes [7], since in convolutional coding the information sequences are not grouped into distinct blocks and encoded so a continuous sequence of information bits is mapped into a continuous sequence of encoder output bits. Convolutional coding can achieve a larger coding gain than can be achieved using a block coding with the same code rate [8]. Convolutional codes have their popularity due to good performance and flexibility to achieve different coding rates.

An important characteristic of convolutional codes [9] is that the encoder has memory. That is the n-tuple output generated by the encoder is a function of not only the input k-tuple but also the previous N-1 input k-tuples. The integer N is called the constraint length [7]. The convolutional code is generated by passing the information sequence through a finite state shift register. In general, the shift register contains N stages and m linear algebraic function generators based on the generator polynomials [9]. The input data k bits is shifted into and along the shift register. The number of output bits for each k input bits is n bits. The code rate is given as R= k / n.

3.3. RS Coding

The mechanism by which Reed-Solomon codes [10]

correct errors is that the encoder adds redundant bits to the digital input data block. The decoder attempts to correct and restore the original data by removing the errors that are introduced in transmission, due to many reasons that might be the noise in the channel or the scratches on a CD. There are different families of Reed-Solomon codes and each has its own abilities to correct the number and type of errors.

These codes are linear and a subsection of BCH codes [11] and can be denoted as RS(n, k) with s-bit (Where s are symbols represented as a bit) symbols.

An n-symbol codeword is made by the k data symbols of s bits each and the encoder adds parity bits. Errors in a codeword can be corrected by the decoder up to t symbols where, 2t=n-k.

4. Research Methodology

Quantitative research methodology (AI-Mahmoud & Zoltowski, 2009) is adopted in this work. MATLAB is used as a tool to implement the OFDM system, error correcting techniques for the OFDM system and after that a performance comparison is made between these techniques. The outcomes of this work will be a BER (Bit Error Rate) vs. SNR (Signal to Noise Ratio) comparison, which will tell the behavior of all these three codes (Linear block codes, Convolutional codes and Reed-Solomon codes) under different SNR. The assessment will help us to analyze the performance of the three coding techniques in combination with OFDM. The finding of this work may help the OFDM system designers to choose the error correcting codes that match their requirements.

4.1. Code Modification

A code C has three fundamental parameters. Its length n, its dimension k, and its redundancy r = n-k. Each of these parameters has a natural interpretation for linear codes, and although the six basic modification techniques [3], are not restricted to linear codes it will be easy initially to describe them in these terms. Each of these one parameter and increases or decreases the other two parameters accordingly. We have:

1. Augmenting. Fix n; increase k; decrease r
2. Expurgating. Fix n; decrease k; increase r
3. Extending. Fix k; increase n; increase r
4. Puncturing. Fix k; decrease n; decrease r
5. Lengthening. Fix r; increase n; increase k
6. Shortening. Fix r; decrease n; decrease k

The six techniques fall naturally into three pairs, each member of a pair the inverse process to the other. Since the redundancy of a code is its "dual dimension" each technique also has a natural dual technique. In this workwe used the extending feature of code modification. We have modified our OFDM input & output code correction techniques according to the extended features and performed the simulation accordingly.

4.2. Extending & Puncturing

In extending or puncturing [5] a code we keep its dimension fixed but vary its length and redundancy. These techniques are exceptional in that they are one-to-one. Issues related to the extending and puncturing of GRS codes will be discussed in the next two sections.

When extending a code we add extra redundancy symbols to it. The inverse is puncturing, in which we delete redundancy symbols. Puncturing may cause the minimum distance to decrease, but extending will not decrease the minimum distance and may, in fact, increase it. To extend a linear code we add columns to its generator matrix, and to puncture the code we delete columns from its generator.

Let us call the [n + 1, k] linear code C_+ a coordinate extension of C if it results from the addition of a single new redundancy symbol to the [n, k] linear code C over the field F. Each codeword $c_+ = (c_1,\ldots, c_n, c_{n+1})$ of the extended code C+ is constructed by adding to the codeword $c = (c_1,\ldots,c_n)$ of C a new coordinate $C_{n+1} = \sum_{i=1}^{n} a_i c_i = a.c$, for some fixed $a = (a_1,\ldots,a_n) \in F_n$. Here I imply that the new coordinate is the last one, but this is not necessary. A coordinate extension can add a new position at any place within the original code.

The most typical method of extending a code is the appending of an overall parity check symbol, a null symbol chosen so that the entries of each new codeword sum to zero. This corresponds to a coordinate extension in which 'a' has all of its entries equal to -1.

5. Simulation & Results

To keep the conditions same for all the simulations the input data, the fading channel and the noise is once generated and saved. The input data is kept same for testing of all coding schemes. The total number of carriers [12] chosen in the OFDM system is 128 and used is 104. Eb/No (Energy per bit to noise power ratio) is used in testing. Eb/No is useful when comparing BER in digital modulation schemes without taking the bandwidth into account.

Rayleigh fading channel is once generated and saved. The same fading is used for all the simulations. The fading is defined by the following MATLAB equation:

$$\text{rayleigh} = \text{sqrt}(0.5 * (\text{randn}(N,1) + i * \text{randn}(N,1)))$$

Where 'randn()' generates values from the standard normal distribution function and 'N' defines the length of fading signal.

The results of simulation for code rate of 1/3 as per MATLAB simulation are shown in Figure 2. It is found that convolution code after the extending & puncturing features applied gives 2 dB improvement than Reed Solomon Code and 4 dB improvement than flat Block Codes at BER of 10^{-3}.

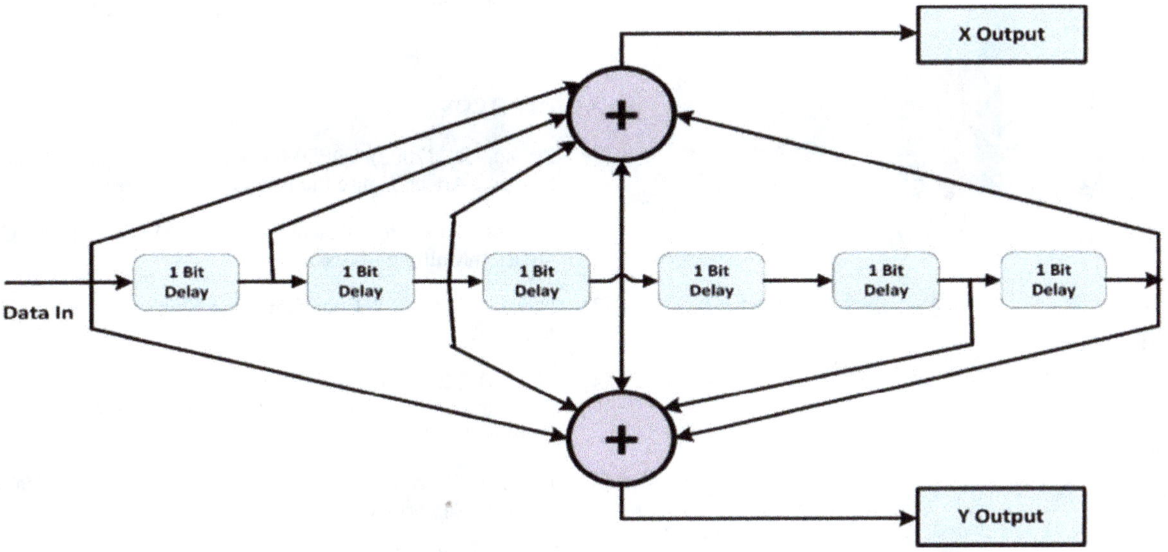

Figure 2. *Convolutional encoder with length 7 & rate ½.*

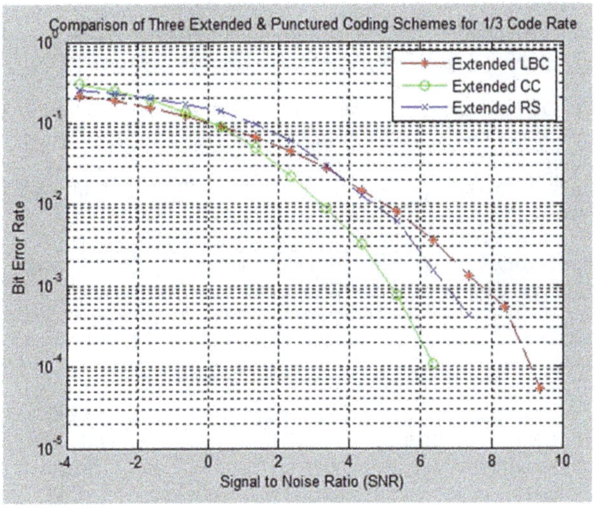

Figure 3. *Comparison of Three Extended & Punctured Coding Schemes for 1/3 Code Rate.*

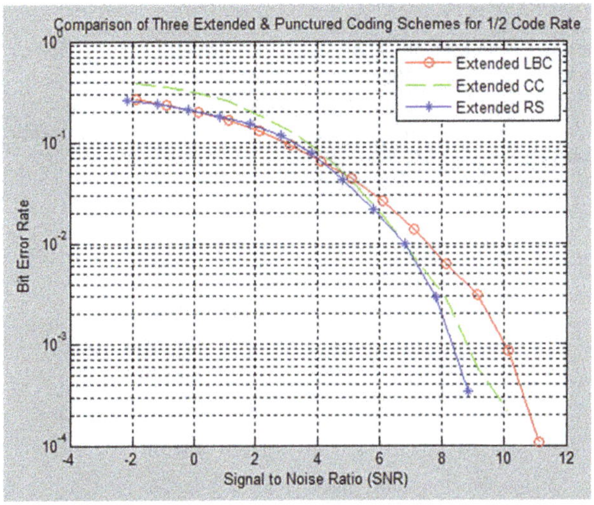

Figure 4. *Comparison of Three Extended & Punctured Coding Schemes for 1/2 Code Rate.*

The results of simulation for code rate of 1/2 are shown in Figure 3. It is found that RS code after the extending & puncturing features applied gives 1 dB improvement than Convolution Code and 2 dB improvement than Block Codes at BER of 10^{-3}.

The final results of simulation for code rate of 2/3 are shown in Figure 4. It is found that Convolution code after the extending & puncturing features applied gives.5 dB improvement than RS Code and 1.5 dB improvement than Block Codes at BER of 10^{-3}.

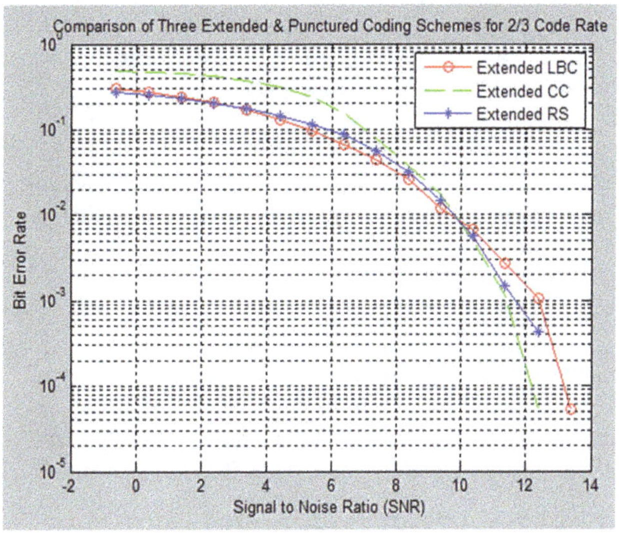

Figure 5. *Comparison of Three Extended & Punctured Coding Schemes for 2/3 Code Rate.*

Hence from all the performance analysis made above for different scenarios are shown in the following graphs:

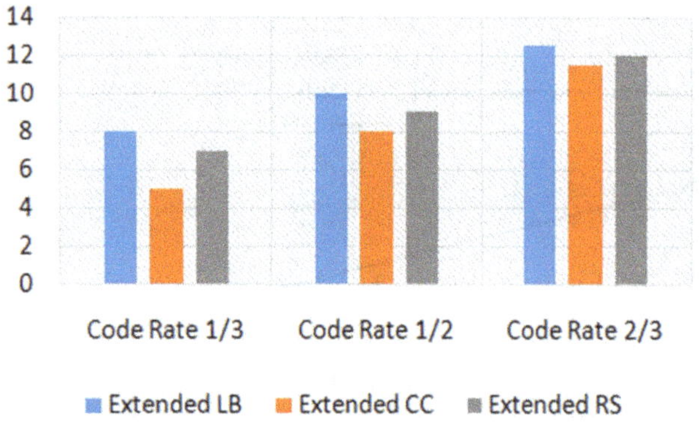

Figure 6. *Comparison of all three error rectification techniques in terms of extended & punctured coding scheme for OFDM systems.*

6. Conclusion

In this work it is found that Convolutional codes are very good in performance at lower code rates but due to complex decoding structure they are difficult to implement. Another problem that may arise in them is error propagation while decoding. Convolutional codes have problem of very complex decoding so if the length of input data is increased, the trellis used will be complex and the decoding will become even more complicated.

Linear block codes are very simple to implement. They also show good performance and are also good for lower code rates and highly suitable for application with less complexity requirement and not so high performance requirement.

Reed-Solomon codes in comparison to linear block codes are difficult to implement but their performance is much better and consistent than linear block codes since they can handle long bursts of errors. These codes showed good performance on all three tested code rates.

Hence with the system that requires good performance and along with that can also handle complexity, there is no better than convolution code. But if the system can't handle complex architecture then RS codes might be more suitable for the purpose.

7. Limitations

Limitation of this work certainly includes the use of a model noise channel as this result is yet to be tested in physical condition. If someone changes the generator matrix then it will also change the results. Thus it is not possible to generalize these results by having different environment other than described in Section V in simulation environment.

References

[1] Prasad, R. (2000), OFDM for Wireless Communications Systems, Artech House Publishers, Boston, London.

[2] Hughes Software Systems (2002), Multi Carrier Code Division Multiple Access.

[3] Haykin, S. (2006), Communication Systems, John Wiley & Sons, Inc.

[4] Sklar, B. (2001), Digital Communications: Fundamentals and Applications, Second Edition, Upper Saddle River, NJ: Prentice-Hall.

[5] Ryan, W. E., & Lin, S. (2009), Channel Codes: Classical and Modern. Cambridge University Press.

[6] Agarwal, Arun, & Patra, S. K. (2011). Performance Prediction of OFDM based DAB System using Block coding technique", Proceedings of ICETECT, India.

[7] MacKay, David J. C. (2003), Information Theory, Inference, and Learning Algorithms, Cambridge University Press.

[8] Witrisal, K., Kim, Y., & Prasad, R. (1999), A Novel Approach for Performance Evaluation of OFDM with Error Correction Coding and Interleaving, Centre for Wireless Personal Communications (CEWPC), IRCTR Delft University of Technology.

[9] Chatzigeorgiou, I. A. (2012), A Comparison of Convolutional and Turbo Coding Schemes For Broadband FWA Systems.

[10] Koetter, R. (2005), Reed–Solomon Codes, MIT Lecture Notes.

[11] AI-Mahmoud, M., & Zoltowski, Michael D. (2009), Performance Evaluation of CODE-SPREAD OFDM with error control coding, School of Electrical and Computer Engineering Purdue University West Lafayette, IN 47907-2035.

[12] Cho, Y. S., Kim, J. (2010), MIMO-OFDM Wireless Communications with MATLAB, John Wiley & Sons, Inc.

[13] Haque, D., Enayet Ullah, S., & Ahmed, R. (2008), Performance evaluation of a wireless Orthogonal Frequency Division Multiplexing system under various concatenated FEC channel-coding schemes, Proceedings of 11th International Conference on Computer and Information Technology, Khulna, Bangladesh, 25-27 December.

FPGA Based Packet Classification Using Multi-Pipeline Architecture

R. Sathesh Raaj, J. Kumarnath

Department of Electronics and Communication Engineering, PSNA College of Engineering and Technology, Dindigul, India

Email address:

Sathesh2311psna@gmail.com (R. S. Raaj), jkumarnath@gmail.com (J. Kumarnath)

Abstract: This paper proposes a decision-tree-based linear multi-pipeline architecture on FPGA's for packet sorting. We reflect on the next-generation packet classification problems where more than 5-tuple packet header fields has been classified. From traditional fixed 5-tuple matching, Multi-field packet classification has been evolved for flexible matching with arbitrary combination of numerous packet header fields. The recently proposed Open Flow switching requires classifying each packet using up to 12-tuple packet header fields. It become a great task to develop scalable solutions for next-generation packet classification that support larger rule sets, additional packet header fields and higher throughput. This paper proposes a 2-D multi-pipeline decision-tree-based architecture for next-generation packet classification which exploits the abundant parallelism and other desirable features such as current field-programmable gate arrays (FPGAs),. We propose several optimization techniques for the state-of-the-art decision-tree-based algorithm by examine the various traditional 5-tuple packet classification methods. By using set of 12-tuple rules, the framework has been developed to partition the rule set into multiple subsets each of which is built into an optimized decision tree. To maximize the memory utilization. a tree-to-pipeline mapping scheme is carefully designed while underneath high throughput. Our proposed architecture can store up to 1K synthetic 12-tuple rules or 10K real-life 5-tuple rules in on-chip memory of a single up to date FPGA, and maintain 80 or 40 Gbps throughput for least packets of size (40 bytes) respectively. To utilize the memory properly and to sustaining high throughput, a mapping scheme based on tree-to-pipeline is designed carefully. This paper deal with the profuse parallelism and other preferred features provided by present field-programmable gate arrays and propose a 2-D multi-pipeline decision tree based architecture for next-generation packet sorting. The Verilog Hardware description languages (VHDL) are used to design the proposed architecture and synthesized using Xilinx Software.

Keywords: Field Programmable Gate Array (FPGA), Multi-Pipeline Architecture, Multi-Field Packet Classification, Open Flow Switching, 2-D Multi-Pipeline Decision-Tree-Based Architecture, 12-Tuple Rules, 5-Tuple Rules, Verilog Hardware Description Languages (VHDL)

1. Introduction

The development of the next-generation internet demand routers to support a several value added services such as firewall processing ,network functionalities, quality of service (QoS) differentiation, traffic billing, virtual private networks, policy routing, and other value added services. To proffer these services, the router needs to organize the packets into diverse categories based on a set of rules which are predefined, which specify the value ranges of the several fields in the packet header. Such kind of function is called multi-field packet classification. In previous network applications, problems based on packet classification frequently consider the fixed 5-tuple fields: 32 -bit source/destination IP addresses, 8-bit trans-port layer protocol and 16-bit source/destination port numbers. In recent times network virtualization emerges as an needed features for next-generation enterprise, cloud computing networks and data center. This entails the underlying data plane be flexible and offer clean interface for control plane. Such efforts can be seen in Open Flow switch which handle explicitly the network laws by a rule set with rich definition as the hardware- software interface [2]. In Open Flow, up to 12-tuple header fields are considered such as Open-Flow-like packet classification and the *next-generation packet classification* problems. To design a high speed router, we need rule set size and multi -field packet classification. It has become one of the fundamental challenges. For example, the present link rate has been pushed above the OC-768 rate that is 40 Gbps, which requires meting out a packet every 8 ns in the most awful case (where the packets are of minimum size that is 40 bytes). Such high throughput is not possible using present software-based solutions [4]. Forth coming generation packet classification on more header fields poses an even higher challenge. Most of the present work in done by variety of hashing schemes such as Blooms filters and ternary content addressable memory .However, TCAMs are not scalable with respect to clock rate when compared to SRAMs, power consumption, or circuit area Most of TCAM-based solutions also suffer from range expansion when converting ranges into prefixes .Hashing-based solutions like Bloom Filters have become popular due to their $O(1)$ time performance and high memory proficiency. It is to be noted that, hashing cannot provide deterministic performance due to potential collision and is incompetent in handling wildcard or prefix matching [13].In Blooms filter, the secondary module is always needed to resolve false positives inherent, which may be slow and can limit the overall performance [14]. As an alternative, our work focuses on optimizing and mapping state-of-the- art packet classification algorithms onto SRAM-based parallel architectures such as field-programmable gate array (FPGA). FPGA technology has become an attractive option for implementing real-time network processing engines [7], [10], [15] due to its ability to reconfigure and to offer abundant parallelism. State- of-the-art SRAM-based FPGA devices such as Xilinx Virtex -6 [16] and Altera Stratix-IV [17] provide high clock rate, low power dissipation and large amounts of on-chip dual- port memory with configurable word width. We make use of these desirable features in current FPGAs for scheming high-performance next-generation packet classification engines.

Table 1. *Shows the Header field supported in current open flow.*

Header field	Notation	# of bits
Ingress port		Variable
Source Ethernet addresses	Eth src	48
Destination Ethernet address	Eth dst	48
Ethernet type	Eth type	16
VLAN ID		12
VLAN Priority		3
Source IP address	SA	32
Destination IP address	DA	32
IP Protocol	Prtl	8
IP Type of Service	ToS	6
Source port	SP	16
Destination port	DP	16

Ingrained node-to-stage mapping scheme is used for mapping the tree structure onto the pipeline architecture, which allows imposing the bounds on the memory size as well as the number of nodes in each stage. As a result, the memory consumption of the architecture is increased. Using external SRAM the memory allocation scheme is enabled to handle even larger rule sets. We make use of the dual-port high-speed Block RAMs provided in modern FPGAs to achieve a high throughput of two packets per clock cycle (PPC). Service interruption become possible due to memory-based linear architecture. All the routing path are located to avoid large routing delay hence high clock frequency is obtained,

2. Related Work

Traditional 5-tuple packet classification is considered a drenched research area, few work has been done on FPGAs. Decomposition based packet classifications algorithms such as DCFL{22} and BV{19}are commonly used in most of the readily available FPGA. Lakshman *et al.* [19] propose the Parallel Bit Vector (BV) algorithm, which is a decomposition-based algorithm targeting hardware implementation. It performs the parallel lookups on each individual field first. The lookup on each field returns a bit vector with each bit representing a rule. Taylor *et al.* [22] introduce Distributed Cross producting of Field Labels (DCFL), which is also a decomposition-based algorithm leveraging several observations of the structure of real filter sets. They putrefy the multi-field searching problem and use independent search engines, which can function in parallel to find the alike conditions for each filter field. Jedhe et al. [15] realize the DCFL architecture in their entire firewall implementation on a Xilinx Virtex 2 Pro FPGA, using a memory intensive approach, as opposed to the logic intensive one, so that on-the-fly update is possible. Two recent works [24], [25] discuss several issues in implementing decision-tree-based packet classification algorithms on FPGA, with different motivations. Luo *et al.* [24] propose a method called *explicit range search* to allow more cuts per node than the Hyper Cuts algorithm. Based on the cost of increased memory utilization, the tree height is radically reduced.At each internal node in order to find which child node to traverse it is needed to determine the a varying number of memory accesses, which may be infeasible for pipelining. Since the authors do not implement their FPGA design, the actual performance results are undecided.

3. Proposed Architecture Design

To achieve line-rate throughput, we map the decision forest including trees onto a parallel multi-pipeline architecture with P linear pipelines, as shown in Fig. 6, where P=2. Each pipeline is used for traversing a decision tree as well as matching the rule lists attached to the leaf nodes of that tree. The tree stages are the pipeline stages for tree traversal which is known as tree stages while those used for rule list matching are called the rule stages. Each tree stage contains a memory block storing the tree nodes and the cutting logic which generates the memory access address based on the input packet header values. At the end of tree traversal, the index of the consequent leaf node is regain to access the rule stages. Since a leaf node contains a list of list Size rules, we need list Size rule stages for matching these rules. All the leaf nodes of a tree have their rule lists mapped onto these list Size rule stages. Each rule stage includes a memory block storing the full content of rules and the matching logic which performs parallel matching on all header fields .Each incoming packet goes through all the pipelines in parallel. A different subset of header fields of the packet may be used to traverse the trees in different pipelines. Each pipeline outputs the rule ID or its corresponding action. The priority resolver picks the result with the highest priority among the outputs from the pipelines.

Table 2. *Formation of Rules.*

Header Filed	Bits Allocated	Rule Type
Source Address	32	Rule1
Destination Address	32	Rule2
Source Port Number	16	Rule3
Destination Port Number	16	Rule4
Transport layer protocol	3	Rule5
Ethernet Type	16	Rule6
VLAN ID	12	Rule7
VLAN Priority	3	Rule8

3.1. Pipeline

Like the HyperCuts with the *push common rule upwards* heuristic enabled, our algorithm may reduce the memory consumption at the cost of increased search time, if the process to match the rules in the *internal rule list* of each tree node is placed in the same critical path of decision tree traversal. Any packet traversing the decision tree must perform: 1) matching the rules in the internal rule list of the existing node and 2) branching to the child nodes, in series. The number of memory accesses all along the grave path can be very large in the worst cases. Though the throughput can be improved by using a deep pipeline, the large delay transitory the packet classification engine need the router to use a large buffer to store the payload of all packets being classified. Moreover, since the search in the rule list and the traversal in the decision tree have different structures, a heterogeneous pipeline is needed, which complicates the hardware design.

3.2. Tree-to-Pipeline Mapping

Before the FPGA implementation, the size of the memory in the pipeline stages should be known. However, when simply mapping each level of the decision tree onto a separate stage, the memory distribution across stages can vary extensively. Allocating memory with the maximum size for each stage results in large memory wastage. This propose a Ring pipeline architecture which employs TCAMs to achieve balanced memory distribution at the cost of halving the throughput to one packet per two clock cycles, i.e., 0.5 PPC, due to its non-linear structure. Our task is to map the decision tree onto a pipeline (i.e., Tree Pipeline in our architecture) to achieve balanced memory distribution over stages, while sustaining a throughput of one packet per clock cycle (which can be further improved to 2 PPC by employing dual-port RAMs). The memory distribution across stages should be balanced not only for the Tree Pipeline, but also for all the Rule Pipelines. Note that the number of words in each stage of a Rule Pipelines depends on the number of tree nodes rather than the number of words in the corresponding stage of Tree Pipeline, as shown in Fig. 8. The challenge comes from the various number of words needed for tree nodes. As a result, the tree-to-pipeline mapping scheme requires not only balanced memory distribution, but also balanced node distribution across stages. Moreover, to maximize the memory utilization in each stage, the sum of the number of words of all nodes in a stage should approach some power of 2. Otherwise, for example, we need to allocate 2048 words for a stage consuming only 1025 words. The above problem is a variant of bin packing problems, and can be proved to be NP-complete. We use a heuristic similar to our previous study of trie-based IP lookup, which allows the nodes on the same level of the tree to be mapped onto different stages. This provides more flexibility to map the tree nodes, and helps achieve a balanced memory and node distribution across the stages in a pipeline, as shown in Fig. 3. Only one constraint must be followed.

Constraint 1: If node is an ancestor of node in the tree, then must be mapped to a stage preceding the stage to which is mapped.

We impose two bounds, namely and for the memory and node distribution, respectively. The values of the bounds are some power of 2. The criteria to set the bounds is to minimize the number of pipeline stages while achieving balanced distribution over stages. The complete tree-to-pipeline mapping algorithm, where denotes a tree node, the number of stages, the set of remaining nodes to be mapped onto stages, the number of words of the the stage, and the number of nodes mapped onto the Nth stage. We manage two lists, Ready List and Next Ready List. The former stores the nodes that are available for filling the current stage, while the latter stores the nodes for filling the next stage. We start with mapping the nodes that are children of the root onto Stage 1.When filling a stage, the nodes in Ready List are popped out and mapped onto the stage, in the decreasing order of their heights.2 After a node is assigned to a stage, its children are pushed into Next Ready List. When a stage is full or Ready List becomes empty, we move on to the next stage. At that time, Next Ready List is merged into Ready List. By these means, Constraint 1 is met. The complexity of this mapping algorithm is , where denotes the total number of tree nodes. Our tree-to-pipeline mapping algorithm allows two nodes on the same tree level to be mapped to different stages. We implement this feature by using a simple method. Each node stored in the local memory of a pipeline stage has one extra field: the distance to the pipeline stage where the child node is stored. When a packet is passed through the pipeline, the distance value is decremented by 1 when it goes through a stage. When the distance value becomes 0, the child node's address is used to access the memory in that stage. External SRAMs are usually needed to handle very large rule sets, while the number of external SRAMs is constrained by the number of I/O pins in our architecture. By assigning large values of and for one or two specific stages, our mapping algorithm can be extended to allocate a large number of tree nodes onto few external SRAMs which consume controllable number of I/O pins.

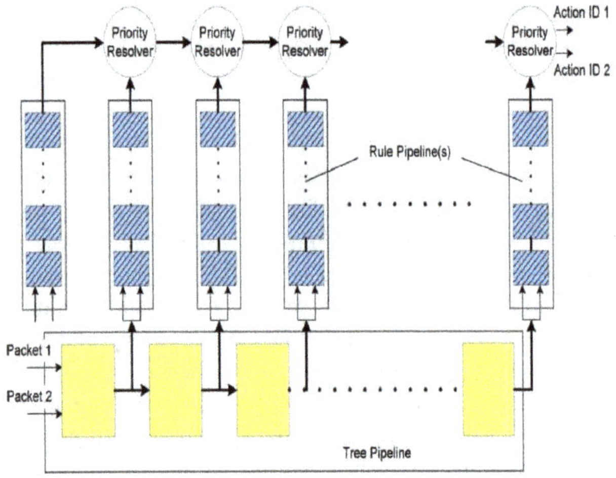

Figure 1. *2-D Multi-Pipeline Architecture.*

3.3. Pipeline for Rule Lists

When a packet accesses the memory in a Tree Pipeline stage, it will obtain the pointer to the rule list associated with the current tree node being accessed. The packet uses this pointer to access all stages of the Rule Pipeline attached to the current Tree Pipeline stage. Each rule is stored as one word in a Rule Pipeline stage, benefiting from the largeword width provided by FPGA. Within a stage of the Rule Pipeline, the packet uses the pointer to retrieve one rule and compare its header fields to find a match. When a match is found in the current Rule Pipeline stage, the packet will carry the corresponding action information with the rule priority along the Rule Pipeline until it finds another match where the matching rule has higher priority than the one the packet is carrying.

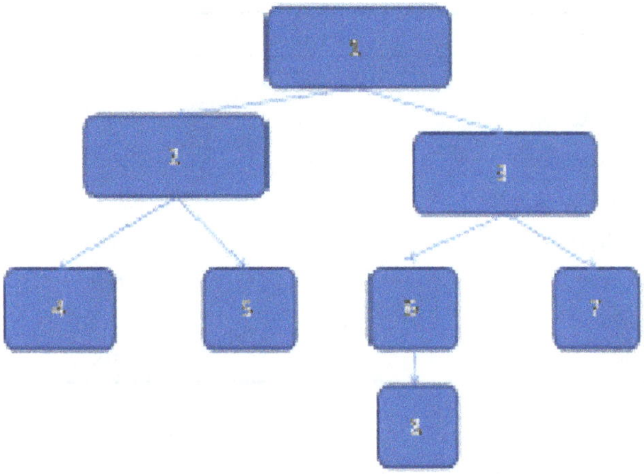

Figure 2. *Decision Tree.*

4. Simulation Results

Table 3. *Shows the Device Utilization Summary.*

Device Utilization Summary				
Logic Utilization	**Used**	**Available**	**Utilization**	**Note(s)**
Number of Slice Flip Flops	682	9,312	7%	
Number of 4 input LUTs	1,987	9,312	21%	
Logic Distribution				
Number of occupied Slices	1,034	4,656	22%	
Number of Slices containing only related logic	1,034	1,034	100%	
Number of Slices containing unrelated logics	0	1,034	0%	
Total Number of 4 input LUTs	2,008	9,312	21%	
Number used as logic	1,987			
Number used as a route-thru	21			
Number of bonded IOBs	18	232	7%	
IOB Flip Flops	8			
Number of GCLKs	1	24	4%	
Total equivalent gate count for design	22,689			
Additional JTAG gate count for IOBs	864			

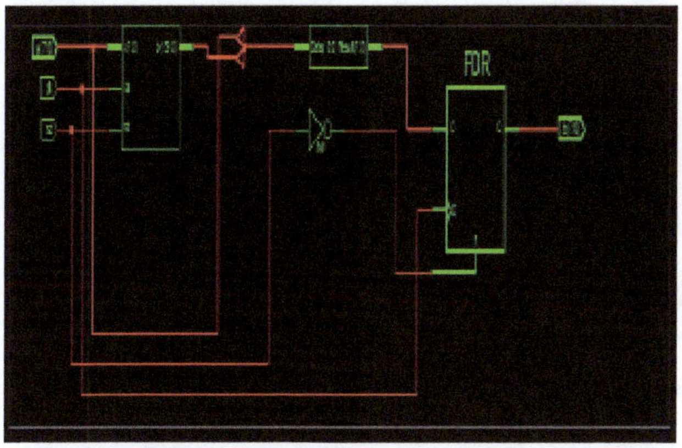

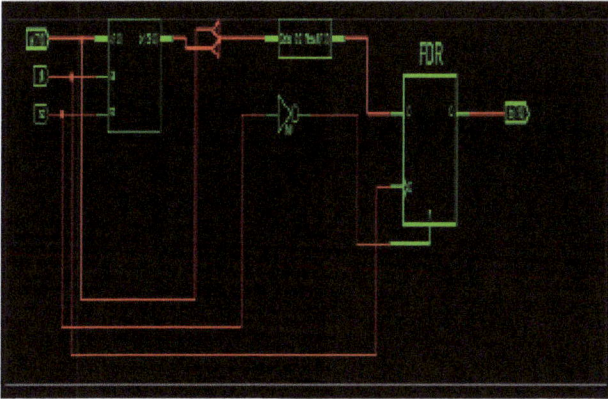

Figure 3. *RTL View of Proposed Architecture.*

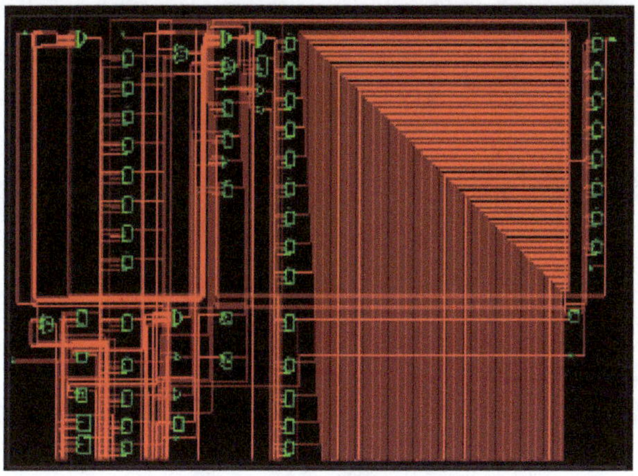

Figure 4. *Technology Schematic View.*

References

[1] M. Casado, T. Koponen, D. Moon, and S. Shenker, "Rethinking packet forwarding hardware," in *Proc. Hot Nets—VII*, 2008, pp. 1–6.

[2] N. Mc Keown, T. Anderson, H. Balakrishnan, G. Parulkar, L. Peterson, J. Rexford, S. Shenker, and J. Turner, "Open Flow: Enabling innovation in campus networks," *SIGCOMM Comput. Commun. Rev.*, vol. 38, no. 2, pp. 69–74, 2008.

[3] Open Flow Foundation, "Open Flow Switch Specification, Version 1.0.0," 2009. [Online]. Available: http://www.openflowswitch.org/documents/openflow-spec-v1.0.0.pdf

[4] P. Gupta and N. Mc Keown, "Algorithms for packet classification," *IEEE Network*, vol. 15, no. 2, pp. 24–32, 2001.

[5] F. Yu, R. H. Katz, and T. V. Lakshman, "Efficient multimatch packet classification and lookup with TCAM," *IEEE Micro*, vol. 25, no. 1, pp. 50–59, Jan. 2005.

[6] K. Lakshminarayanan, A. Rangarajan, and S. Venkatachary, "Algorithms for advanced packet classification with ternary CAMs," in *Proc. SIGCOMM*, 2005, pp. 193–204.

[7] H. Song and J. W. Lockwood, "Efficient packet classification for network intrusion detection using FPGA," in *Proc. FPGA*, 2005, pp. 238–245.

[8] S. Dharmapurikar, H. Song, J. S. Turner, and J. W. Lockwood, "Fast packet classification using bloom filters," in *Proc. ANCS*, 2006, pp. 61–70.

[9] I. Papaefstathiou and V. Papaefstathiou, "Memory-efficient 5D packet classification at 40 Gbps," in *Proc. INFOCOM*, 2007, pp. 1370–1378.

[10] A. Nikitakis and I. Papaefstathiou, "A memory-efficient FPGA-based classification engine," in *Proc. FCCM*, 2008, pp. 53–62.

[11] W. Jiang and V. K. Prasanna, "Sequence-preserving parallel IP lookup using multiple SRAM-based pipelines," *J. Parallel Distrib. Comput.*, vol. 69, no. 9, pp. 778–789, 2009.

[12] H. Yu and R. Mahapatra, "A power- and throughput-efficient packet classifier with n bloom filters," *IEEE Trans. Comput.*, vol. 60, no. 8, pp. 1182–1193, Aug. 2011.

[13] W. Jiang and V. K. Prasanna, "Large-scale wire-speed packet classification on FPGAs," in *Proc. FPGA*, 2009, pp. 219–228.

[14] I. Sourdis, "Designs & algorithms for packet and content inspection" Ph.D. dissertation, Comput. Eng. Div., Delft Univ. Technol., Delft, The Netherlands, 2007. [Online]. Available: http://ce.et.tudelft.nl/publicationfiles/ 1464_564 sourdis phdthesis.pdf

[15] G. S. Jedhe, A. Ramamoorthy, and K. Varghese, "A scalable high throughput firewall in FPGA," in *Proc. FCCM*, 2008, pp. 43–52.

[16] Xilinx, Inc., San Jose, CA, "Xilinx Virtex-6 FPGA family,"2009.[Online].Available:www.xilinx.com/products/virtex6/

[17] Altera Corp., San Jose, CA, "Altera Stratix IV FPGA," 2009.[Online].Available:http://www.altera.com/products/devices/stratix-fpgas/ stratix-iv/

[18] D. E. Taylor, "Survey and taxonomy of packet classification techniques," *ACM Comput. Surv.*, vol. 37, no. 3, pp. 238–275, 2005.

[19] T. V. Lakshman and D. Stiliadis, "High-speed policy-based packet forwarding using efficient multi-dimensional range matching," in *Proc. SIGCOMM*, 1998, pp. 203–214.

[20] S. Singh, F. Baboescu, G. Varghese, and J. Wang, "Packet classification using multidimensional cutting," in *Proc. SIGCOMM*, 2003, pp. 213–224.

[21] P. Gupta and N. McKeown, "Classifying packets with hierarchical intelligent cuttings," *IEEE Micro*, vol. 20, no. 1, pp. 34–41, 2000.

[22] D. E. Taylor and J. S. Turner, "Scalable packet classification using distributed crossproducing of field labels," in *Proc. INFOCOM*, 2005, pp. 269–280.

[23] W. Eatherton, G. Varghese, and Z. Dittia, "Tree bitmap: Hardware/ software IP lookups with incremental updates," *SIGCOMM Comput. Commun. Rev.*, vol. 34, no. 2, pp. 97–122, 2004.

[24] Y. Luo, K. Xiang, and S. Li, "Acceleration of decision tree searching for IP traffic classification," in *Proc. ANCS*, 2008, pp. 40–49.

[25] A. Kennedy, X.Wang, Z. Liu, and B. Liu, "Low power architecture for high speed packet classification," in *Proc. ANCS*, 2008, pp. 131–140.

[26] Y. Luo, P. Cascon, E. Murray, and J. Ortega, "Accelerating Open Flow switching with network processors," in *Proc. ANCS*, 2009, pp. 70–71.

[27] J. Naous, D. Erickson, G. A. Covington, G. Appenzeller, and N. Mc Keown, "Implementing an Open Flow switch on the Net FPGA platform," in *Proc. ANCS*, 2008, pp

28

Localization through compressive sensing: A survey

A. Ali

TRENDS Lab, ITU University, Lahore, Pakistan

Email address:
anum.ali@live.com

Abstract: User mobile device or for wireless node detection localization is a primary concern not only in normal days but especially during emergency situations. There is variety of useful and necessary applications related to localization and it is an important technology playing critical role in wireless communication. The conceptual point of view is to sense the localization (coordinates of the user) from a specific region of interest (ROI). For reducing the complexity and increasing efficiency, the data samples for location sensing is limited in a term of taking sparsity of the detected signal in known transformed domain by taking fewer data samples. This whole phenomenon is called compressive sensing. This paper introduces this technology especially in location-sensing and discusses the present techniques.

Keywords: Cognitive Radio, Localization, Mobile Networks, Wireless Networks, Sparsity, Compressive Sensing, Signal Detection

1. Introduction

In mobile and wireless network architecture location and mobility management have been an important factor for many good reasons such as giving on time rescue services, in case of GPS; in none availability of satellite plane of view directing users to their destination and many other useful applications. Location-sensing or localization is an automatic means of position determination for the user through signal detection from their devices. Efficient location-sensing require sampling of fewer data blocks from received signal and in many cases continuous signal is not received, fewer or interrupted signal has been detected. From these few samples based on sparsity technique location estimation is performed.

This paper discusses different present techniques for localization of user through compressed sensing. Localization has gained its popularity in many domains including mobile ad-hoc and vehicular networks, robotics and Public Protection and Disaster Relief (PPDR) communication system. There are surveys [1][37] purely based on location-sensing techniques through trilateration methods, none at the moment were related to compressive-sensing for localization of user nodes.

The outline of this survey is as follows: Section 2 discusses the main challenges and parameters for accurate localization of the user. Also what were the drawbacks of non-compressive techniques previously used for location-sensing. Later in the section 3, compressive sensing is explained and the re-formulation of the location parameters in form of sparse values is explained. In section 4 focus will be on the effective algorithms for localization explaining the sparse techniques and recent developments to perform efficient localizations.

2. Localization Issues and Parameters

2.1. Linearizing Vs Non- Linear

For location estimation usually parameters are taken into 2 or 3D dimensional coordinates. In cellular network location parameters are taken within the network without the aid from external resources such as GPS. Mostly UE is known in normal days and if not known the parameters are detected from within the network. Not like GPS, cellular network localization parameters are detected from limited region of interest (ROI). While detecting signals the nodes may be moving generating time-stamped measurements. These parameters may be in non-linear coordinates. These parameters should be combined to form a trajectory leading to the user location.

For accurate and sparse calculations (discussed in next section 3) all parameters are converted to linear parameters. Geometric methods can locate an object distance and measurements. Before sparsity was not introduced multiple dimensional coordinates were used. Following kinds of nonlinear parameters exists

1) Lateration-Single Dimension
2) Trilateration-2D
3) Multi-lateration-3D

Every UE exhibits three or more parameters. For deploying compressive sensing, an efficient and less complex computation requires to convert all parameters in lateration to mere approximation values. The description is illustrated in the following figure 1.

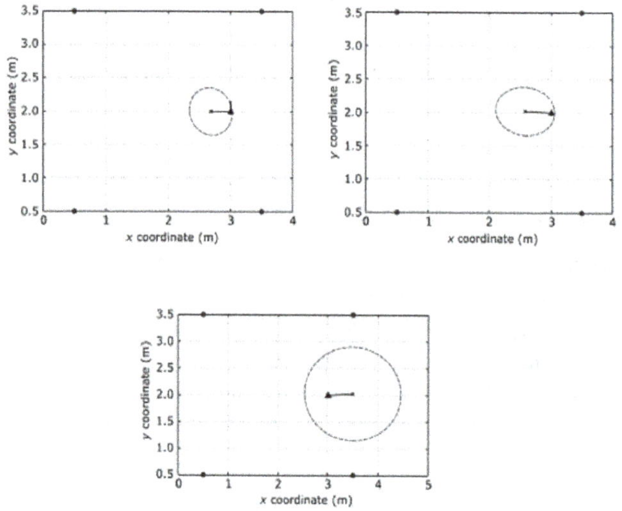

Figure 1. *Location-sensing coordinates measurement*

Location estimate in cellular/wireless networks are mostly processed through "RF finger prints" terminology. It is very similar to human finger printing. UE location-dependent signal parameters are extracted accordingly with their time-averages. There are two types of RF finger printing either reference or target [2]. The parameters obtained are unique set of geographic coordinates. A simple set of data matrix is shown below.

$$A = \begin{bmatrix} ID_1 & RSS_1 & RTD_1 \\ \ldots & \ldots\ldots & \ldots\ldots \\ ID_{n-1} & RSS_{n-1} & RTD_{n-1} \end{bmatrix} \quad (2.1)$$

By using number of mathematical techniques such as Euclidean distance or Sum of Absolute Difference (SAD) these three dimensional values compute the distances with reference with the adjacent or known coordinate as shown in following equation

$$d_{i,j} = \sqrt{\sum_{m=1}^{N} \left(\left\lfloor \frac{S_{i,j}'(n_m, 2) - A(m, 2)}{\delta} \right\rfloor \right)^2} \quad (2.2)$$

Here in above equation $S_{i,j}$ is the N-dimensional RSS space. As discussed above for localization many non-linear parameters were considered and computed. As the paper concentrate on compressive sensing effectiveness for localization rather than using trilateral coordinate system, further in next section sparsity in compressive sensing methodology is discussed in details.

3. Fundamentals of Compressive Sensing

3.1. Sparse Representation

Taking the location parameters from region of interest (ROI) and re-formulating it in *l*-minimization matrix for compressive sensing on the data is termed as sparse representation. The reason to apply sparse transformation for location estimation is due to the in-efficiency of location-sensing technology those require computation on large amount of data that cost overhead to its management and require high budget for hardware and software. Re-formulation in compressive sensing provides fundamentally advance approach for cost-effective and time-consuming solutions.

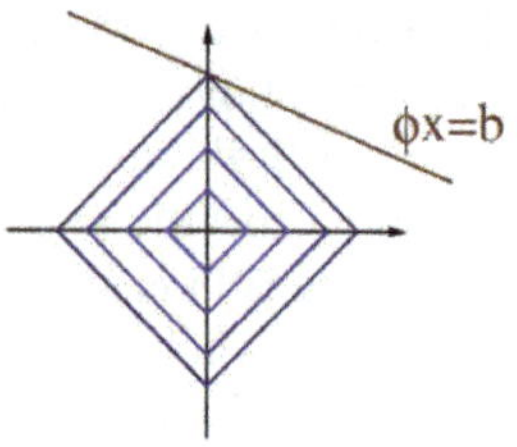

Figure 2. *L-minimization [3]*

By using fewer samples in linear domain compressive sensing implies sparsity. The explanation can be put forwarded as having a unknown signal vector $£^N$, it is sampled using n functions for linearizing and later reconstruct it, where n < N when signal space is bigger than measurements. Nearly from mid-eighteen it has been researched that minimization on l-norm can recover sparse measurements as illustrated in above figure. Usually a sparse matrix derived from discrete-time domain signal is represented as follow

$$\Theta = \{\Omega_i\}_{i=1}^{M} \quad (3.1)$$

$$\Psi = \Theta e \quad (3.2)$$

The discrete signal is represented as S^M and e is M x1 column vector of weighted product of co-efficient

$$e_i = \langle \Psi, \Omega_i \rangle = \Omega_i^T \Psi \quad (3.3)$$

The "T" symbol denotes transpose of the sparse vector. The above equation is the sparse representation of signal. Only the basis linear combination of *k* vectors are considered, meaning their values are the most significant such as

$$\begin{aligned} &\text{if } e_i \Rightarrow k \neq 0 \\ &\text{then } M - k = 0 \quad (k < M) \end{aligned} \quad (3.4)$$

$$M = (k \log N / k) \quad (3.5)$$

Where e_i is the linear projection of *M* signal, having *N* as intermediate acquiring samples. In a matrix if most of the elements are non-zero then the matrix or vector is considered dense not sparse matrix. Transform coding is successfully processed on the data samples those are k-sparse signals through compressive sensing. This framework is considered incoherent and represented as sparse representation. There exists number of different techniques for sparsing the data such as wavelet transformation, Logan phenomena, Lasso, the matching pursuit and least absolute shrinkage. Using not all signal samples but only few intervals is actually sparsity of signal where the sample is most weighted one.

3.2. Compressive Sensing vs Data Compression

There are two types of compression lossless and lossy. Compression sensing and data compression are two very different technologies. Before discussing compressive sensing in detail, the difference between two techniques should be well cleared. Data compression is a methodology of discarding and reducing data for increasing bit storage. There are number of different models and coding techniques for performing data compression.

Compressive sensing (CS) is very similar to transform coding, involving large amount of data. Transforming code process input signals into dense form of high dimensional space. The signal is sampled into sparsity form in a known transform domain. By sparsity it is meant, the matrix having samples of most weighted coefficients of a received signal that through transformation becomes zero.

3.3. Spatial Sparsity in ROI

Incoherency and sparsity are the two main pillars on which CS relies. High-dimensional signals especially trilateral coordinates for localization can easily be presented using few small set of variables and co-efficient through sparsity as shown in the figure 3.

4. Present Techniques

This section discusses present algorithms and methodologies for localization of user nodes through compressive sensing. As discussed in above sections the significance of sparsity theory over certain old techniques like FFT and Nyquist sampling theorem. In the following algorithm [5], a pair-wise distance measured matrix is derived by using sparsity. The central node only transmit small noisy compressive signal and a pair-wise matrix is constructed from those samples. CS uses l-minimization matrix to find pair-wise matrix through sparsing. By applying l-minimization algorithm, a sparse pair-wise distance matrix is reconstructed for learning locations of nodes. Suppose $S_k \in ROI^n$ is a sparse matrix S having pair-wise distance values. In the matrix each value is a two dimensional location vector as expressed in equation 4.1.

$$S_k = [S_{k1}, S_{k2}, \ldots\ldots S_{kn}] \qquad (4.1)$$
$$\|S_{ki}\|_0 \leq k, i \in \{1, 2, \ldots.., n\}$$

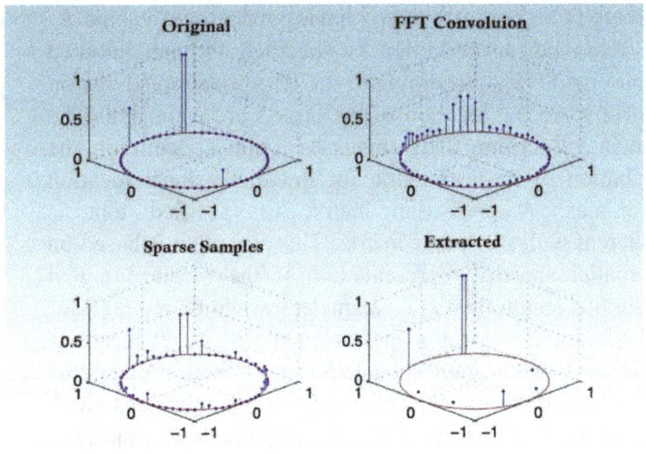

Figure 3. *Sparse Data Samples*

Further on three steps are performed on the matrix, in step 1, Floyd or Dijkstra path algorithm is applied on the values to recover sparse pair-wise values. In step 2, MDS algorithm is implied on the resultant matrix S`. The output from step 1 and 2 gives 3D relative coordinates of the nodes. Since this technique derive 3D coordinates for single node, the next techniques uses compressive sensing to derive the location of multiple points. The next popular technique [8] was evolved for missile launch system not for PPDR or public service schemes. The algorithm is very simple and straight forward by approaching the problem through Received Signal Strength (RSS) parameters. The RSS values are stored in a sparse matrix for pin-pointing the multiple location targets. The locations are then extracted from the sparse values through l-minimization matrix technique. Like previously discussed techniques that measured the k-sparse representations, instead RSS measurements in M-dimensional coordinates are measured accurately by convoluting with original received signal according to below equation

$$b = \Omega\Theta\xi + \varepsilon \qquad (4.2)$$

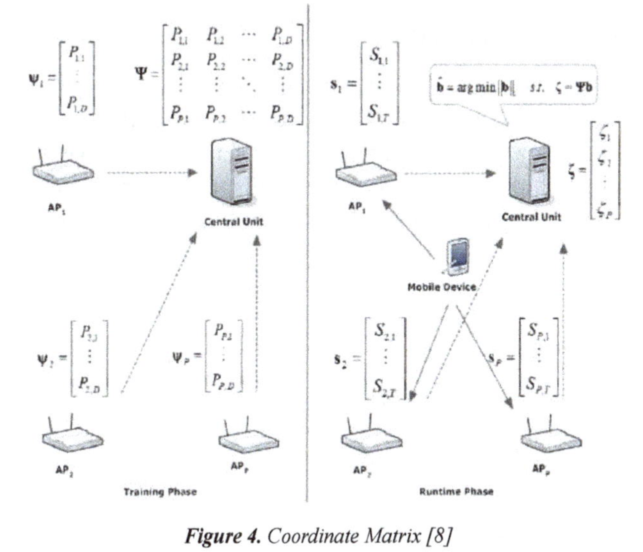

Figure 4. *Coordinate Matrix [8]*

Where Ω is the sparse matrix having sparse coefficients ε

A fixed power definition is specified through adopted channel model and according to the RSS matrix readings on the grid scale the location of the targets are estimated. The following algorithm [10] emphasize main concern on the distribution of the estimation for low-dimensional location coordinates. A projection matrix is specified that is incoherent with the sparse matrix. This algorithm is based on the spatial sparsity representation. If the received signal having of l length then $l+2$ parameters would be required for the estimation of location points on the M target locations having k sparse samples containing amplitudes of source signals referred as "Localization via spatial sparsity". Algorithms [7][9], [10] is based on both previous techniques [4][5] discussed above. Extracting the received signal strength and plotting over k sparse spree, a simple illustration is shown in figure 4.

A new technique is proposed in algorithm [12] defined as Greedy Matching Pursuit "GMP". GMP is an algorithm similar to OMP and CoSAMP [39] algorithms that could offer much better performance in regard to the unknown target locations from a measured signal. By adopting target energy decay model [40], [41] the states of signal energy received at certain location for pointed target from another location j is approximated as:

$$C_{ij} = \frac{J_0 G_{ij}}{d_{ij}^{\alpha}} \tag{4.3}$$

Here J_0 is the received signal intensity at i, d_{ij} is the derived distance from Euclidean formula between the know target location i with the required / estimated target location j, G_{ij} holds the Raleigh fading for the received target signal. Sparsity is implied on the resultant matrix, after getting sparse representation points on the grid, energy of the target signal will be highest where there are most of the targets resides.

Acknowledgements

I would like to acknowledge TRENDS Lab of ITU University, Lahore Pakistan for their support and guidance in writing in this paper. Also I am thankful for ITU University administration to give access to their library resources for further literature review and allow me to use the material in research work.

References

[1] J. F. Jiang, G. J. Han, C. Zhu, Y. H. Dong, N. Zhang, "Secure localization in wireless sensor networks: A survey", Journal of Communications, vol.6, no.6, pp.460-470, 2011.

[2] Reza Zekavat, R. Michael Buehrer, "Handbook of Position Location: Theory, Practice and Advances," ISBN: 978-0-470-94342-7

[3] Olga V.Holtz, "Compressive sensing: a paradigm shift in signal processing", Dec, 2008

[4] Richard G. Baraniuk, "More Is Less: Signal Processing and the Data Deluge",DOI: 10.1126/science.1197448 , 717 (2011); 331 Science

[5] Chen Feng, Shahrokh Valaeel, Zhenhui Tan Department of Electrical and Computer Engineering, University of Toronto, State Key Laboratory of Rail Traffic Control and Safety, Beijing Jiaotong University, "Localization of wireless sensors using compressive sensing for manifold learning," IEEE The 20th Personal, Indoor and Mobile Radio Communications Symposium, PIMRC, 2009.

[6] C. Feng, S. Valaee, and Z. Tan, "Multiple target localization using compressive sensing," in GLOBECOM'09: Proceedings of the 28th IEEE conference on Global telecommunications, 2009, pp. 4356-4361.

[7] J. geun Park, E. D. Demaine, and S. Teller, "Moving-baseline localization," in Proceedings of Information Processing in Sensor Networks (IPSN), 2008, pp. 15-26.

[8] S. Nikitaki and P. Tsakalides, ldquo, "Localization in Wireless Networks via Spatial Sparsity," Proc. Conf. Record of the 44th Asilomar Conf. Signals, Systems and Computers (ASILOMAR ',10), pp. 236-239, Nov. 2010.

[9] C. R. Berger , Z. Wang , J. Huang and S. Zhou "Application of compressive sensing to sparse channel estimation", IEEE Commun. Mag., vol. 48, no. 11, pp.164 -174 2010

[10] V. Cevher , M. F. Duarte and R. G. Baraniuk "Distributed target localization via spatial sparsity", 16th Eur. Signal Process. Conf., 2008.

[11] S. Nikitaki and P. Tsakalides, "Localization in wireless networks based on jointly compressed sensing," Proc. of European Signal Proc. Conf. (EUSIPCO), pp. 1809 - 1813, Aug.-Sept. 2011.

[12] B. Zhang, X. Cheng, N. Zhang, Y. Cui, Y. Li, and Q. Liang, "Sparse target counting and localization in sensor networks based on compressive sensing," in Proc. IEEE INFOCOM, pp. 2255-2263, 2011.

[13] Wael Guibène and Dirk Slock, "Cooperative Spectrum Sensing and Localization in Cognitive Radio Systems Using Compressed Sensing" Hindawi Publishing Corporation, Journal of Sensors, Volume 2013, Article ID 606413, 9 pages, http://dx.doi.org/10.1155/2013/606413

[14] W. Guibene and D. Slock, "A combined spectrum sensing and terminals localization technique for cognitive radio networks," in Proceedings of the IEEE 8th International Conference on Wireless and Mobile Computing, Networking and Comm's (WiMob '12), 2012.

[15] Sofia Nikitaki University of Crete & FORTH, Heraklion, Greece, Panagiotis Tsakalides University of Crete & FORTH, Heraklion, Greece, "Decentralized indoor wireless localization using compressed sensing of signal-strength fingerprints", PM2HW2N '12, Pages 37-44, ACM New York, NY, USA ©2012, ISBN: 978-1-4503-1626-2 doi>10.1145/2387191.2387198

[16] Gan, Ming; Guo, Dongning; Dai, Xuchu, "Distributed Ranging and Localization for Wireless Networks via Compressed Sensing", eprint arXiv:1308.3548, Publication Date: 08/2013

[17] Lanchao Liu, Zhu Han, Zhiqiang Wu, Lijun Qian, "Spectrum Sensing and Primary User Localization in Cognitive Radio Networks via Sparsity" , EAI Endorsed Transactions on Wireless Spectrum, Copyright © 2014, doi:10.4108/ws.1.1.e2

[18] R. M. Vaghefi and R. M. Buehrer, "Improving positioning in LTE through collaboration," in Proc. IEEE WPNC, 2014.

[19] Raja Jurdak, X. Rosalind Wang, Oliver Obst, and Philip Valencia, CSIRO ICT Centre, Australia, "Wireless Sensor Network Anomalies: Diagnosis, and Detection Strategies", A. Tolk and L.C. Jain (Eds.): Intelligence-Based Systems Engineering, ISRL 10, pp. 309–325.

[20] Sheenam, Navdeep Kaur, SBSTC, Ferozepur, India, "Improvement of Energy Efficiency of Compressive Sensing in Wireless Sensor Networks", ISSN 2348-5426 International Journal of Advances in Science and Technology (IJAST) Vol 2 Issue 2 (June 2014)

[21] W. Guibene and D. Slock, "Cooperative spectrum sensing and localization in cognitive radio systems using compressed sensing," Journal of Sensors, vol. 2013, Article ID 606413, 9 pages, 2013.

[22] K. Hayashi, M. Nagahara, and T. Tanaka, "A user's guide to compressed sensing for communications systems, " IEICE Trans. on Communications, vol. E96-B, no. 3, pp. 685-712, Mar. 2013.

[23] Joseph Lardies, Hua MA, Marc Berthillier. Source localization using a sparse representation of sensor measurements. Soci´et´e Fran¸caise d'Acoustique. Acoustics 2012, Apr 2012, Nantes, France. <hal-00810912>

[24] D.L Donoho and B. Logan, "Signal recovery and the large sieve," SIAM J. Appl. Math., vol.52, no.2, pp.577-591, April 1992

[25] S.G. Mallat, "A Wavelet Tour of Signal Processing", Third ed. The Sparse Way, Academic Press, 2008.

[26] P.Buhlmann and S. van de Geer, Statistics for High-Dimensional Data: Methods, Theory and Applications, Springer, 2011.

[27] S.G. Mallat and Z.Zhang, "Matching pursuits with time-frequency dictionaries," IEEE Trans. Signal Process., vol.41, no.12, pp.3397-3415, Dec. 1993

[28] R. Tibshirani, "Regression shrinkage and selection via the lasso," J.R. Statist. Soc. B, vol.58, no.1, pp.267-288, 1996

[29] J.L. Starck, F.Murtagh, and J.M. Fadili, Sparse Image and Signal Processing: Wavelets, Curvelets, Morphological Diversity, Cambridge University Press, 2010.

[30] J. Yoo, C. Turnes, E. Nakamura, C. Le, S. Becker, E. Sovero, M. Wakin, M. Grant, J. Romberg, A. Emami-Neyestanak, and E. Cand`es, "A compressed sensing parameter extraction platform for radar pulse signal acquisition," Submitted to IEEE J. Emerg. Sel. Topics Circuits Syst., February 2012.

[31] W. Dai, O. Milenkovic, Subspace pursuit for compressive sensing: Closing the gap between performance and complexity, available at: http://www.dsp.ece.rice.edu/cs/SubspacePursuit.pdf (preprint)

[32] I. F. Gorodnitsky and B. D. Rao, "Sparse signal reconstruction from limited data using FOCUSS: A re-weighted minimum norm algorithm," IEEE Transactions on Signal Processing, vol. 45, no. 3, pp. 600–616, 1997.

[33] V. Cevher, A. C. Gurbuz, J. H. McClellan, and R. Chellappa, "Compressive wireless arrays for bearing estimation," in IEEE Int. Conf. on Acoustics, Speech and Signal Processing (ICASSP), Las Vegas, NV, Apr. 2008.

[34] D. Malioutov, M. Cetin, and A. S. Willsky, "A sparse signal reconstruction perspective for source localization with sensor arrays," IEEE Transactions on Signal Processing, vol. 53, no. 8, pp. 3010–3022, 2005.

[35] D.Model and M. Zibulevsky, "Signal reconstruction in sensor arrays using sparse representations," Signal Processing, vol. 86, no. 3, pp. 624–638, 2006.

[36] A. C. Gurbuz, V. Cevher, and J. H.McClellan, "A compressive beamformer," in IEEE Int. Conf. on Acoustics, Speech and Signal Processing (ICASSP), Las Vegas, NV, 2008.

[37] Isaac Amundson and Xenofon D. Koutsoukos, "A Survey on Localization for Mobile Wireless Sensor Networks", R. Fuller and X.D. Koutsoukos (Eds.): MELT 2009, LNCS 5801, 2009, Pages: 235-254

[38] Guevara, J.; Jiménez, A.R.; Prieto, J.C.; Seco, F. Error Estimation for the Linearized Auto-Localization Algorithm. Sensors 2012, 12, 2561–2581.

[39] S. Foucart and H. Rauhut, A Mathematical Introduction to Compressive Sensing, Applied and Numerical Harmonic Analysis, DOI 10.1007/978-0-8176-4948-7 2, © Springer Science+Business Media New York 2013

[40] M. Ding, F. Liu, A. Thaeler, D. Chen, and X. Cheng, "Fault-tolerant target localization in sensor networks," in EURASIP J. Wirel. Commun. Netw., vol. 2007, no. 1, 2007, pp. 19–28.

[41] T. Clouqueur, K. K. Saluja, and P. Ramanathan, "Fault tolerance in collaborative sensor networks for target detection," in IEEE Transactions on Computer, vol. 53, no. 3, 2004, pp. 320–333.

[42] Marco F. Duarte, "Localization and Bearing Estimation via Structured Sparsity Models," IEEE Statistical Signal Processing Workshop (SSP), 2012, Ann Arbor, MI, pp. 333-336.

[43] Emamnuel J. Candès, "Compressive sampling", Applied and Computational Mathematics, California Institute of Technology, Pasadena, CA 91125, U.S.A

[44] Lei Liu Jin-Song Chong, Xiao-Qing Wang, and Wen Hong, "Adaptive Source Location Estimation Based on Compressed Sensing in Wireless Sensor Networks" International Journal of Distributed Sensor Networks, Volume 2012 (2012), Article ID 592471, 15 pages, http://dx.doi.org/10.1155/2012/592471

29

Implementation aspects in DFT modulated filter bank transceivers for cognitive radio

Nour Mansour, Dirk Dahlhaus

Communications Laboratory, University of Kassel, Kassel, Germany

Email address:
mansour@uni-kassel.de (N. Mansour), dahlhaus@uni-kassel.de (D. Dahlhaus)

Abstract: Discrete Fourier transform (DFT) modulated filter banks (FBs) are considered as strong tools used to implement both dynamic spectrum access and spectrum sensing in cognitive radio (CR) systems. High time-frequency (TF) resolution for spectral estimation and effective spectrum access with low complexity transceivers are the basic objectives in CR systems. However, the limitations of self-interference in DFT FBs as well as a primary user interference increase the overall transceiver complexity. In this paper, we design DFT modulated FBs which take into account the aforementioned contradicting requirements of high resolution capabilities, efficient spectrum access and affordable implementation effort for an additive white Gaussian channel. Four simple designs are presented and their performance are investigated and compared for a CR system with basic transmission parameters resembling those of IEEE 802.11g.

Keywords: Cognitive Radio, Filter Banks, Spectrum Access, Spectrum Sensing, Intersymbol Interference, Gabor System

1. Introduction

Intensive research activities are increasing rapidly in the field of cognitive radio (CR) due to the importance of the wireless spectrum in radio communications. To exploit this resource, unused slots of the spectrum by so-called prioritized primary users (PUs) are detected to be used by so-called secondary users (SUs). The SUs are allowed to use idle spectral resources given that the resulting interference experienced by the PU is limited, e.g. in the framework of underlay, overlay or interweave CR systems [1].

Spectrum sensing (SSE) [2], [3] requires good time-frequency (TF) resolution of the employed spectral estimation scheme to allow for subsequent spectrum access (SAC). The latter in turn benefits from good TF resolution of the employed waveforms due to limited availability of TF resources to be used by the SUs. The PU traffic patterns and the bit-error rate (BER) specifications of SU transmission together with spectral masks can be used to specify the conditions under which SSE and SAC schemes are designed. An efficient transceiver architecture for implementing SSE and SAC simultaneously should thus employ a common signal processing approach for both tasks. To this end, one option is to use discrete Fourier transform (DFT) modulated filter banks (FBs) where the spectrum under consideration is sensed by SU receivers (RXs) and potentially accessed by SU transmitters (TXs) in small portions of the TF plane.

A critical issue in the design of FBs is the TF correlation of FB output signals. Mutually orthogonal pulses in the TF plane used for inner products in the FBs can be constructed using DFT FBs which represent implementations of Gabor systems [4]. In case of mutually orthogonal pulses, zero intra-band and cross-band intersymbol interference (ISI) of SU signals as well as high accuracy for spectrum estimation can be achieved under certain conditions. Here, we can distinguish two cases. Firstly, one can consider so-called critically sampled DFT FBs [4]. Riesz bases in the Hilbert space of square-summable time signals $L^2(\mathbb{Z})$ with good TF concentration properties of the resulting pulse to be used in the FB, however, cannot be constructed according to the Balian-Low theorem [5]. Secondly, under-critically sampled FBs are described in [6]. In this case, the pulse TF concentration can be improved at the expense of a potential loss in transmission rate due to the incompleteness of the corresponding Gabor frame and the correspondingly missing perfect reconstruction property. In [6], a high implementation complexity is required to construct a pulse $\mathbf{g}=\left[g[0],\dots,g[L-1]\right]^{\mathrm{T}}$ consisting of L components in the time domain with $L\in\mathbb{N}$ in a so-called paraunitary over-critically sampled FB under the lattice being dual to the sought for Gabor system. Besides, a semidefinite programming to solve a semidefinite relaxation of the original optimization problem with a number of constraints [7] has to be implemented with an additional potential final rank reduction method.

In this paper, an alternative approach to the design of a DFT FB transceiver for SSE and SAC is proposed which aims at a system-specific approach taking into account SU BER specifications and PU interference and simultaneously can be implemented with limited complexity. Here, we consider for simplicity transmission over an additive white Gaussian noise (AWGN) channel. The main idea is to find a pulse $\mathbf{g}$ to be used in both SU TXs and SU RXs that result from constrained optimization. The objective function is the maximization of the pulse TF concentration where different concentration measures are applied. The constraints contain BER specifications of the SU taking into account intra-band and cross-band ISI from SU signals as well as third-party interference including the PU signal modeled as an additive white Gaussian process. Therefore, as long as the constraints are met for given values of the signal-to-noise ratio (SNR) and the interference based on the chosen objective function, no further optimization in the sense of achieving a global optimum of the objective function is required.

Four different constrained optimization approaches for optimizing the SSE performance and meeting a specific BER are presented [8]. The concentration of $\mathbf{g}$ at a certain point in the TF plane is measured using different dispersion metrics. We consider the minimization of a heuristic dispersion measure as well as the minimization of leakage in TF as well as separately in time or frequency based on the Rihaczek distribution. In addition to [8], we show a practical application of the design in the field of wireless local area networks (WLANs).

The paper is organized as follows. Sect. 2 describes the system model, the modulated DFT FBs used in the SU transceiver and the interference experienced by the SU RX. In Sect. 3, we discuss the transceiver design including the pulse optimization based on different leakage metrics for the four approaches. In Sect. 4, an overview of practical implementation aspects of the proposed transceiver is given based on the WLAN standard IEEE 802.11g. Sect. 5 shows pulses resulting from the optimization in Sect. 3 and their corresponding TF energy distribution. Furthermore, simulations are carried out to characterize the BER performance of the DFT FB transceiver in different interference environments. Finally, conclusions are drawn in Sect. 6.

Throughout the paper boldfaced characters are used for vectors and matrices. Furthermore, $\mathbf{E}\{.\}$, $\|\cdot\|$, $\mathbf{X}^{\mathrm{T}}$, a^*, $\langle .,. \rangle$, $|.|$, Diag[**a**] and $\lceil \cdot \rceil$ denote expectation, the Euclidean norm, transposition of matrix $\mathbf{X}$, complex conjugation of a complex number a, the inner product, the absolute value, a diagonal matrix composed of the elements of the vector $\mathbf{a}$ and the ceiling function, respectively.

2. System Model

2.1. Frames and Filter Banks

The authors in [4] derive the equivalence of DFT modulated FBs and Gabor frames in form of a relation between Gabor analysis/synthesis windows and the analysis/synthesis prototype filters of DFT modulated FBs. Here, we make use of this equivalence to describe the FBs.

The properties of a discrete-time signal $x[n]$ can be characterized with respect to time and frequency in the context of a DFT FB which contains a so-called prototype filter that is characterized by its discrete-time impulse response $g[n]$. We define a set $\{g_{\ell,k}[n] = g[n-\ell N]\, e^{j2\pi(n-\ell N)k/K} : (\ell,k) \in \Lambda\}$ in $L^2(\mathbb{Z})$ as a Gabor system which constitutes a set of functions derived from $g[n]$ by time shifts ℓN and frequency shifts k/K with $(\ell,k) \in \Lambda$ and $\Lambda = \mathbb{Z} \times \{0,\ldots,K-1\}$ where K and N are positive integers chosen according to the system specifications. Using $\{g_{\ell,k}[n]\}$, two tasks can be accomplished. Firstly, under the assumption of a complete set $\{g_{\ell,k}[n]\}$, the channel state information can be acquired by SSE, where a received signal $y[n]$ is projected onto $g_{\ell,k}[n]$ by $y_{\ell,k} = \langle y[n], g_{\ell,k}[n] \rangle = \sum_{n\in\mathbb{Z}} y[n] g^*_{\ell,k}[n]$. Secondly, SAC can be done by transmitting data symbols $b^{(\ell,k)}$ at time ℓ and at frequency k in form of a signal $b^{(\ell,k)} g_{\ell,k}[n]$ with $b^{(\ell,k)}$ being drawn from a suitable symbol alphabet. Here, we consider a $\pi/4$-differential quaternary phase-shift keying ($\pi/4-$DQPSK) modulation. Consequently, both sensing and demodulation can be accomplished using the given receiver structure with identical analysis and synthesis windows of the Gabor system. As a result of the aforementioned equivalence, a Gabor system can be implemented using a modulated DFT FB with K channels, where in each channel the signal is filtered and down-sampled by a factor N. Furthermore, the parameters K and N are chosen to be equal in order to exploit the complete capacity offered by the channel of SAC and offer both orthogonal pulses if required as well as perfect reconstruction ability of the resulting DFT FB [8].

2.2. Spectrum Sensing and Access

The two important phases of CR, SSE and SAC, are modeled in Fig. 1 where we assume firstly that the switch is in the SSE position. In this case, the input of the SU RX contains a thermal noise modeled as complex zero-mean AWGN with variance σ_υ^2 as well as potential interference resulting by third parties including the PU signal. Thus the input signal $y[n] = \upsilon[n]$ is represented at the output of the DFT analysis FB whose filters have the $z-$transform

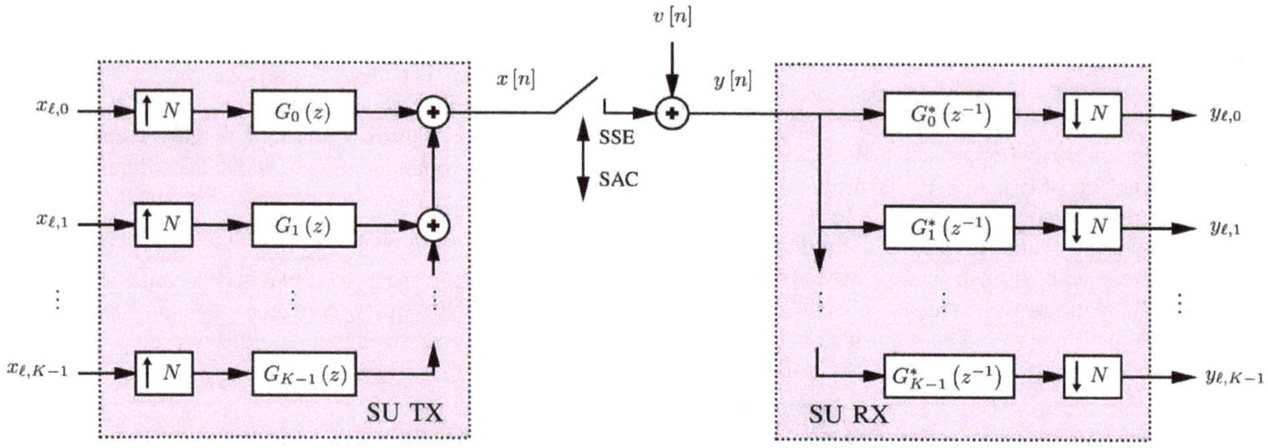

Figure 1. *System model with DFT modulated FBs.*

$G_k^*(z^{-1})$ with

$$G_k(z)=G(zW^k),\ k=0,\ldots,K-1$$

where $W=\mathrm{e}^{-j\frac{2\pi}{K}}$ and $G(z)=\sum_{n=0}^{L-1} g[n]\, z^{-n}$ is the $z-$ transform of $\mathbf{g}$. To this end, the signals are downsampled by a factor N and projected onto the corresponding pulse to provide the TF projections $y_{\ell,k}$ assumed to be jointly Gaussian distributed where the elements of the vector $\mathbf{y}_\ell=[y_{\ell,0},\ldots,y_{\ell,K-1}]$ at the output of the analysis FB are given by

$$y_{\ell,k}=\sum_{n\in\mathbb{Z}} v[n]\, g^*[n-\ell N]\, \mathrm{e}^{-j2\pi(n-\ell N)k/K}$$

$$=\sum_{n\in\mathbb{Z}} v[n] g^*[n-\ell N]\, \mathrm{e}^{-j2\pi nk/K}$$

To determine whether $y_{\ell,k}$ at the kth frequency and ℓth time slots is occupied for transmitting the PU signal or available for the SU to transmit its data, e.g. in interweave CR, a threshold test is performed usually in combination with a suitable prediction. Secondly, we assume for simplicity that the aforementioned SSE has indicated availability of all subbands for SU transmission. In this case, the switch in Fig. 1 is in the SAC position and the SU TX vector $\mathbf{x}_\ell=[x_{\ell,0},\ldots,x_{\ell,K-1}]$ is upsampled by a factor N and the signal $x[n]$ is synthesized by K synthesis filters with $z-$ transforms $G_k(z)$ according to

$$x[n]=\sum_{k=0}^{K-1}\sum_{\ell\in\mathbb{Z}} x_{\ell,k}\, g[n-\ell N]\, \mathrm{e}^{j2\pi(n-\ell N)k/K}$$

$$=\sum_{k=0}^{K-1}\sum_{\ell\in\mathbb{Z}} x_{\ell,k}\, g[n-\ell N]\, \mathrm{e}^{j2\pi nk/K}$$

Upon reception of $y[\ell N+L-1]$, two tasks should be executed at the SU RX, namely detecting $\mathbf{x}_\ell$ and simultaneously taking a decision about the presence of a PU signal in the ℓth time slot [8]. Clearly, both tasks benefit from good TF concentration properties of $\mathbf{g}$ that will be discussed in Sect. 5.

2.3. Self-Interference

If the functions $\{g_{\ell,k}\}$ are not orthogonal in both time and frequency domains, the demodulation of $\mathbf{y}_\ell$ representing sufficient statistics for the detection of $\mathbf{x}_\ell$ in AWGN is subject to self-interference. We define $\hat{x}_{\ell,k}=y_{\ell,k}/\|\mathbf{g}\|^2$ as the estimated value of $x_{\ell,k}$ given by

$$\hat{x}_{\ell,k}=x_{\ell,k}+\rho_{\Omega_{\mathrm{IBI}}^{(\ell,K)}}+\rho_{\Omega_{\mathrm{CBI}}^{(\ell,K)}}+\rho_{\Omega_{\mathrm{RI}}^{(\ell,K)}}+z_{\ell,k} \tag{1}$$

where $\rho_\Omega=\sum_{(\lambda,\kappa)\in\Omega} x_{\lambda,\kappa}\,\rho_{\ell-\lambda,k-\kappa}$. Here, $\rho_{\lambda,\kappa}=\sum_{n=0}^{L-1} g[n]\, g^*[n-\lambda N]\|\mathbf{g}\|^{-2}\, \mathrm{e}^{-j2\pi n\kappa/K}$ denotes a crosscorrelation and $\rho_{\Omega_{\mathrm{IBI}}^{(\ell,K)}}$, $\rho_{\Omega_{\mathrm{CBI}}^{(\ell,K)}}$ and $\rho_{\Omega_{\mathrm{RI}}^{(\ell,K)}}$ are defined as the intra-band interference (IBI), the cross-band interference (CBI) and the residual interference (RI), respectively. Moreover, $z_{\ell,k}=\langle v[n],g_{\ell,k}[n]\rangle\|\mathbf{g}\|^{-2}$ defines the AWGN noise and the third party interference contributions in $\hat{x}_{\ell,k}$ with variance equal to $\mathbf{E}\{|z_{\ell,k}|^2\}=\sigma_v^2\|\mathbf{g}\|^{-2}=\sigma_z^2$. In (1), we define the sets $\Omega_{\mathrm{IBI}}^{(\ell,k)}$, $\Omega_{\mathrm{CBI}}^{(\ell,k)}$ and $\Omega_{\mathrm{RI}}^{(\ell,k)}$ as

$$\Omega_{\mathrm{IBI}}^{(\ell,k)}=\{(\lambda,\kappa):\lambda\in\Omega_{\mathrm{T},\ell},\ \kappa=k\}$$

$$\Omega_{\mathrm{CBI}}^{(\ell,k)}=\{(\lambda,\kappa):\lambda=\ell,\ \kappa\in\Omega_{\mathrm{F},k}\}$$

$$\Omega_{\mathrm{RI}}^{(\ell,k)} = \left\{(\lambda,\kappa) : \lambda \in \Omega_{\mathrm{T},\ell},\ \kappa \in \Omega_{\mathrm{F},k}\right\}$$

with index sets in frequency and time given by

$$\Omega_{\mathrm{F},k} = \{0,\ldots,K-1\} / \{k\}$$

$$\Omega_{\mathrm{T},\ell} = \{\ell-\Delta,\ldots,\ell-1,\ell+1,\ldots,\ell+\Delta\}$$

$$\Delta = \left\lceil \frac{L}{N} \right\rceil - 1.$$

By exploiting the central limit theorem, the IBI, CBI and RI contributions are modeled as zero-mean Gaussian random variables with variance [8]

$$\sigma_\rho^2 = \mathbf{E}\left\{\left|\rho_{\Omega_{\mathrm{IBI}}^{(\ell,k)}} + \rho_{\Omega_{\mathrm{CBI}}^{(\ell,k)}} + \rho_{\Omega_{\mathrm{RI}}^{(\ell,k)}}\right|^2\right\} = \sum_{(\lambda,\kappa)\in\Omega_{\mathrm{IBI}}^{(\ell,k)} \cup \Omega_{\mathrm{CBI}}^{(\ell,k)} \cup \Omega_{\mathrm{RI}}^{(\ell,k)}} \left|\rho_{\ell-\lambda,k-\kappa}\right|^2$$

Here, we consider $x_{\ell,k}$ to be mutually independent DQPSK symbols with unit bit energy. Fig. 2 illustrates the self-interference affecting on demodulating the signal $x_{0,3}$ and the corresponding sets for $\hat{x}_{0,3}$ [8].

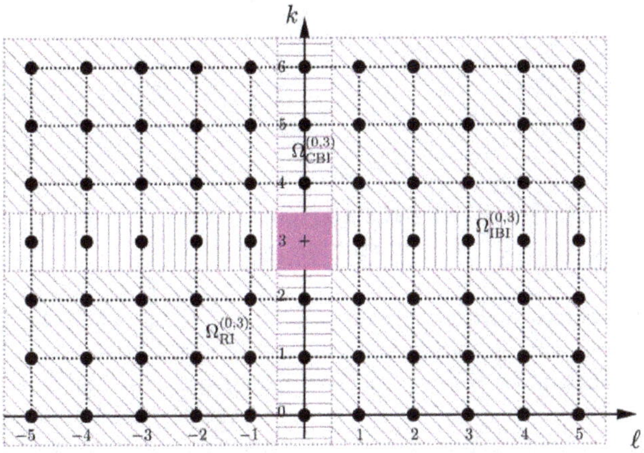

Figure 2. *Self-interference to* $x_{0,3}$ *with* $K=7$ and $\Delta=5$: *IBI, CBI and RI hatched vertical, horizontal and diagonal, respectively.*

3. Transceiver Design

3.1. Optimization Constraints

The main task in designing the transceiver in Fig. 1 is optimizing $\mathbf{g}$ subject to certain constraints. Here, two types of constraints are distinguished.

Firstly, we consider the constraints arising from system specifications formulated as a lower bound R_{min} on the data transmission rate R in bit/s and an upper bound $\bar{P}_{\mathrm{b,max}}$ on the average BER $\bar{P}_{\mathrm{b}}$ of the SU in form of

$$R \geq R_{\mathrm{min}} \tag{2}$$

$$\bar{P}_{\mathrm{b}} \leq \bar{P}_{\mathrm{b,max}}. \tag{3}$$

The constraint in (2) is met upon choosing an appropriate value K as explained in Sect. 4 independently of the form of $\mathbf{g}$. On the contrary, the constraint (3) depends on $\mathbf{g}$ and has therefore to be taken into account in the aforementioned optimization. In view of the Gaussian interference approximation in Sect. 2.3, the average BER reads [9]

$$\bar{P}_{\mathrm{b}}(\zeta) = \mathrm{Q}_1(a_-,a_+) - \frac{1}{2} I_0(a_- a_+) \exp^{\left[-\frac{1}{2}(a_-^2+a_+^2)\right]} \tag{4}$$

with $a_\pm = a_\pm(\zeta) = \sqrt{\left(2\pm\sqrt{2}\right)\zeta}$ and $\zeta = \dfrac{1}{\sigma_v^2+\sigma_\rho^2}$. Furthermore, $I_0(x)$ and $\mathrm{Q}_1(a,b)$ are the zeroth order modified Bessel function and the Marcum Q function [9], respectively. For a given value $\bar{P}_{\mathrm{b,max}}$ in (3), the constraint can be formulated as

$$\mathrm{C}_1 \quad : \zeta \geq \bar{P}_{\mathrm{b}}^{-1}(\bar{P}_{\mathrm{b,max}}) \tag{5}$$

The second type of constraints arises from implementation aspects. The norm of $\mathbf{g}$ is limited to one, i.e. we have

$$\mathrm{C}_2 \quad : \left\|\mathbf{g}\right\| = 1. \tag{6}$$

Finally, in order to obtain a $\mathbf{g}$ providing a system $\left\{g_{\ell,k}[n]\right\}$ of pulses being orthogonal in both time and frequency, a necessary condition reads

$$\mathrm{C}_3 \ : g[n] = g[L-n-1] \ \text{ for } n \in \{0,\ldots,L-1\} \tag{7}$$

in conjunction with $\mathbf{g} \in \mathbb{R}^L$ [10]. For later use, we collect the three constraints in (5), (6) and (7) symbolically in the vector

$$\mathbf{C}: \quad [\mathrm{C}_1, \mathrm{C}_2, \mathrm{C}_3].$$

3.2. Objective Functions

Our objective is the minimization of a dispersion metric of the pulse $\mathbf{g}$ subject to the constraints $\mathbf{C}$. Four different metrics are considered below.

3.2.1. Heuristic Dispersion Measure

A heuristic dispersion measure (HDM) is considered as the first approach [8]. Here, the time index n in $\mathbf{g}$ is interpreted as a random variable whose probability $\Pr(\theta = n)$ is defined by $\Pr(\theta = n) = g[n]^2 / \left\|\mathbf{g}\right\|^2$. In view of (6), we have $\Pr(\theta = n) = g[n]^2$ and choose the variance of θ as the dispersion metric, i.e.

$$\eta_{\mathrm{HDM}}(\mathbf{g}) = \sum_{n=0}^{L-1} n^2 \Pr(\theta = n) - \left[\sum_{n=0}^{L-1} n \Pr(\theta = n)\right]^2$$

$$= \mathbf{g}^{\mathrm{T}}\mathbf{D}\mathbf{g} - \left(\mathbf{g}^{\mathrm{T}}\mathbf{Z}\mathbf{g}\right)^2$$

with $\mathbf{D} = \mathrm{Diag}\left[0,\ldots,(L-1)^2\right]$ and $\mathbf{Z} = \mathrm{Diag}\left[0,\ldots,L-1\right]$. Thus the optimization problem based on the heuristic dispersion measure reads

$$\mathbf{g}_{\mathrm{HDM}} = \arg\min_{g\in L^2(\mathbb{Z})} \eta_{\mathrm{HDM}}(\mathbf{g}) \quad \text{s.t. } \mathbf{C}.$$

3.2.2. Time-Frequency Concentration

In order to obtain a pulse with TF concentration (TFC), firstly, we measure the energy of the pulse $\psi_{\Omega_{\Delta,\mathrm{T}},\Omega_{\Delta,\mathrm{F}}}(\mathbf{g})$ inside a given TF window $\Omega_{\Delta,\mathrm{T}} \times \Omega_{\Delta,\mathrm{F}}$ based on the Rihaczek distribution [6]. Secondly, the leakage of the pulse energy outside the TF window is to be minimized. Here, we choose

$$\Omega_{\Delta,\mathrm{T}} \times \Omega_{\Delta,\mathrm{F}} = \left\{\frac{L-\Delta_\mathrm{T}}{2}+1,\ldots,\frac{L+\Delta_\mathrm{T}}{2}\right\} \times \left[-\frac{\Delta_\mathrm{F}}{2},\frac{\Delta_\mathrm{F}}{2}\right),$$

where $L+\Delta_\mathrm{T}$ is assumed even, Δ_T is a positive integer chosen as $\Delta_\mathrm{T} = N$ and $\Delta_\mathrm{F} = 1/K$. The objective function now becomes

$$\eta_{\mathrm{TFC}}(\mathbf{g}) = 1 - \frac{\psi_{\Omega_{\Delta,\mathrm{T}},\Omega_{\Delta,\mathrm{F}}}(\mathbf{g})}{\|\mathbf{g}\|^2} = 1 - \psi_{\Omega_{\Delta,\mathrm{T}},\Omega_{\Delta,\mathrm{F}}}(\mathbf{g}),$$

where we define [6]

$$\psi_{\Omega_{\Delta,\mathrm{T}},\Omega_{\Delta,\mathrm{F}}}(\mathbf{g}) = \sum_{n\in\Omega_{\Delta,\mathrm{T}}} \int_{f\in\Omega_{\Delta,\mathrm{F}}} R_\mathbf{g}(n,f)\,\mathrm{d}f$$

with $\mathrm{R}_g(n,f) = g[n]\sum_{v\in\mathbb{Z}} g[v]\mathrm{e}^{j2\pi f(v-n)}$ known as the Rihaczek distribution [11]. The energy of the pulse can be formulated in a matrix form as $\psi_{\Omega_{\Delta,\mathrm{T}},\Omega_{\Delta,\mathrm{F}}}(\mathbf{g}) = \bar{\mathbf{g}}^\mathrm{T}\mathbf{S}\bar{\mathbf{g}}$ where $S = \{S_{n,v}\} = \{\Delta_\mathrm{F}\,\mathrm{sinc}(\Delta_\mathrm{F}\pi(n-v))\}$ is a $(L\times L)$- dimensional matrix and the reduced pulse $\bar{\mathbf{g}}$ is formulated as

$$\bar{\mathbf{g}} = \left\{0,\ldots,0,g\left[\frac{L-\Delta_\mathrm{T}}{2}+1\right],\ldots,g\left[\frac{L+\Delta_\mathrm{T}}{2}\right],0,\ldots,0\right\}^\mathrm{T}.$$

Therefore, the optimization problem based on pulse energy in TF domain is given by

$$\mathbf{g}_{\mathrm{TFC}} = \arg\min_{g\in L^2(\mathbb{Z})} \eta_{\mathrm{TFC}}(\mathbf{g}) \quad \text{s.t. } \mathbf{C}.$$

where $\eta_{\mathrm{TFC}}(\mathbf{g}) = 1 - \bar{\mathbf{g}}^\mathrm{T}\mathbf{S}\bar{\mathbf{g}}$.

3.2.3. Time Concentration

In a corresponding time concentration (TC) approach, the concentration is only considered in time while the frequency window $\Omega_{\Delta,\mathrm{F}}$ is extended to the interval $\Omega_\mathrm{F} = [0,1)$ [8]. Consequently, the optimization function for TC reads

$$\eta_{\mathrm{TC}}(\mathbf{g}) = 1 - \frac{\psi_{\Omega_{\Delta,\mathrm{T}},\Omega_\mathrm{F}}(\mathbf{g})}{\|\mathbf{g}\|^2} = 1 - \psi_{\Omega_{\Delta,\mathrm{T}},\Omega_\mathrm{F}}(\mathbf{g})$$

with $\psi_{\Omega_{\Delta,\mathrm{T}},\Omega_\mathrm{F}}(\mathbf{g}) = \sum_{n\in\Omega_{\Delta,\mathrm{T}}} \int_{f\in\Omega_\mathrm{F}} R_\mathbf{g}(n,f)\,\mathrm{d}f = \sum_{n\in\Omega_{\Delta,\mathrm{T}}} g[n]^2.$

The optimized problem is formulated as

$$\mathbf{g}_{\mathrm{TC}} = \arg\min_{g\in L^2(\mathbb{Z})} \eta_{\mathrm{TC}}(\mathbf{g}) \quad \text{s.t. } \mathbf{C}.$$

where $\eta_{\mathrm{TC}}(\mathbf{g}) = 1 - \|\bar{\mathbf{g}}\|^2$.

3.2.4. Frequency Concentration

To formulate the frequency concentration (FC) metric, we extend the time window interval $\Omega_{\Delta,\mathrm{T}}$ to $\Omega_\mathrm{T} = \{0,\ldots,L-1\}$. The corresponding optimized problem becomes

$$\mathbf{g}_{\mathrm{FC}} = \arg\min_{g\in L^2(\mathbb{Z})} \eta_{\mathrm{FC}}(\mathbf{g}) \quad \text{s.t. } \mathbf{C}.$$

with

$$\eta_{\mathrm{FC}}(\mathbf{g}) = 1 - \frac{\psi_{\Omega_\mathrm{T},\Omega_{\Delta\mathrm{F}}}(\mathbf{g})}{\|\mathbf{g}\|^2} = 1 - \psi_{\Omega_\mathrm{T},\Omega_{\Delta\mathrm{F}}}(\mathbf{g}) = 1 - \mathbf{g}^\mathrm{T}\mathbf{S}\mathbf{g}.$$

4. Practical Implementation Based on the WLAN Standard IEEE802.11g

In this section we use the aforementioned DFT modulated FB transceiver designs to consider practical CR implementations with physical layer parameters resembling IEEE 802.11g. That is, we define the system parameters including the modulation scheme such that the required data transmission rate of IEEE 802.11g can be reached. For complexity reasons, DQPSK is used to avoid the need for channel estimation in differential detection of the received data symbols. In IEEE 802.11g, the minimum data rate in (2) to be supported by the CR system is $R_{\min} = 54$ Mbit/s. We assume a sampling rate $\alpha = 30$ Msample/s and choose $K = 512$. Clearly, in view of the 2-bits carried by one DQPSK symbol, the maximum data rate R results to 60 Mbit/s. For the aforementioned parameters, we can thus leave subcarriers at the frequency band boundaries unloaded in order to have a certain separation to neighboring bands. More accurately, we use the 462 subcarriers in the band center and leave 25 subcarriers at both band edges unloaded as shown in Fig. 3. The resulting duration of a symbol carrying 512 DQPSK symbols is about $17\,\mu s$.

Furthermore, we choose $L = 1536 = \bar{L}K$ as the length of our optimized pulse $\mathbf{g}$ with $\bar{L} = 3$. Note that the TF concentration of $\mathbf{g}$ being important for the SSE performance depends mainly on the TF resolution parameters $\Delta_T = 512$ and $\Delta_F = 1/512$. Concerning possible approaches for solving the optimization problems in Sect. 3, one has to realize that these problems are highly nonlinear. Thus, using interior-point (IP) and sequential quadratic programming (SQP) are powerful methods which can effectively solve our optimization problems easily and very reliably with a small number of iterations required for convergence [12], [13].

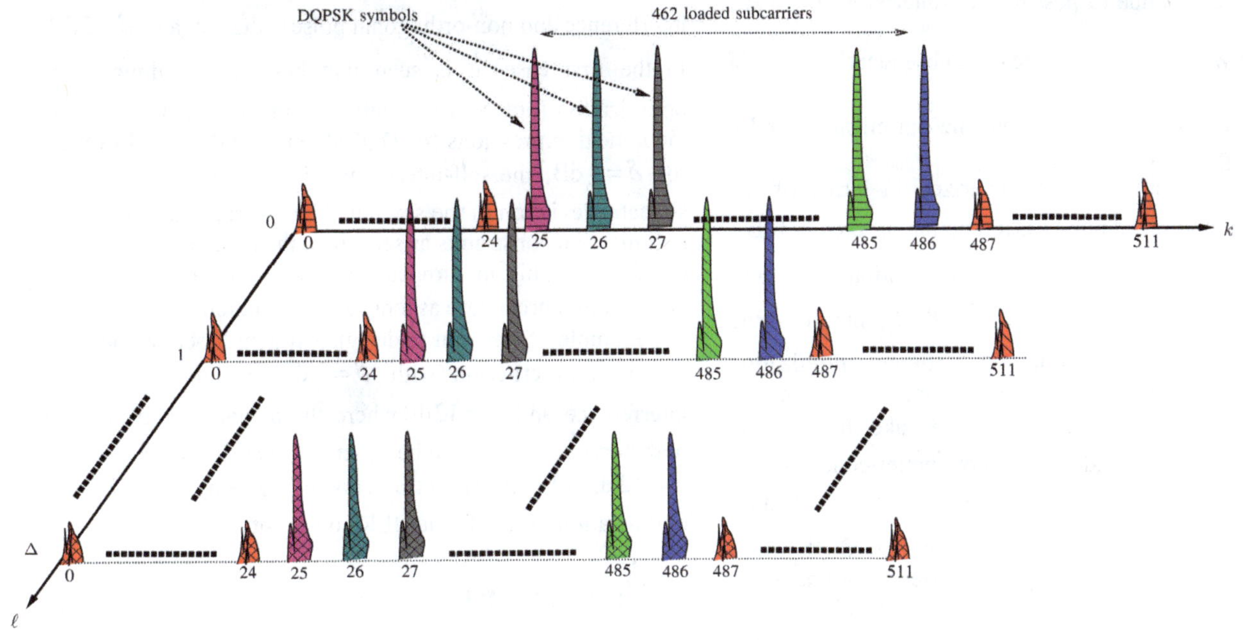

Figure 3. The loaded and unloaded subcarriers for reaching the data transmission rate of IEEE 802.11g.

Due to the large value of K, a polyphase structure for implementing the FBs at both transmitter and receiver provides a considerable saving in complexity [10]. To be more specific, we choose the kth analysis polyphase subfilter as $\mathbf{e}_k = \left[e_k[0],\ldots,e_k[\bar{L}-1]\right]^T$ being obtained from $\mathbf{g}$ by an $N-$fold downsampling with phase shift k, i.e.

$$e_k[n] = g[Nn+k].$$

If the transmitter and the receiver use a common hardware architecture, we can make use of the fact that the kth synthesis polyphase subfilter $r_k[n]$ satisfies $r_k[n] = e_{K-1-k}[n]$, so that the synthesis filter is the permutation of the corresponding analysis filter which leads to corresponding saving in filter implementation. Note that each of the polyphase subfilters has a number of coefficients $\bar{L} = 3$ which is a factor of K less than the number of coefficients in the original pulse $\mathbf{g}$.

Now we compare the computational complexity C measured in units of filter operations per second (FOPS) [14] required using the original FB in Fig. 1 with the complexity $\bar{C}$ due to applying the aforementioned polyphase structure. In the original FB implementation, the complex input signal is sampled at a rate α and three real multipliers are required to perform the multiplication of two complex numbers. Thus we have for the filtering

$$C = 3\alpha K L = 3\cdot 30\cdot 10^6\cdot 512\cdot 1536 \text{ FOPS } \approx 71\cdot 10^{12}\text{FOPS}.$$

By applying the polyphase structure, two real multipliers are required to perform the multiplication of the complex input signal and the real-valued coefficients of $e_k[n]$. Furthermore, the polyphase subfilters are computed at a sampling rate $\frac{\alpha}{N}$. Thus,

$$\bar{C} = \frac{\alpha}{N}\left(2K\bar{L} + \frac{K}{2}\,\mathrm{ld}\,K\right) \approx 315\cdot 10^6\,\text{FOPS},$$

where $\frac{K}{2}\,\mathrm{ld}\,K$ is the DFT implementation complexity. The complexity reduction is thus $\frac{C}{\bar{C}} = 225\cdot 10^3$.

5. Performance Analysis

The performance analysis is based on two parts. First, the energy concentration of the pulses governing the SSE performance is considered. Secondly, we study the SAC performance in terms of the achievable BER for a given pulse resulting from the constrained optimization in Sect. 3. Before starting the analysis, we introduce standard parameters, namely the signal-to-interference-plus-noise ratio (SINR)

$\zeta = \frac{1}{\sigma_\upsilon^2 + \sigma_\rho^2}$, the signal-to-noise ratio (SNR) $\gamma = \frac{1}{\sigma_\upsilon^2} = \frac{1}{\sigma_z^2}$, the signal-to-interference ratio (SIR) $\xi = \frac{1}{\sigma_\rho^2}$ and finally the loss due to possible self-interference and PU interference $\delta = \frac{\gamma}{\zeta} = 1 + \frac{\sigma_\rho^2}{\sigma_\upsilon^2}$. Note that subsequently, a certain value δ is assumed given for both the optimization and a corresponding simulation scenario.

The TF concentration is characterized in terms of the leakage of the optimized pulses as a function of $\delta \in [0, 3\text{dB}]$ and γ, where $\delta = 0$ dB represents a situation with zero self-interference, i.e. $\sigma_\rho^2 = 0$. For $\gamma = 8\text{dB}$, the resulting pulses are shown in Fig. 4 for the four different optimization approaches in Sect. 3. As can be seen, in time domain and for $\delta = 0\,\text{dB}$, both HDM and TC based pulses take shapes being close to a rectangular window and show better concentration within Δ_T than TFC and FC based $\mathbf{g}$. A coarse inspection of Fig. 4 suggests furthermore that for increasing δ, the pulse concentration decreases and the shapes differ according to the type and quantity of the interference term. To judge the TF concentration in greater detail, the absolute value of the frequency-discrete Rihaczek distribution $\left|\text{R}_{g,d}(n,f)\right| = \left|\int_{\frac{k}{K}-\frac{\Delta_\text{F}}{2}}^{\frac{k}{K}+\frac{\Delta_\text{F}}{2}} \text{R}_{g,d}(n,f)\,\text{d}f\right|$ in [dB] is illustrated in Fig. 5 for $\delta = 3\text{dB}$ and $\gamma = 8\text{dB}$. For the HDM and TC cases, the latter shows the best concentration in time domain and both of them have almost zero IBI and RI, but non-zero ICI due to the aforementioned allowed loss. Correspondingly, the FC case experiences zero ICI while IBI and RI are about the same as for the TFC case. Next, we consider the leakage values $\eta_\text{TC}(.)$, $\eta_\text{FC}(.)$ and $\eta_\text{TFC}(.)$ for $\mathbf{g}_\text{HDM}$, $\mathbf{g}_\text{TFC}$, $\mathbf{g}_\text{TC}$ and $\mathbf{g}_\text{FC}$ as a function of $\xi = \gamma$, i.e. $\delta = 3\text{dB}$, in Fig. 6. Clearly, all leakage measures decrease for increasing values of ξ due to a decreasing interference term. Apparently, as a direct consequence of the large value of K, the leakage $\eta_\text{FC}(.)$ of any pulse is very close to zero and according to the pulse design, the leakage $\eta_\text{TFC}(.)$ is the largest among all measured leakages for all pulses. Obviously, the objective function affects the leakage measure as can be seen from $\eta_\text{TC}(\mathbf{g}_\text{TC})$, $\eta_\text{TC}(\mathbf{g}_\text{HDM})$, and $\eta_\text{FC}(\mathbf{g}_\text{FC})$ for which the corresponding leakages are zero. The relevant leakage measure for SSE in the TF domain is clearly $\eta_\text{TFC}(.)$. Here, the least leakage metric is achieved by $\eta_\text{TFC}(\mathbf{g}_\text{TFC})$ which is almost identical to $\eta_\text{FC}(\mathbf{g}_\text{FC})$.

The BER performance is considered for two cases in Fig. 7, namely interference-free transmission with bi-orthogonal pulses and thus $\delta = 0\,\text{dB}$ as well as transmission with interference and non-orthogonal pulses where again $\delta = 3\text{dB}$. In the first case, it is seen that both $\bar{P}_\text{b}$ resulting from optimization (opt) and simulations (sim) are identical to the theoretical expressions for DQPSK in AWGN (4). However, for $\delta = 3\,\text{dB}$, the self-interference is not Gaussian anymore so that a deviation of the simulated BERs from the ones in the optimization constraints arises where the latter are higher than the former. This in turn can be taken into account in the optimization procedure as long as the deviation is less than δ. For example, if we want to design a system based on the TFC optimization criterion with $\delta = 3\text{dB}$ maximum allowable interference and $\gamma = 12\text{dB}$ where the deviation between the both curves in Fig. 7 can be observed to be 1.7 dB, a value of $\zeta = 12\,\text{dB} - 1.3\,\text{dB} = 10.7\,\text{dB}$ should be employed in the optimization to satisfy the BER constraint.

6. Conclusions

Trading implementation complexity in a cognitive radio transceiver against high-resolution spectrum sensing and minimum bit-error rate performance in spectrum access can be taken into account in the design of suitable DFT modulated filter banks. The system specifications translate into a constrained optimization procedure for finding corresponding prototype filter coefficients. If the design is applied to a system with characteristics similar to IEEE802.11g, the system specifications can be met by properly treating interference phenomena arising from both third parties as well as from self-interference in the filter bank. The transceiver complexity can benefit from the polyphase implementation of DFT modulated filter banks.

The approach is currently extended to the case of time-/frequency selective fading channels [16] where the channel parameterization translates into properties of the corresponding TF signal representations, and the resulting pulses will be applied to indoor cognitive radio environments.

Acknowledgments

This work has been supported by the German Federal Ministry of Education and Research project *Entwurf einer Cognitive Radio-ARchitektur basierend auf Optimierten Zeit-FreqUenz-SignALdarstellungen* (CAROUSAL).

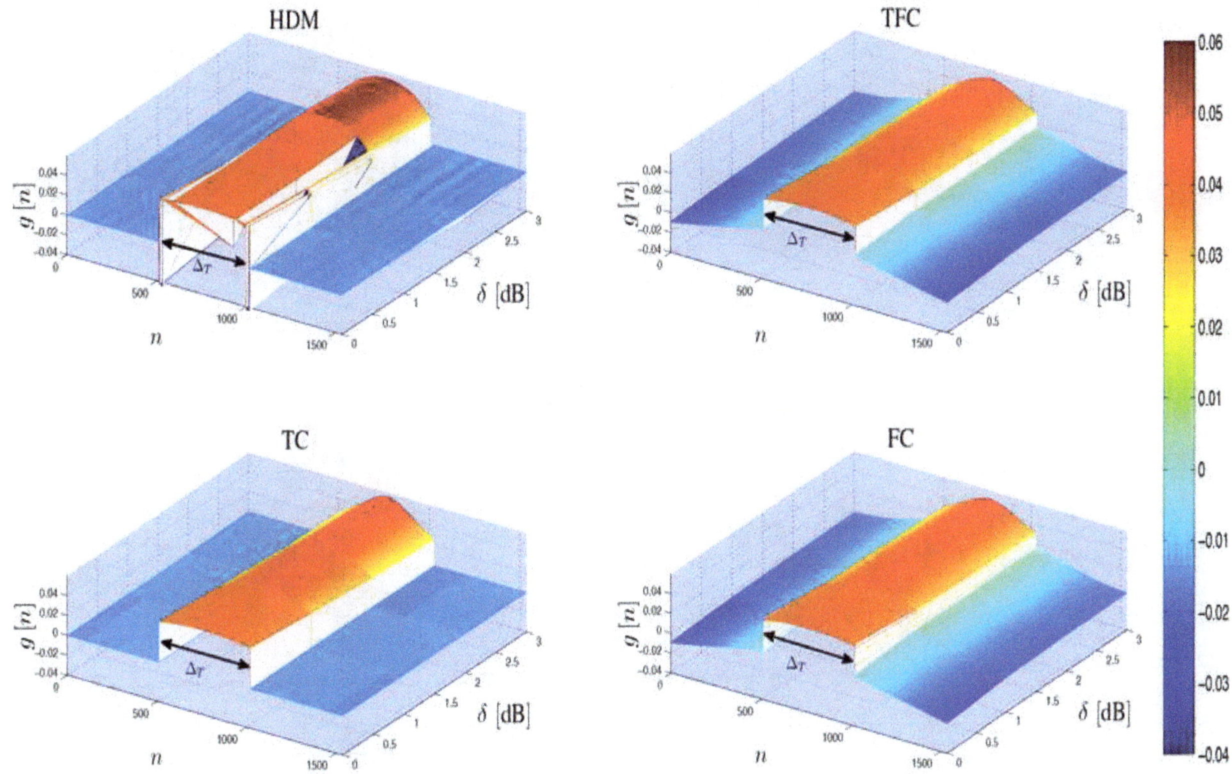

Figure 4. The optimized pulses in case of different values of δ *using HDM, TFC, TC and FC and* $\gamma = 8\,\mathrm{dB}$.

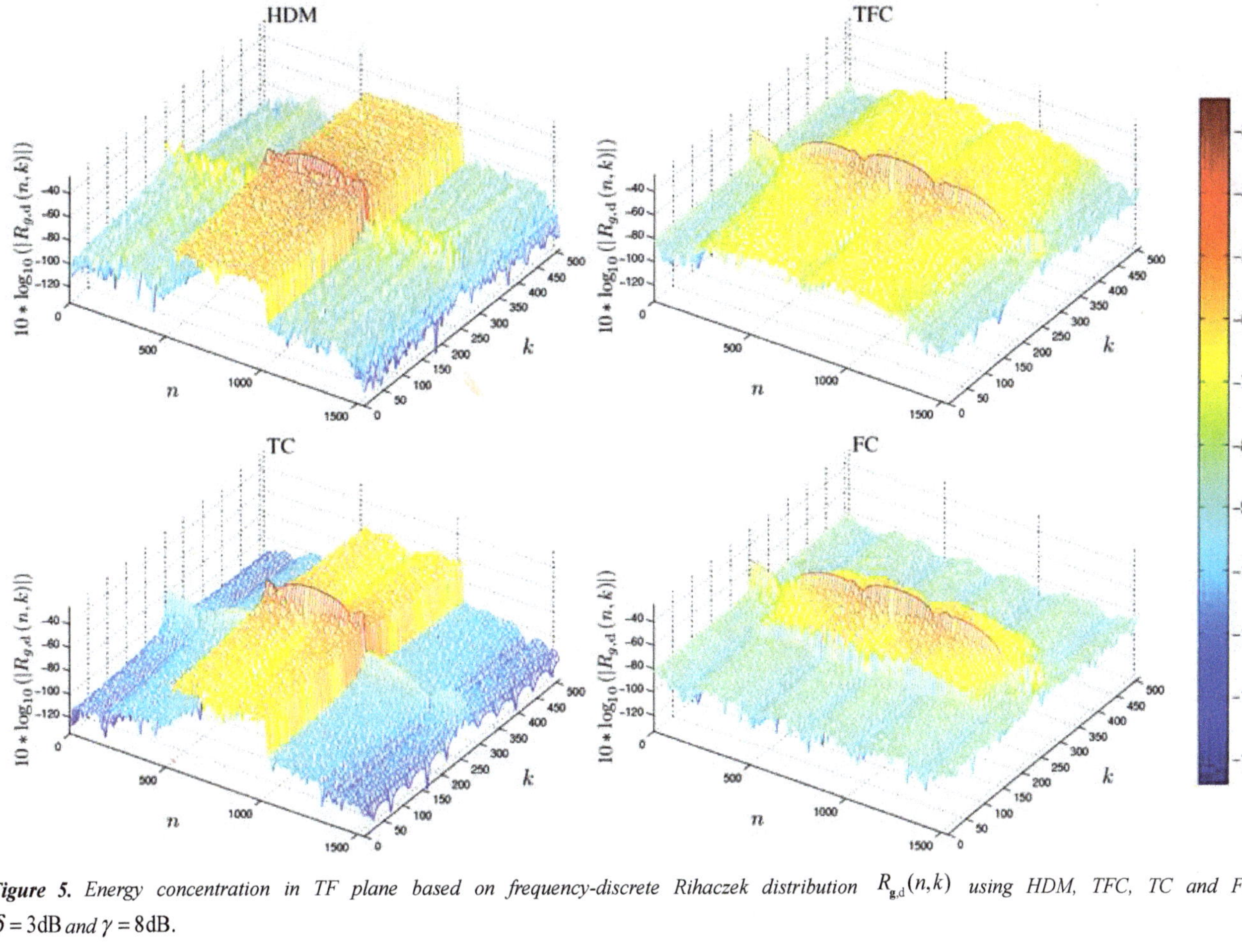

Figure 5. Energy concentration in TF plane based on frequency-discrete Rihaczek distribution $R_{g,d}(n,k)$ *using HDM, TFC, TC and FC for* $\delta = 3\mathrm{dB}$ *and* $\gamma = 8\mathrm{dB}$.

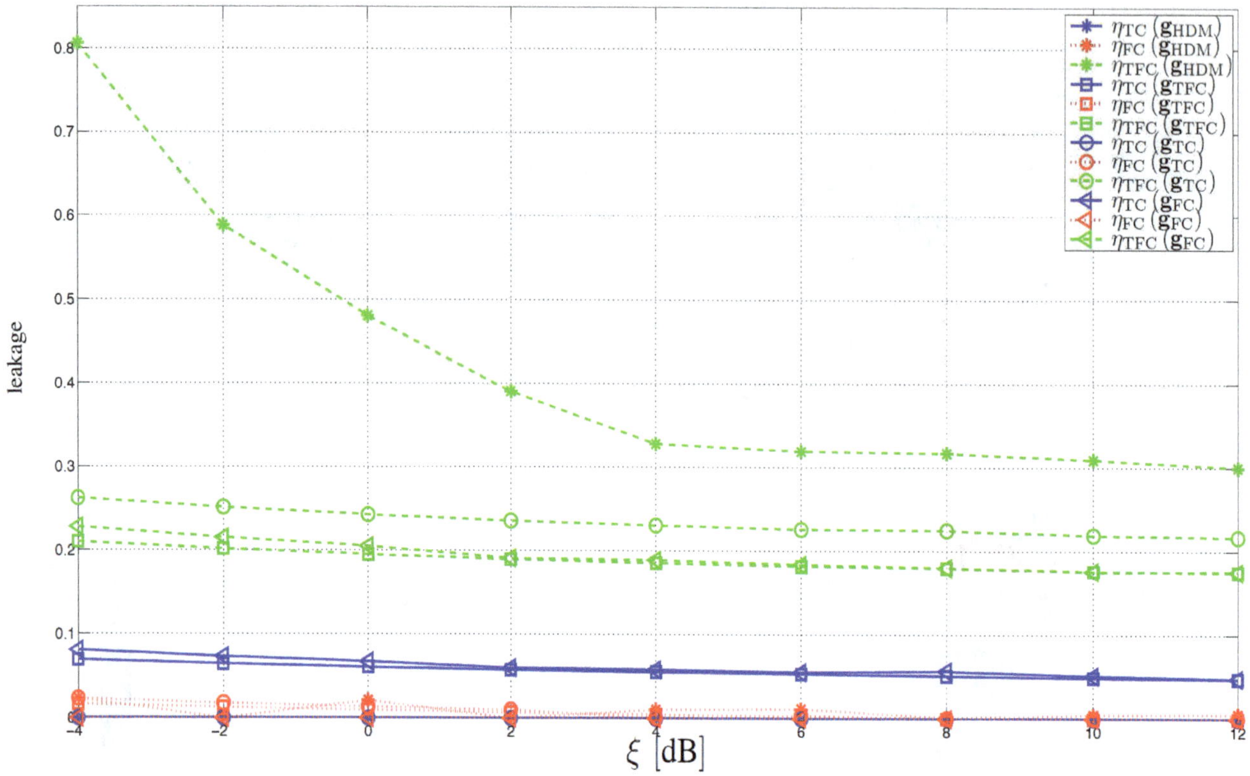

Figure 6. *Leakage values of* $\eta_{TC}(.)$, $\eta_{FC}(.)$ and $\eta_{TFC}(.)$ *for* $\mathbf{g}_{HDM}$, $\mathbf{g}_{TFC}$, $\mathbf{g}_{TC}$ and $\mathbf{g}_{FC}$ *as a function of* $\xi = \gamma$ *and* $\delta = 3\text{dB}$.

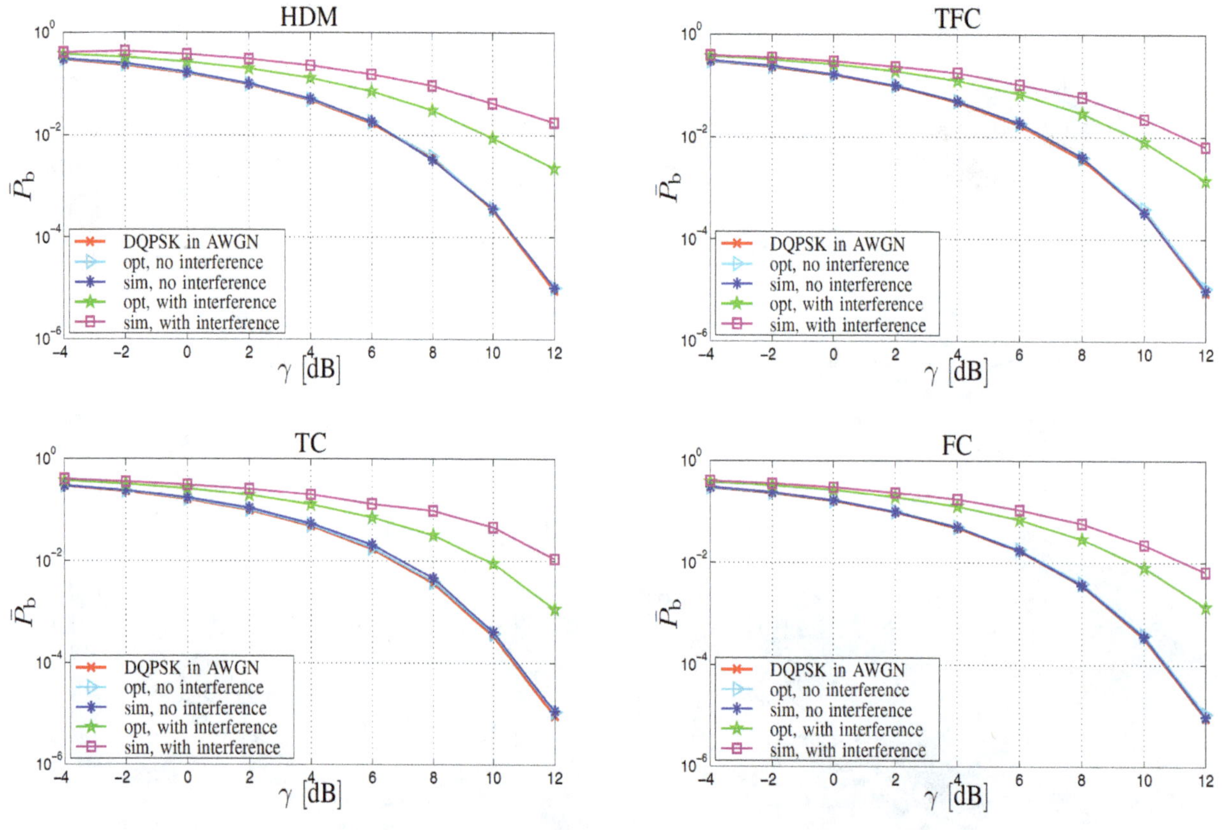

Figure 7. *BER performance in AWGN with DQPSK signaling in case of* $\delta = 0\,\text{dB}$ *(no interference) and* $\delta = 3\text{dB}$ *(with interference).*

References

[1] E. Biglieri, A. J. Goldsmith, L. J. Greenstein, N. B. Mandayam, and H. V. Poor, Principles of cognitive radio. Cambridge University press: Cambridge University Press, New York, 2013.

[2] W.-B. Chien, C.-K. Yang, and Y.-H. Huang, "Energy-saving cooperative spectrum sensing processor for cognitive radio system," circuits and systems I: Regular Papers, IEEE Transactions on, vol. 58, no. 4, pp. 711–723, April 2011.

[3] D. Joshi, D. Popescu, and O. Dobre, "Adaptive spectrum sensing with noise variance estimation for dynamic cognitive radio systems," in Information Sciences and Systems (CISS), 2010 44th Annual Conference on, March 2010, pp. 1–5.

[4] H. Bölcskei, F. Hlawatsch, and H. G. Feichtinger, "Equivalence of DFT filter banks and Gabor expansions," in Proc. of SPIE: wavelet applications in signal and image processing III, vol. 2569, July 1995, pp. 128–139.

[5] H. Feichtinger and T. Strohmer, Gabor analysis and algorithms: theory and applications, ser. Applied and numerical harmonic analysis. Birkhäuser Verlag GmbH, 2012.

[6] T. Hunziker, U. Rehman, and D. Dahlhaus, "Spectrum sensing in cognitive radios: Design of DFT filter banks achieving maximal time frequency resolution," in 8th International Conference on Information, Communications and Signal Processing (ICICS) 2011, Dec 2011, pp. 1–5.

[7] Z. Ju, T. Hunziker, and D. Dahlhaus, "Optimized paraunitary filter banks for time-frequency channel diagonalization," in EURASIP journal on advances in signal processing, vol. 2010, Dec. 2010.

[8] N. Mansour and D. Dahlhaus, "Interference in DFT modulated filter bank transceivers for cognitive radio," in European Wireless 2014; 20th European Wireless Conference; Proceedings of, May 2014, pp. 1–7.

[9] J. Proakis, Digital communications. McGraw Hill, New York, 4th ed., 2000.

[10] P.P. Vaidyanathan, Multirate systems and filter banks. Prentice Hall Signal Processing Series, New Jersey, 1993.

[11] B. Boashash, "Time-frequency signal analysis and processing," Queensland university, Brisbane, Australia.

[12] S. Boyd and L. Vandenberghe, Convex optimization. Cambridge UK: Cambridge University Press, New York, 2004.

[13] P. Gill, W. Murray, and M. Saunders, "SNOPT: An SQP algorithm for large-scale constrained optimization," SIAM Journal on Optimization, vol. 12, no. 4, pp. 979–1006, 2002.

[14] N.J. Fliege, Multirate Digital Signal Processing: Multirate systems, filter banks, wavelets, John Wiley & Sons, Chichester, 1994.

[15] S. Verdu, Multiuser detection. Cambridge University Press: Press Syndicate of the University of Cambridge, 1998.

[16] T. Hunziker, Z. Ju, and D. Dahlhaus, "Time-frequency channel parameterization with application to multi-mode receivers," IEICE Trans. Commun., pp. 3717–3725, Dec 2009.

Problem of Recognition of Hamiltonian Graph

Kochkarev Bagram Sibgatullovich

Department of Mathematics and Mathematical Modeling, Institute of Mathematics and Mechanics Named After Nikolai Ivanovich Lobachevsky, Kazan (Volga Region) Federal University, Kazan, Russia

Email address:
bkochkar@gmail.com

Abstract: In this article the author introduces the notions of combinatorial and of polynomial combinatorial sets in enumerative combinatorics. Formulates the problem of finding in combinatorial set of element with an easily recognizable property. The author proposes an efficient algorithm for solving this problem, which cancels known in the theory of algorithms abstract Turing, Church and Markov. We prove the criterion of polynomiality of the formulated problem. As a special case of this problem considers the problem of recognition of a Hamiltonian cycle in an undirected graph. We prove non-polynomiality this problem, which implies in particular the hypothesis of Jacques Edmonds *P≠NP*.

Keywords: Polynomial Problem, NP-problem, NPC-problem, Hamiltonian Cycle, Hamiltonian Graph

1. Introduction

In 1964 A. Cobham [1] and, independently, in 1965 J. Edmonds [2] introduced the concept of complexity class P.

Definition [1-2]. A problem (language) L belongs to P if there exists an algorithm A that decides (recognizes) L in a polynomial number of steps ($< O(n^k)$), where n is the length of the input, and k is some constant. The class of problem P is called polynomial (practical).

According to [3] J. Edmonds also introduced the complexity class NP. This class of problems (languages) that can be checked by polynomial algorithms.

Definition 2 [3]. A problem (language) L belongs to NP if there exists a two-input polynomial-time algorithm A and a polynomial $p(x)$ with integer coefficients that

$$L = \{x \in \{0,1\}^* : \textit{there exists a certificate } y \textit{ with } |y| \le p(|x|) \textit{ and } A(x,y) = 1\}$$

In this case we say that algorithm A verifies the solution x in polynomial number of steps of the length x.

According to definition 2, if L belongs to P and $|y| \le p(|x|)$, then $L \in NP$. But, if $L \in P$ and the length of certificate is not bounded from above by a polynomial of the length of x then $L \notin NP$ and P is not subset of NP. Note that in [3-5] $P \subseteq NP$, that is the erroneous statement. According to [3] J. Edmonds also expressed the hypothesis $P \neq NP$. In [6-10] we have constructed classes of polynomial problems with not polynomial certificates. According to the reasoning above this implies the positive solution of problem S. A. Cook [11] and the hypothesis of J. Edmonds $P \neq NP$. In definition 2 of the class NP there is a limit to the length of the certificate generated by the inspection algorithm, apparently caused by the algorithm for checking the correctness of solution of the classical algebraic problem of solving algebraic equation. It is well known that if x_0 is a solution of the equation $p(x) = 0$, where $p(x)$ is a polynomial from x with integer coefficients, then the verification of the correctness of the decision x_0 is the equality $p(x_0) = 0$. Obviously, if the solution is sought in a polynomial number of steps, then the validation also is carried out in a polynomial number of steps. Of course, the way to check the correctness of the solution of the problem depends on the nature of the problem. There are problems that have no solutions, and hence can't be checked, although the certificates to verify the correctness of the decision and there are. For example, in the thirties of the 19 th century Galois was able to prove [12] that for any $n \ge 5$ you can specify unsolvable by radicals of equation n degree with integer coefficients.

Naturally, according to definition 2 of the class NP, any polynomial problem belongs to NP, since the composition of polynomials is a polynomial. But if to remove restriction on length of the certificate that is explainable because the way of check of correctness of the decision depends both on a problem, and from properties of the decision which correctness is checked. Therefore the verification of decision may not be polynomial.

If we are interested not only a solution, but also check of its correctness, usually believe that the decision is a direct task, and check of its correctness – inverse. So finding of the cipher in a cryptosystem is a direct problem, and the decryption is the inverse problem. Therefore in cryptography it is very important that the transcript was algorithmically more difficult than finding the cipher. In this interpretation, a cryptographic system is the solution S. A. Cook's problem, proposed by the Institute Clay among the seven Millennium problems whose solution was presented in [6-10]. Although this problem was resolved already, we can say, seven years ago [7] and published on International conferences [6, 9] and in licensed journals [13-14] and notified to the Institute Clay, but the result is still the Clay Institute not recognized and therefore in [15] this problem is considered not solved. Note that in [16-17] there are positive feedbacks on our work on this issue.

2. NP-complete Problems and Their Complexity

In 1971 S. A. Cook [18] in the class of *NP* allocated a subclass of the most difficult problems of the so-called *NPC* (*NP*-complete) problem and proved *NP* completeness of two specific problems. Independently the concept of *NP*-completeness has also introduced L. A. Levin [19], who also proved the *NP* completeness of several problems. Since then many authors have proved the *NP* completeness thousands of problems, but the nature of their complexity not officially clarified.

Definition 3 [3]. We will say that a language $L_1 \subseteq \{0,1\}^*$ polynomial-time reducible to language $L_2 \subseteq \{0,1\}^*, (L_1 \leq_P L_2)$, if there exists a computable in polynomial number of steps the function $f:\{0,1\}^* \to \{0,1\}^*$ such that for all $x \in \{0,1\}^*$ $x \in L_1$ if and only if $f(x) \in L_2$. We call the function f the reducing function, and a polynomial-time algorithm F that computes f is called a reducing algorithm.

Definition 4 [3]. A language $L \subseteq \{0,1\}^*$ is called *NP* complete, if

1. $L \in NP$, and
2. $L' \leq_P L$ for any $L' \in NP$.

According to the definition 3, 4 if some *NP* complete problem is polynomial, then all *NP* complete problems are polynomial, and Vice Versa, if some *NP* complete problem is not polynomial, then all *NP* complete problems are not polynomial. All attempts to prove, that some *NP* complete problem is polynomial were in vain. In [20] we built *NP* complete problem that is not polynomial.

In this paper we propose one method of proving this fact using enumerative combinatorics and we prove that the recognition problem undirected Hamiltonian graph, *NP* completeness of which was proven in 1972 R. M. Karp [21], is not polynomial.

Mathematical Institute Clay explains the problem of S. A. Cook [11] by the following example: let's say someone in a big company wants to make sure that in this company there is one of his friends. If he is told that his friend is sitting in the corner, then only a fraction of a second to make sure that it is so. However, if no such information, he will be forced to circumvent the entire room, examining visitors. Note that this example fails because in modern conditions this problem is easily solved, e.g., using a microphone. Another thing, if we want to find some object with easily verifiable (in polynomial time of the length of the input) a symptom among the many combinatorial objects.

Enumerative combinatorics [22] deal with counting the number of elements in the finite set S. In enumerative combinatorics, the combinatorial elements of set S have a simple combinatorial definition, and some additional structure. Shows that the set S contains many elements and the main question is to determine (estimate) their combinatorial number, not a search, for example, some special item. The problem, the solution of which we intend to present in this work is, precisely, to build an effective search algorithm is a special element among the elements of some combinatorial sets.

In enumerative combinatorics [22] is usually given an infinite class of finite sets S_i, where i runs over some set of indices I (the set of nonnegative integers N), and we want to count combinatorial number $|S_i|$ of elements in each S_i "simultaneously". The set S_i we will call combinatorial sets.

However, for our purposes it is sufficient to know the particular answer to the question: "$|S_i|$ polynomial or not from the i?" In the case of an affirmative answer to the question we have to prove that $|S_i| < O(i^k)$, where k is some constant.

Definition 5 [23]. Combinatorial set S_i is called polynomial if $|S_i| \leq O(i^k)$, where k is a constant; otherwise it is called non-polynomial.

Problem statement: let S_i combinatorial set and x some object from S_i with easy to verify (in a polynomial number of steps of $|x|$) characteristic (property) α. It is necessary to construct an algorithm that for $\geq |S_i|$ the number of steps finds the object (element) x. The following algorithm is proposed to solve the problem: we assume that a finite set S_i of combinatorial objects is in some capacity Ω. From Ω we derive successively without returning all objects to the last, inclusive. For the latter object we check that the properties α. If the last retrieved object has the property α, then the problem is solved. If this object does not have property α, then all of the extracted of Ω objects we returned in Ω and renewable the extraction process of Ω objects, but with each subsequent extraction of the object check whether the extracted object is α property or not. Obviously, for some $|S_i|+m, m\leq|S_i|$ step, the object will have the property α and the problem will be solved. From the constructed algorithm follows theorem.

Theorem [23]. If the combinatorial set S_i is polynomial, the formulated problem can be solved in polynomial number of steps, namely, the number of steps t of the constructed algorithm satisfies the inequalities $|S_i|\leq t\leq C|S_i|$, where C is some constant ≥ 1. If combinatorial set S_i is not a polynomial, then the problem is not polynomial. Obviously, if $|S_i|\geq k^i, k\geq 2$, then the combinatorial set S_i is not a polynomial.

The algorithm, which implies the above theorem obviously satisfies all the requirements of the intuitive notion of algorithm, but does not fit [23] the formal definition of the algorithm (for example, "Turing machine"). Therefore, the algorithm cancels famous Turing thesis: "any algorithm can build a Turing machine that is equivalent to a given algorithm". Due to the equivalence of a Turing other well-known formal algorithms (recursive function, normal algorithms) relevant abstracts are also voided.

The problem of finding a Hamiltonian cycle in a undirected graph were studied [3] for over a hundred years. Some Hamiltonian cycle of an undirected graph $G=(V,E)$ is a simple cycle that contains each vertex of V. A graph which contains some Hamiltonian cycle is called Hamiltonian; otherwise, the count is not Hamiltonian. However, not all graphs are Hamiltonian. Every bibartite graph with odd number of vertices is not Hamiltonian [3]. We [3] to formulate the Hamiltonian cycle problem, "does some given graph G Hamiltonian cycle? As a formal language: HAM-CYCLE={$G: G$ *is a Hamiltonian graph*}".

Let $G=(V=V_1\cup V_2, E)$ be a bipartite graph such that $|V_1|=\left\lfloor\frac{i}{2}\right\rfloor, |V_2|=\left\lceil\frac{i}{2}\right\rceil, i$ is an odd natural number, $|E|=\left\lfloor\frac{i}{2}\right\rfloor\left\lceil\frac{i}{2}\right\rceil$. Thus, in the graph G every vertex from V_1 is connected with each vertex of V_2. Furthermore, let $G'=(V,E')$ is the graph obtained from graph G by adding one edge u, connecting two vertices $v\in V_2, v'\in V_2$. As the combinatorial set S_i, we consider the set of all simple paths from a vertex v in the graph G that contains each vertex in V. Obviously, all simple paths from vertex v to vertex v' in graph G, that contains each vertex in V, adding edge u in graph G' are transformed into Hamiltonian cycles. Thus, combinatorial number $|S_i|$ for graph G and G' are the same. It is obvious that $|S_i|=(\left\lfloor\frac{n}{2}\right\rfloor!)^2$. Since $|S_i|$ is not a polynomial, then according to the theorem, the Hamiltonian cycle problem is not polynomial. It also follows the results.

Corollary 1. All NP complete problems are not polynomial.

Corollary 2. Hypothesis J. Edmonds $P\neq NP$ is correct.

3. Conclusions

Certainly, there are polynomial problems, check of which correctness of the decision demands the polynomial number of steps (for example, problems of sorting [24]), however it doesn't mean that the class of polynomial problems enters the class NP, i.e. $P\subseteq NP$ as it was noted in introduction, is the wrong statement.

Thus, the problem $P=NP$, it is trying to solve the authors [25-26] does not exist in nature. Therefore, it is impossible to solve by any means, and not only natural [24]. Once again we repeat: on the one hand, we have shown [6-10] the existence of polynomial problems the verifying the correctness of the decision which requires a non-polynomial number of steps that need cryptography on the other hand, we have proved that NP complete problems are not polynomial.

References

[1] Cobham A. The intrinsic computational difficulty of functions //In Procedings of the 1964 Congress for Logic, Methodology, and the Philosophy of Science.-North-Holland, 1964. - P. 24-30.

[2] Edmonds J. Paths, trees and flowers //Canadian Journal of Mathematics.-1965-Vol. 17. - P. 449-467.

[3] Cormen T. H., Leiserson Ch. E., Rivest R. L., Introduction to Algorithms, MIT Press, 1990. 2002. P. 955.

[4] Cook S. A., The P versus NP Problem //Manuscript prepared for the Clay Mathematics Institute for the Millennium, April, 2000, www.cs.toronto.edusacook.

[5] Razborov A. A. Theoretical Computer Science: vzglyad mathematica, http://old.Computerra.ru/offline/2001/379/6782.

[6] Kochkarev B. S. Prilogenie monotonnykh funktsij algebry logiki k probleme Kuka, Nauka v Vuzakh: matematika, fizika, informatika, Tezisy dokladov Mejdunarodnoj nauchno-obrazovatelnoj konferentsii, 2009, pp. 274-275. (in Russian).

[7] Kochkarev B. S., On Cook's problem, http://www.math.nsc.ru/conference/malmeet/08/Abs.

[8] Kochkarev B. S. About one class polynomial problems with not polynomial certificates //arxiv: 1210.7591v1 [math. CO] 29 Oct 2012/.

[9] Kochkarev B. S., About one class polynomial problems with not polynomial certificates //Second International Conference "Claster Computing"CC2013 (Ukraine, Lviv, June 3-5, 2013) P. 99-100.

[10] Kochkarev B. S., K probleme Kuka //Matematicheskoe obrazovanie v shkole I v Vuze v usloviyakh perekhoda na novye obrazovatelnye standarty. Materialy Vserossiyskoy nauchno-practicheskoy konferentsii s mejdunarodnym uchastiem. TGGPU, Kazan, 2010. s. 133-136 (in Russian).

[11] The Clay Mathematics Institute www.Claymath.org.Millennium Prize Problems.

[12] Kurosh A. G. Kurs vysshey algebry. Izd. F. -M. Literatury, M., 1962, 432. (in Russian).

[13] Kochkarev B. S., Vzaimootnosheniya mejdu slojnostnymi klassami P, NP i NPC. Problems of modern science and education. 2015, 11 (41), s. 6-8 (in Russian).

[14] Kochkarev B. S., Dokazatelstvo gipotezy Edmondsa i reshenie problemy Kuka. Nauka, Tekhnika i Obrazovanie, 2014, 2 (2), s. 6-9. (in Russian).

[15] Ravenstvo klassov P i NP https://ru.wikipedia.org/wiki.

[16] En waarmee toverde het internet vandaag een lach op Uwge https://www.uscki.nl.

[17] Gipoteza J. Edmondsa i problema S. A. Kuka http://www.refereed.ru.

[18] Cook S. A. The complexity of theorem proving procedures. //In Proceedings of the Third Annual ACM Symposium on Theory of Computing. P. 151-158, 1971.

[19] Levin L. A. Universal sorting problems. //Problemy Peredachi Informatsii, 9(3), s. 265-266, 1973.

[20] Kochkarev B. S. Proof of the hypothesis Edmonds's, not polynomial of NPC problems and classification of the problems with polynomial certificates.//arxiv: 1303.2580v1 [cs. CC] 7 Mar 2013.

[21] Karp R. M. Reducibility among combinatorial problems. //In Raymond E. Miller and James W. Thatcher, editors, Complexity of computer Computations, P. 85-103, Plenum Press, 1972.

[22] Stanley R. P. Enumerative Combinatorics v. 1, 1986.

[23] Kochkarev B. S. Ob odnom algoritme, ne soglasuyutchemsya s tezisami Turinga, Churcha i Markova, Problems of modern science and education, M., 2014, 3(21) c. 23-25 (in Russian).

[24] Kochkarev B. S. Gipoteza J. Edmondsa i problema S. A. Kuka. //Vestnik TGGPU, 2011 2 (24) s. 23-24 (in Russian).

[25] Razborov A. A., Rudich S. Natural proof. Proceedings of the 26 th Annual ACM Symposium on the Theory of Computing. P. 204-213 DOI: 10.1145/195058.195134.

[26] Babay priblizilsya k resheniyu pronlemy tysyacheletiya. http://lenta.ru/news/2015/11/20/graphtheory/.

A novel high spectral efficiency waveform coding-OVTDM

Li Daoben

School of Information and Communication, Beijing Univ. of Posts & Telecomm, Beijing, China

Email address:
Lidaoben2014@163.com

Abstract: Different from other Coding, based on the Overlapped Multiplexing Principle discovered by author， a novel OVTDM (Overlapped Time Division Multiplexing) Waveform Coding is proposed. Instead of the encoding matrix and mapped signal constellation, any engineering sense band-limited Multiplexing Waveform can be employed. By its data weighted shift overlapped versions, the coding gain and spectral efficiency are both achieved. The heavier the overlap of the data weighted Multiplexing Waveform, the higher the coding gain and spectral efficiency as well as the closer the output to the optimum complex Gaussian distribution. The encoder structures, parameters, optimum and fast decoding algorithms, pre-coding, some implementation problems as well as the bit error performance are estimated and discussed. Simulations show that OVTDM is suitable for high spectral efficiency applications and its spectral efficiency is roughly proportional to SNR.

Keywords: Overlapped Multiplexing Principle, Waveform Encoding, Spectral Efficiency, Shannon Capacity

1. Preface

Shannon Theory is well known the guidance of communications. If let a signaling symbol (pulse) of received signal to carry more information, multi-levels called level division should be employed. Along with the code constraint length increasing, the number of distinguish levels of a received "pulse" approaches to $\sqrt{1+P_S/P_N}$ carrying $\max 0.5\log_2(1+P_S/P_N)$ bits/symbol information, where P_S/P_N is the signal noise ratio (SNR). Later by level division and assuming channel obey Nyquist criterion with bandwidth strictly limited to B. A continues channel was easily transformed to a discrete memory-less signaling symbol sequence with rate 2B symboles/s, i.e. a discrete memory-less channel. The "Shannon capacity" $C = B\log_2(1+P_S/P_N)\ bps/Hz$ was easily obtained in this way. Obviously channel capacity may have different even better form if level division and Nyquist criterion are no longer employed.

There are 2K combinations of K bits with total duration KT_b needing 2K one to one mapping represented symbols. If the channel is regarded as a memory-less symbol sequence, surely level division, i.e. a signal constellation of 2K levels is the only choice. However the received signal is continuous one, why don't employ 2K waveforms? It is a general knowledge that in a very noisy environment, people can still distinguish a huge number of weak voices by their waveforms rather than levels.

The fatal weakness of Nuquist Criterion is that it violates the uncertainty principle and the no ISI (Inter-symbol Interference) Nyquist Channel is physically unrealizable. In fact, in any field X (X denotes time T, frequency F, space S, code C as well as their hybrid H) system, the overlapping between adjacent data is unavoidable, the higher the data rate the heavier the ISI. Why don't to utilize ISI adroitly? The Overlapped Multiplexing Principle discovered by [12] reveals that the overlapping between adjacent and neighboring data in any system is never interference but a beneficial coding constraint relation offering benefit coding gain. The destroy fact coming outside the system is only the interference. The channel capacity will be reduced by brutally force equalizing a channel with code constraint into a Nyquist channel of thoroughly losing coding constraint relation.

Leaving from Nyquist criterion and level division, based on the Overlapped Multiplexing Principle and waveform division, a novel OVTDM (Overlapped Time Division Multiplexing) waveform coding scheme is proposed in the paper. By the shift data weighted overlapped version of an engineering sense band-limited Multiplexing Waveform, there appears a OVTDM coding with high spectral efficiency, high coding gain, no coding redundancy, relative low decoding complexity.

It is well known that the transmitted signal should be not only in complex field, but also in complex Gaussian distribution under additive complex Gaussian noise environment. Unfortunately all nowadays coding need mapping to a signal constellation in complex field. Though most of the sequence to sequence coding is blameless, their final outputs can never be in complex Gaussian distribution, due to the mapped signal constellations are all in uniformly distribution. Even "shaping" scheme may centralize the signal constellation a little. Such modification can never solve their fatal weakness and gives at most 1.53 dB gain.

Although there are non-finite field coding, e.g. [10] and partial response coding, as well as the superimposed coding [9]. However such "superimposition" is not the "shifted overlapping with ISI" and the key point is that they never leave the uniform distributed signal constellation.

FTN (faster than-Nyquist) criticizes Nyquist signaling rate but insists on a strictly band-limited Nyquist channel (Sinc pulse is an exception?). On the other hand, so far FTN is only a little faster than Nyquist rather than much faster than Nyquist like OVTDM, and its drawback is still treats symbol overlapping as an interference rather than a beneficial coding gain.

OVTDM belongs to a novel waveform coding, it is based on waveform division rather than level division and its output automatically approaches to optimum complex Gaussian distribution. OVTDM employs an engineering sense band-limited multiplexing waveform. By its shift data weighted overlapped version, OVTDM will have least output levels and maximum Euclidean branch distance as well as an approaching to optimum complex Gaussian output.

Except Nyquist criterion another obstacle of limiting spectral efficiency η is the signal levels, i.e. the number of points in a signal constellation. People use to put 2K levels without coding and at least 2K+1 levels with coding for η=Kbits/symbol. Therefore the signal levels will be increased exponentially with η, even shaping may shrink the levels a little. For average power limited channel, the more the signal levels the smaller the distance between them and the lower the noise immunity.

OVTDM essentially is a convolutional waveform coding scheme. Except the near complex Gaussian distribution outputs, OVTDM also has least number of output levels. For binary (+1,-1) data input, the output of a K folds OVTDM only has K+1 levels with spectral efficiency η=Kbits/symbol and 2K distinguished output sequences within the code constraint length KT_b. Since there are only K+1 levels for each code node, their Euclidean distance between code nodes can be increased at most. Surely relative higher noise immunity can be achieved.

The output distribution of a K folds OVTDM with binary (+1,-1) input is the Kth order binomial distribution approaching to optimum Gaussian distribution with K. It is well known that polynomial and binomial distribution can all approach to Gaussian distribution with K. Two stage concatenate OVTDM structure parallel putting in orthogonal I, Q channels is proposed in the paper, where the 1st stage is a K1th order pure OVTDM (no relative shift) changing binary (+1,-1) input into multilevel real input and the 2nd stage is a K2th order shifted OVTDM making output polynomial distribution approach to Gaussian. The total spectral efficiency η of such I, Q parallel concatenate OVTDM structure is η= 2K1K2bits/symbol, and I, Q real distribution outputs together approach to complex Gaussian.

Instead of encoding matrix and mapped signal constellation, any band-limited Multiplexing Waveform can be employed in OVTDM. By its shift data weighed overlapped version, coding gain and spectral efficiency η are both achieved. Then what is the Optimum Multiplexing Waveform? What effect is the channel filtering? Such problems will be discussed in the paper. The channel capacity of OVTDM is roughly linear to SNR rather than logarithm of SNR. The reason is it utilizes physically realizable engineering sense rather than Nyquist sense band-limited waveform. No matter how fast decay with tail spectrum outside filter's bandwidth, physically realizable spectrum tail always extends to infinite. The system capacity should be linearly to SNR (see [15] and appendix C of the paper). The performance of OVTDM can go far beyond the Shannon capacity when spectral efficiency η is high enough, the reasons are as follows:

Leaving obstacle from Nyquist criterion, recovering discrete memory-less sample sequences with no code constraint to waveforms with strong code constraint relation;

Employing waveform division instead of level division;

The code outputs no longer have exponentially increased but algebraically increased levels with spectral efficiency η. The Euclidean distance between nodes of OVTDM increased at most.

No matter how fast decay with tail spectrum outside filter's bandwidth, spectrum tail always extends to infinite rather than strictly no tail.

Even only field time T coding OVTDM is discussed in the paper. It is easily expanded to other field X (OVXDM).

2. OVTDM System Model

A *OVTDM Model*

The complex envelop (center frequency removed) model of an OVTDM system is shown in Fig.1.

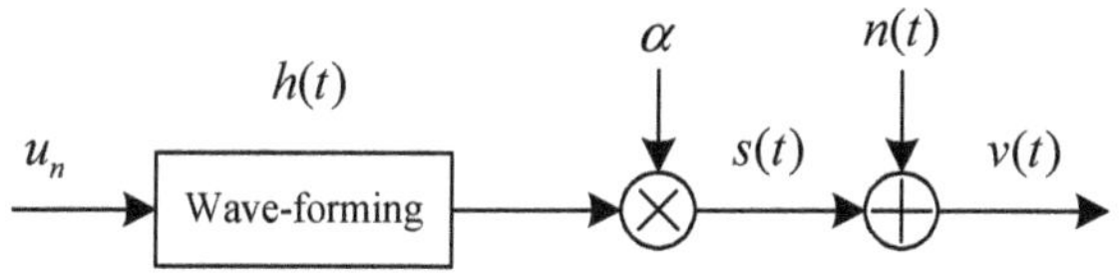

Fig 1. *OVTDM model*

Assuming that $U \triangleq [u_0, u_1, \cdots]^T$ be the transmitted i.i.d. (identical, independent, distribute) data sequence; $h(t)$ be the so called Band-limited Multiplexing Waveform, which is an impulse response function, including all the filters in the system. e.g. wave-forming, transmitter, receiver, channel, equalizer etc. u_n carrying Q bits be the nth transmitted data with duration T; E_0 be the transmitted data energy; α be the attenuation factor of the channel; $E = \alpha E_0$ be the received data energy, $n(t)$ be the complex envelope of additive white

Gaussian noise with power spectrum density N_0. Then the received signal's complex envelope is

$$v(t) = \sqrt{2E}\sum_n u_n h(t-nT) + n(t) = s(t) + n(t), \quad (1)$$

Where $h(t)=0, t \notin (0,\Delta)$ is the employed any physical realizable band-limited multiplexing waveform;

$(K-1)T < \Delta \le KT$; $K \triangleq \lfloor \Delta / T + 1 \rfloor$ is the number of overlapped folds of $h(t)$; $\lfloor \bullet \rfloor$ is the least integer of $\bullet$.

When $t \in [nT,(n+1)T]$, $n = 0,1,2,\ldots$, the received signal can also be represented as

$$v_n(t) = \sqrt{2E}\sum_{i=0}^{K-1} u_{n-i} h_i(t) + n_n(t) = s_n(t) + n_n(t), \quad (2)$$

Where:

$$\begin{cases} s_n(t) \triangleq s(t)[U(t-nT) - U(t-(n+1)T)] \\ n_n(t) \triangleq n(t)[U(t-nT) - U(t-(n+1)T)] \\ v_n(t) \triangleq v(t)[U(t-nT) - U(t-(n+1)T)] \\ h_i(t) \triangleq h(t+iT)[U(t) - U(t-T)] \end{cases} \quad (3)$$

$U(t)$ is the unite step function.

Therefore $v_n(t)$ is just the complex convolution of transmitted data sequence $\mathrm{U} \triangleq [u_0, u_1, \cdots]^T$ with Multiplexing Waveform sequence $\mathrm{h}(t) \triangleq [h_0(t), h_1(t), \cdots, h_{K-1}(t)]^T$.

Fig.2 is its model of waveform convolutional encoder with rate 1 and constraint length K. The spectral efficiency reaches $\eta = KQ$ bits/symbol. For the sake of simplicity in Fig.2 tape coefficients $h_0, h_1, \cdots, h_{K-1}$ are not number sequence but waveform sequence $h_0(t), h_1(t), \cdots, h_{K-1}(t)$.

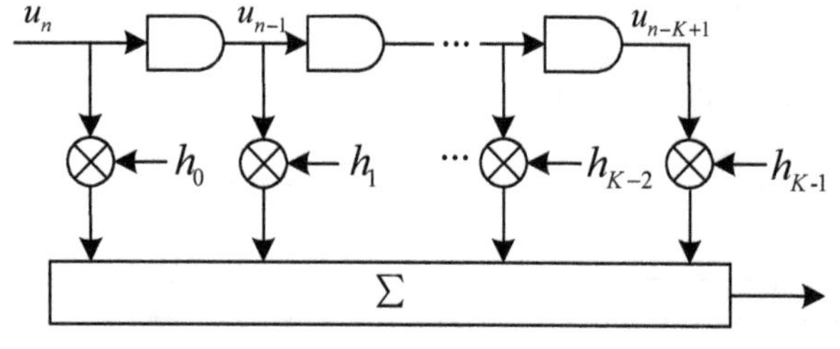

Fig 2. *The complex waveform convolutional encoder model of OVTDM with shift unit T and overlapping folds K*

For 2Q-nary real data, the output of a K folds OVTDM of real $h(t)$ will have $K(2^Q - 1)+1$ levels with spectral efficiency KQ bits/symbol. The reason is that since the width of symbol (multiplexing waveform $h(t)$) expanded K times, the original signal bandwidth B should be shrunk to B/K. In order to keep the same data rate there have to be K symbols overlapped together. Let the frame length be L occupying $[K+(L-1)]T$ sec. carrying LQ bits. Then its spectral efficiency η becomes

$$\eta = \frac{LQ}{[K+(L-1)]TB/K} \xrightarrow{L \gg K} \frac{KQ}{BT} \text{bits/s/Hz} \quad (4)$$

When $L \gg K$, the spectral efficiency η and the capacity of such K folds OVTDM system will increase K times. Its output is binomial distribution of order K, approaching Gaussian distribution when $K \gg 1$.

Similarly, when $h(t)$ is real, for quaternary (+1, -1, +j, -j) independent data steam. Any K folds OVTDM output has $(K+1)^2$ levels with spectral efficiency η=2K bits/symbol. Its output is orthogonal two binomial distributions of order K, approaching complex Gaussian distribution when $K \gg 1$.

OVTDM does destroy the one to one mapping relation between the input and output symbols but keeps the one to one mapping relation between the input sequence and the output sequence [2][12]. For binary data within constraint length K, there are 2K input binary data sequences, but also 2K output waveform sequences. They are absolutely in one to one mapping relation. Similarly, for quaternary data within constraint length K, there are 2K input binary data sequences and 2K output waveform sequences both in orthogonal I and Q channels. One to one mapping relation still kept. Surely MLSD (Maximum Likelihood Sequence Detection) should be employed. From the total 2K possible waveform sequences in each I and Q channel, to select the most possible coded waveform sequence that is nearest to the received signal waveform sequence [2][12].

B *Power Spectrum of OVTDM*

Let the Multiplexing waveform and its spectrum be $h(t)$ and $H(f)$ respectively. $h(t) \leftrightarrow H(f)$

They are a pair of Fourier transform.

The output waveform of OVTDM and the corresponding spectrum are

$$\sum_n u_n h(t - n/T) \leftrightarrow H(f)\sum_n u_n e^{j2\pi fn/T} \quad (5)$$

Then output power spectrum of OVTDM is

$$E\left\{\left|H(f)\sum_n u_n e^{j2\pi fn/T}\right|^2\right\} = E\left\{\sum_n\sum_{n'} u_n u_{n'}^* e^{j2\pi f(n-n')/T}\right\}|H(f)|^2, \quad (6)$$

Since input data is i.i.d. $E(u_n u_{n'}^*) = E(|u_n|^2)\delta_{n,n'}$, formula (6) becomes

$$|H(f)|^2 \sum_n E(|u_n|^2) \quad (7)$$

For i.i.d. data, since only K adjacent data overlapped together, there are only K terms in summation. Finally the power spectrum of a K folds OVTDM is $K|H(f)|^2$. That is completely same as its band-limited multiplexing waveform $h(t)$.

C *The Difference of OVTDM from others*

OVTDM employ not level division but waveform division. OVTDM belong to waveform coding. Instead of the encoding matrix and mapping constellation, any engineering sense band-limited Multiplexing Waveform $h(t)$ can be employed. By its data weighted shift overlapped versions, the coding gain and spectral efficiency η are both achieved. The system performance is only determined by $h(t)$. Then what is the optimum $h(t)$? If the restraint condition is the coding constraint length K, the answer is very simple. Due to the symmetry principle, rectangular $h(t)$ is the optimum, since its coding constraint relation is equal and maximum. However the problem will become complicated for other restraint condition, e.g. spectral efficiency η. Since $h(t)$ should be related to the "time–bandwidth production" of it. First what is required "time–bandwidth production"? 2nd there is so many definitions of "bandwidth" and "time duration" [12]. Which definition is suitable? Or need to re-define? According to the uncertainty principle strictly limited bandwidth and time duration signal is physically unrealizable. The optimum $h(t)$ is no longer the rectangular one. For non-flat $h(t)$, Code constraint length and relation should all be changed. That is a problem of that when moving a little part the whole situation will be complete different. Such question can't be solved by nowadays coding theory. Several $h(t)$ are given for simulation in the paper. They are rectangular, raised cosine, raised cosine spectrum waveform as well as their compounds. Since employing rectangular $h(t)$ has some theoretical meaning, and employing other $h(t)$ can be approached by filtering in engineering. It is lucky that their performance difference is not huge relative to their gain. Which tell us that even the optimum $h(t)$ is found, the performance improvement may not huge. Any way finding optimum $h(t)$ is still an open problem.

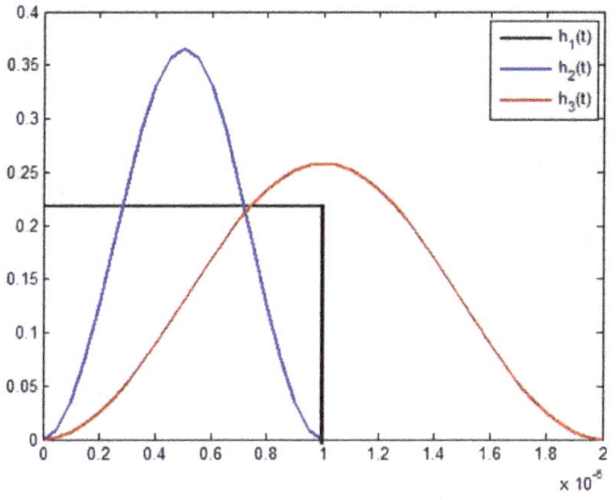

Fig 3. *Some basic Multiplexing waveforms*

Some tested $h(t)$ in the paper as in [12] are as follows:

$h_1(t) \sim$ Rectangular wave with duration T,

$$B_e = 1/T, B_0 = 1/T.$$

$h_2(t) \sim$ Raised Cosine wave 1 with duration T,

$$B_e = 1/T, B_0 = 2/T.;$$

$h_3(t) \sim$ Raised Cosine wave 2 with duration $2T$,

$$B_e = 1/2T, B_0 = 1/T.$$

$h_8(t) = \frac{1}{2}\sin c(\pi t/T)\frac{Cos(\pi t/T)}{(1-t^2/T^2)} \sim$ Raised Cosine spectrum wave with most energy within $(-T,T)$, $B_e = 1/T, B_0 = 2/T$.

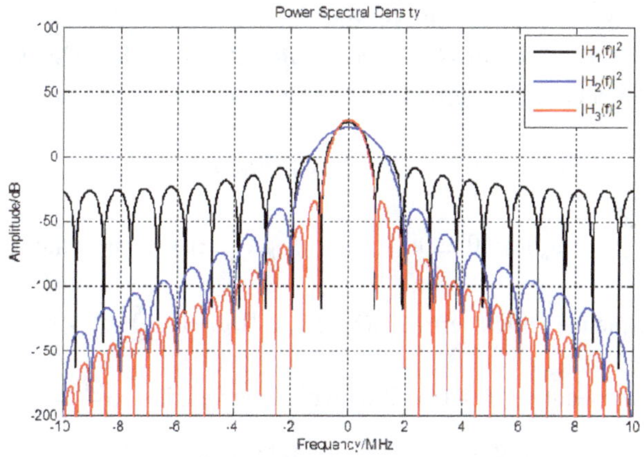

Fig 4. *Power spectrum of the basic Multiplexing waveforms*

$$h_4(t) = h_1(t) \otimes h_2(t) \sim \ 2T, B_e = 0.6/T, B_0 = 1/T.$$

$$h_5(t) = h_1(t) \otimes h_3(t) \sim 3T, B_e = 0.4/T, B_0 = 1/T.$$

$$h_6(t) = h_2(t) \otimes h_2(t) \sim 2T, B_e = 0.75/T, B_0 = 1/T.$$

$$h_7(t) = h_2(t) \otimes h_3(t) \sim 3T, B_e = 0.45/T, B_0 = 1/T.$$

The power spectrum of $h_l(t)$ $(l = 4,5,6,7)$ are the production of their components.

Where: $\sim aT, a = 1,2,3$ denotes the time duration of $h_l(t), l = 1,2,...,7$; B_0, B_e respectively denote the first zero bandwidth and the equivalent noise bandwidth, the last is a fictive rectangular bandwidth of high $H_l(0)$ $(l = 1,2,...,8)$ with the same filtered power in white Gaussian noise.

OVTDM is a waveform coding scheme. The system performance is controlled by $h(t)$. People may worry about $h(t)$ would be destroyed by multipath Rayleigh fading channel. However the random time dispersion of the channel (which causes frequency selective fading) may put additional overlapping to $h(t)$ and has no effect to spectral efficiency η. On the contrary, it is a beneficial factor of improving system performance. Duo to that the additional overlapping increases the coding and implicit diversity gain simultaneously.

3. A Bit Error Probability upper Bound of OVTDM

Just like the convolutional codes, Evaluating on accurate OVTDM's bit error probability is very difficult. That is still an open problem. There were many error probability upper bound of ISI channel, but all gave a pessimistic and no general result. This chapter will give an optimistic and general result by a "Modified Minimum Euclidean Distance Sphere Bound" [2] and a non-normalized masked distribution.

A *Node error event of OVTDM*

An error event begins at $t = jT$, ends at

$$t = (j+k+K)T, k = 0,1,2,...$$

Suppose the correct data sequence be $\mathrm{u} \triangleq [u_0, u_1, ..., u_{j-1}, u_j, u_{j+1}, ...u_{j+k+K}, u_{j+k+1}, ..., u_{N-1}]$, One error sequence with S errors from $\mathbf{u}$ be

$$\mathrm{u}_S^{'} \triangleq [u_0, u_1, ..., u_{j-1}, u_j^{'}, u_{j+1}^{'}, ...u_{j+k+K}^{'}, u_{j+k+K+1}, ..., u_{N-1}],$$

$\mathrm{e}_S \triangleq \frac{1}{2}(\mathrm{u} - \mathrm{u}_S^{'}) = [0,0,....,e_j, e_{j+1}, ..., e_{j+k}, 0, ..., 0]$, be the error sequence with S errors, where $e_n \in \{1,0,-1\}$, $\sum_n e_n^2 = s$, and no consecutive $K-1$ position with no errors, i.e. no $\underbrace{00...0}_{\geq K-1}$ between adjacent $e_n \neq 0$. There are totally $2^S (K-1)^{S-1}$ e_S event with shortest length $S+K-1$ and longest length $(K-1)(S-1)+K$.

B *Node error probability and Bit error probability of OVTDM*

Let u be the correct data sequence, $\mathrm{u}_S^{'}$ be its only alternative, than their pairwise probability is

$$P(\mathrm{u} \to \mathrm{u}_S^{'}) \triangleq P_e(\mathrm{e}_S) = erfc\{[\frac{1}{2N_0} \int_0^{NT+\Delta} |s(t) - s'(t)|^2 dt]^{1/2}\}, \quad (8)$$

Where: $erfc(x) \triangleq \frac{1}{\sqrt{2\pi}} \int_x^{\infty} e^{\omega^2/2} d\omega,$

$s(t) = \frac{\sqrt{2E_S}}{2} \sum_n u_n h(t-nT) \sim$ is the correct signal,

$s'(t) = \frac{\sqrt{2E_S}}{2} \sum_n u_n^{'} h(t-nT) \sim$ is its only alternative.

The Euclidean distance between $s(t)$ and $s'(t)$ is

$$\int_0^{NT+\Delta} |s(t) - s'(t)|^2 dt = 2E_S \sum_n \sum_m e_n e_m h_{n-m}^0, \quad (n, m = 0,1,..., N-1) \quad (9)$$

Where:

$$\tilde{h}_{n-m}^0 \triangleq \int_0^{NT+\Delta} h(t-nT) h^*(t-mT) dt = \tilde{h}_{m-n}^{0\,*} \quad (|n-m| = 0,1,..., K-1) \quad (10)$$

Since $\sum_n e_n^2 = s,\ h_0^0 = 2,$

Let: $h_l^0 \triangleq Re\, \tilde{h}_l^0 =$

$$Re \int_0^{\Delta - lT} h(t) h^*(t+lT) dt,\ l = 0,1,..., K-1), \quad (11\text{-}1)$$

$$\sigma_l \triangleq \frac{1}{S} \sum_n e_n e_{n-l}, \quad (11\text{-}2)$$

$$\varepsilon_S \triangleq \sum_{l=1}^{K-1} \sigma_l h_l^0, \quad (11\text{-}3)$$

$$d^2 \triangleq E / N_0, \quad (11\text{-}4)$$

(11-4) is so called the normalized SNR. Thus (8) becomes

$$P_e(\mathrm{e}_S) = erfc\{[2sd^2(1+\varepsilon_S)]^{\frac{1}{2}}\} < \frac{1}{2} \exp\{-sd^2(1+\varepsilon_S)\}, \quad (12)$$

People are interested in when a node error event occurs, the probability of average S bits in error, i.e.

$$P_e(s) = P_e(\cup \mathrm{e}_S), \quad (13)$$

Since in Trellis diagram of OVTDM, each node represents one bit entering into the channel, the bit error probability P_b is

$$P_b = E\{n_b\} = \sum_{s=1}^{\infty} s P_e(s) = P_e(\cup \mathrm{e}_s) \leq \sum_{\mathrm{e}_s} P_e(\mathrm{e}_S), \quad (14)$$

Since there are $2^S (K-1)^{S-1}$ different e_s with S errors, the union bound on $P_e(s)$ becomes

$$P_e(s) \leq 2^S (K-1)^{S-1} E\{P_e(\mathrm{e}_S)\}, \quad (15)$$

Such a bound only suitable for small S and K. when $S \gg 1$ or $K \gg 1$, (15) may greater than 1 and becomes useless. A best way is to use the "Modified Minimum Euclidean Distance Sphere Bound" [2], that gives

$$P_e(s / \mathrm{u}) \leq (s-1)(K-1) E\{e^{-sd^2(1+\varepsilon_S)} / \mathrm{u}\},\ (s \geq 2, K > 2), \quad (16)$$

There is equal probable $(s-1)(K-1)$ $\mathrm{u}_S^{'}$ surrounding u , then

$$E\{e^{-sd^2(1+\varepsilon_S)} / \mathrm{u}\} \geq \frac{1}{(s-1)K-1)} e^{-sd^2[1+Min\, \varepsilon_s(\mathbf{u})]},\ (K > 2) \quad (17)$$

Where: $\varepsilon_{s,i}(\mathrm{u})$ is the conditional ε_s under u and its only alternative is $\mathrm{u}_{s,i}^{'}$. Thus

$$\begin{aligned} P_e(s) &= \sum_{\mathrm{u}} P(\mathrm{u}) P_e(s/\mathrm{u}) \le (s-1)(K-1) \sum_{\mathrm{u}} P(\mathrm{u}) E\{exp\text{-}sd^2(1+\varepsilon_S)\} \\ &= (s-1)(K-1)E\{exp - sd^2(1+\varepsilon_S)\}, \quad (s \ge 2, K > 2), \end{aligned} \tag{18}$$

Note: When $K = 2$, u'_S with S bits error from u is only consecutive S bits different from u, therefore (17) is hold only for $K > 2$. There should be no $(s-1)(K-1)$ coefficient for $K = 2$.

When $S = 1, \varepsilon_S = 0$, we have

$$P_e(1) = erfc\ (2d^2) < \mathrm{e}^{-d^2}, \tag{19}$$

An upper bound on P_b is

$$\begin{aligned} P_b &= \sum_{s=1}^{\infty} sP_e(s) \le e^{-d^2} + \\ &+ (K-1)\sum_{s=2}^{\infty}(s^2-s)e^{-sd^2}E\{e^{-sd^2\varepsilon_S}\}, \ (K>2) \end{aligned} \tag{20}$$

To find an upper bound on P_b, the evaluation of $E\{e^{-sd^2\varepsilon_S}\}$ or its upper bound is of importance.

C *One upper bound on Bit error probability of OVTDM*

$K = 2$

At such case $\varepsilon_s = \sigma_1 h_1$, and according to appendix B, σ_l only exists even moments as

$$\begin{aligned} \mathrm{E}\{\sigma_l^{2k}\} &= \frac{1}{2^{s-1}s^{2k}} \sum_{i=0}^{s-1} \binom{s-1}{i} (s-2i-1)^{2k}, \\ k &= 0,1,2,\cdots, l = 1,2,\cdots,K-1, \end{aligned} \tag{21}$$

$$\begin{aligned} P_e(s) &\leqslant \mathrm{e}^{-sd^2}\mathrm{E}\{\mathrm{e}^{-sd^2\sigma_1 h_1}\} \\ &= \mathrm{e}^{-sd^2}\sum_{k=0}^{\infty}\frac{(sd^2h_1)^{2k}}{(2k)!}\bullet\frac{1}{2^{s-1}s^{2k}}\sum_{i=0}^{s-1}\binom{s-1}{i}(s-2i-1)^{2k} \\ &= \frac{1}{2^s}\mathrm{e}^{-sd^2}\sum_{i=0}^{s-1}\binom{s-1}{i}\left[\mathrm{e}^{(s-2i-1)h_1d^2}+\mathrm{e}^{-(s-2i-1)h_1d^2}\right] \\ &= \frac{1}{2^{s-1}}\mathrm{e}^{-sd^2}\left[\mathrm{e}^{h_1d^2}+\mathrm{e}^{-h_1d^2}\right]^{s-1}, \end{aligned} \tag{22}$$

Finally

We have

$$\begin{aligned} P_b &\leqslant \mathrm{e}^{-d^2}\sum_{s=1}^{\infty}s\left[\frac{1}{2}\mathrm{e}^{-d^2}\left(\mathrm{e}^{h_1d^2}+\mathrm{e}^{-h_1d^2}\right)\right]^{s-1} \\ &= \mathrm{e}^{-d^2}\left[1-\frac{1}{2}\mathrm{e}^{-d^2}\left(\mathrm{e}^{h_1d^2}+\mathrm{e}^{-h_1d^2}\right)\right]^{-2}, \end{aligned} \tag{23}$$

(23) is identical to the bound given by Viterbi and Omura [1].

For any physical realizable channel, $h_1^0 < 1$, then for $d^2 \gg 1$

$$P_b < \mathrm{e}^{-d^2}, \quad d^2 \gg 1; K = 2, \tag{24}$$

The least favorite multiplexing waveform is the rectangular one with $h_1^0 = 1$,

$$h(t) = \sqrt{\frac{2}{KT}}[u(t) - u(t-KT)], \quad K = 2, \tag{25}$$

$$P_b \leqslant 4\mathrm{e}^{-d^2} \quad d^2 \gg 1; K = 2,. \tag{26}$$

$K>2$

Although the distribution of $\sigma_l (l = 1,2,\cdots,K-1)$ is known, due to the strong dependence among σ_l, to find the distribution of $P_r(\varepsilon_s)$ is still difficult. So far most of people employ numerate method, but it is unreality when $s \gg 1$ or $K \gg 1$.

Since $E(\varepsilon_s) = 0$, $\mathrm{E}\{\mathrm{e}^{-sd^2\varepsilon_s}\}$ is controlled by $\varepsilon_s < 0$, especially its terminal of $\varepsilon_s < 0$ when $d^2 \gg 1$. That requires finding the minimum energy of $\sum_n e_n h(t-nT)$, such question had been studied by too many people in the past, there is impossible to list them all. On the other hand $\min\varepsilon_s$ must be related to a least favorite node error sequence, meaning

$$P_r\{\varepsilon_s = \min\varepsilon_s\} = P_r(\sigma_1 = \min\sigma_l). \tag{27}$$

Node: If σ_1 is not the minimum, need reordering.

At the terminal of $\varepsilon_s < 0$, Let $\varepsilon_s \cong a(K,s)\sigma_1$,

Where:

$$a(K,s) \triangleq \frac{\min\varepsilon_s}{\min\sigma_l} = \frac{s}{s-1}|\min\varepsilon_s|, \tag{28}$$

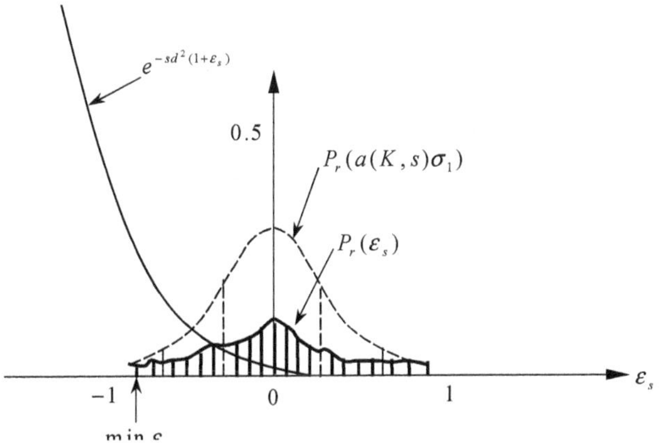

Fig 5. The difference between $P_r(\varepsilon_s)$ *and* $P_r(a(K,s)\sigma_1)$.

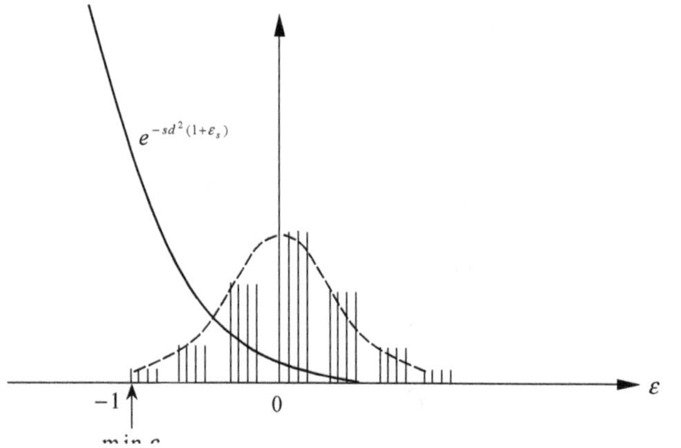

Fig 6. Non-normalized $P_r\,'[a(K,s)\sigma_1]$ with $K-1$ times spectrum line

Fig.5 shows the difference between the numerical evaluated real $P_r(\varepsilon_s)$ and $P_r(a(K,s)\sigma_1)$. ε_s , $a(K,s)s_1$ are all discrete random variables, $P_r(a(K,s)\sigma_1)$ is always above $P_r(\varepsilon_s)$, due to $P_r(\varepsilon_s)$ has $K-1$ times more spectrum lines than $P_r(a(K,s)\sigma_1)$, and at the terminal of $\varepsilon_s < 0$, this two probabilities are equal.

Since ε_s is consisted of the summation of $K-1$ weighted identical distributed dependent $\sigma_l (l = 1,2,\cdots,K-1)$. Between adjacent two spectrum lines of $a(K,s)\sigma_1$, there are no more than $K-1$ spectrum lines of ε_s .

Since $P_r(\varepsilon_s)$ is unknown, $P_r(a(K,s)\sigma_1)$ is known. If adding K-1 lines between two adjacent lines of $a(K,s)\sigma_1$, a new non-normalized distribution $P_r\,'[a(K,s)\sigma_1]$ will be obtained, that has total probability greater than 1, and $P_r(\varepsilon_s)$ will be masked by it (Fig.6), except in very few case, at the terminal of $\varepsilon_s > 0$, However it is no effect to the final integration, duo to at $\varepsilon_s > 0$, $e^{-sd^2(1+\varepsilon_s)} \ll 1$, the final integration is controlled only by $\varepsilon_s < 0$ (Fig.5).

If replacing $P_r(\varepsilon_s)$ by $P_r\,'[a(K,s)\sigma_1]$, the expectation operation for $e^{-sd^2\varepsilon_s}$ will be increased $A(K,d,s)$ times.

Where:

$$A(K,d,s) = \sum_{i=0}^{K-1} e^{-i\frac{a_s d^2}{K-1}} = \frac{1-e^{-a_s d^2}}{1-e^{-i\frac{a_s d^2}{K-1}}} \qquad (29)$$

$$0 < a_s < \left|Min\ \varepsilon_S\right| / s.$$

When $d^2 \gg K$, $A(K,d,s) \cong 1$;

$$K \gg d^2,\ A(K,d,s) \cong K-1;$$

Otherwise $1 \le A(K,d,s) \le K-1$.

After such modification, we have

$$P_e(s) \le A(K,d,s)(K-1)(s-1)e^{-sd^2}E\{e^{-sd^2 a(K,s)\sigma_1}\}, \qquad (30)$$

Since

$$E\{e^{-sd^2 a(K,s)\sigma_1}\} = \sum_{k=0}^{\infty}\frac{(sd^2 a)^{2k}}{(2k)!}E\{\sigma_1^{2k}\}, \qquad (31)$$

Substitute $E\{\sigma_1^{2k}\}$ into (31), we have

$$P_e(s) < e^{-d^2}\frac{A(K,d,s)(s-1)}{(s-1)^{s-2}}B^{s-1}(K,d,s), \qquad (32)$$

Where

$$B(K,d,s) \triangleq (K-2)e^{-d^2} + \frac{1}{2}e^{-d^2}[e^{-a(K,s)d^2} + e^{a(K,s)d^2}], \qquad (33)$$

Thus $P_b < e^{-d^2}\{1+(K-1)\sum_{s=2}^{\infty}(s^2-s)[\frac{1}{K-1}B(K,d,s)]^{s-1}\}$,

$$K > 2, \qquad (34)$$

Especially $P_b < e^{-d^2}\{1+(K-1)^2\sum_{s=2}^{\infty}(s^2-s)[\frac{1}{K-1}B(K,d,s)]^{s-1}\}$,

$$d^2 \gg K, K > 2, \qquad (35)$$

$$P_b < e^{-d^2}\{1+(K-1)\sum_{s=2}^{\infty}(s^2-s)A(K,d,s)[\frac{1}{K-1}B(K,d,s)]^{s-1}\},$$

$$K \gg d^2, K > 2, \qquad (36)$$

Now the key point becomes to evaluate $a(K,s)$ or equivalently the minimum Euclidean distance between any two received signals. Such task seems very difficult, duo to the fact when $K \gg 1$ *or* $s \gg 1$, the total number of the node error sequences is very huge. However, there is no need to evaluate all $a(K,s)$, only evaluate several small s's $a(K,s)$ is enough, Then, in general such evaluations is not difficult but easy. Even most physical $a(K,s)$ can be found by intuition way (appendix A). The reason is that, for the physical realizable $h(t)$, series $|h_1^0|,|h_2^0|,\ldots,|h_{K-1}^0|$ is strictly monotonic decreasing with attenuation coefficient less than any arithmetic one. If the evaluation from $s = 1,2,\ldots,$ *until* s_M, $a(K,s_M) \le 1$, then just let $a(K,s) = a(K,s_M), \forall s > s_M$, and s_M usually is not large.

Thus for $|h_1^0| \le 1$ channel, a loser upper bound on P_b can be obtained by let all $a(K,s) = 1$ as

$$P_b < e^{-d^2}[1+(K-1)\sum_{s=2}^{\infty}\frac{(s^2-s)}{(K-1)^{s-1}}A(K,d,s)B^{s-1}(K,d)],$$

$$(K > 2, |h_1^0| \ge 1, \qquad (37)$$

Where

$$B(K,d) \triangleq (K-2)e^{-d^2} + \frac{1}{2}(1+e^{-2d^2}), \qquad (38)$$

When $d^2 \gg 1$, we have

$$\begin{aligned} P_b &< e^{-d^2}[1+(K-1)\sum_{s=2}^{\infty}\frac{(s^2-s)}{2^{s-1}(K-1)^{s-1}} \\ &= e^{-d^2}\{1+[1-\frac{1}{2(K-1)}]^3\},\ (d^2 \gg K, K>2, |h_1^0| \le 1), \end{aligned} \tag{39}$$

It is a pity that $|h_1^0| \le 1$ exists only for small K. However, no matter in what situation, in $\sum_n e_n \tilde{h}(t-nT)$ at least one whole $h(t-n'T)$ should be survived for odd s. Therefore for odd s, $|h_1^0|>1$ is impossible, $|h_1^0|>1$ only exists for even s. Since $a(K,s)$ is monotonic decreasing function of s, there must exist a s_M, for $s \ge s_M, a(K,s) \le 1$, and let $a(K,s)=1\ (s \ge s_M)$ .Thus only the term with $a(K,s)>1$ should be remained. Therefore a loser upper bound on P_b can be obtained as

$$\begin{aligned} P_b < e^{-d^2}\{1+(K-1)\sum_{s=2}^{\infty}\frac{(s^2-s)}{(K-1)^{s-1}}A(K,d,s)B^{s-1}(K,d)] \\ +(K-1)\sum_{s=2,4,\dots,s_M}\frac{s^2-s}{(K-1)^{s-1}}A(K,d,s) \\ \times[B^{s-1}(K,d,s)-B^{s-1}(K,d)]\},\ (K>2). \end{aligned} \tag{40}$$

When $d^2 \gg K$,

$$\begin{aligned} P_b < e^{-d^2}\{1+[1-\frac{1}{2(K-1)}]^{-3} \\ +(K-1)\sum_{s=2,4,\dots,s_M}\frac{s^2-s}{2^{s-1}(K-1)^{s-1}}[e^{(s-1)[a(K,s)-1]d^2}-1]\} \\ (K>2, d^2 \gg \mathrm{K}), \end{aligned} \tag{41}$$

When $K \gg d^2$,

$$\begin{aligned} P_b < e^{-d^2}\{[1+2(K-1)B(K,d)(1-\frac{1}{K-1}B(K,d))^{-3}] \\ +(K-1)^2\sum_{s=2,4,\dots,s_M}\frac{s^2-s}{(K-1)^{s-1}}[B^{s-1}(K,d,s)-B^{s-1}(K,d)]\} \\ (K>2, K \gg d^2). \end{aligned} \tag{42}$$

and (42) are the final results, In fact (42) is suitable for any SNR d^2, duo to it uses the largest upper bound (K-1) on $A(K,d,s)$.

D *Bit error probability upper bound of OVTDM for special multiplexing waveform* $h(t)$ $(K \gg 1, d^2 \gg K)$

Two special multiplexing waveforms $h(t)$ are considered in this section, they are:

1） *Rectangular waveform*

$$h(t) \triangleq \sqrt{\frac{2}{KT}}[u(t)-u(t-KT)];$$

2） *Truncated exponential waveform*

$$h(t) \triangleq 2\sqrt{\alpha}e^{-\alpha t}[u(t)-u(t-KT)],\ \alpha = b/KT,\ b \ge 7$$

Note: Larger truncate number b will cause smaller cut off power that becomes interference, b has no effect to the conclusion!

Due to the following reasons to study above two waveforms have both theoretical and engineer importance:

1. According to the symmetry principle, rectangular $h(t)$ uniformly distributes signal's energy to all time delay, it belongs to the least favorite waveform, but has the simplest decoding complexity at condition of all $h(t)$ have same η.
2. Physical realizable $h(t)$ can be looked upon as the different exponential waveforms' linear combination.
3. By employing a special "Perfect Complete Complementary Orthogonal Code Pairs Mate" [12], rectangular $h(t)$ can be working well in OFDM system without losing performance.

1. Rectangular $h(t)$

$$\text{Its } h_l^0 = 2(1-l/K),\ (l=1,2,\dots,K-1), \tag{43}$$

Its least favorite node error sequence is

$$xx\underbrace{00\dots0}_{K-2}xx\underbrace{00\dots0}_{K-2}\dots\dots xx\underbrace{00\dots0}_{K-2}$$

Where: xx denotes polar alternative + − errors. The least favorite waveform of $\sum_n e_n h(t-nT)$ is shown in Fig.7 as

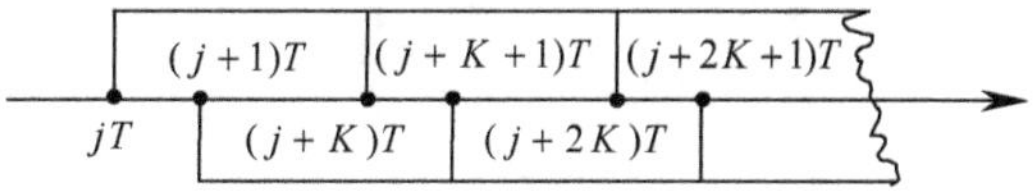

Fig. 7. *Least favorite* $\sum e_n h(t-nT)$ *for rectangular* $h(t)$

It can be found:

$$Min\ \varepsilon_s = \begin{cases} -1+2/sK, & (s=2,4,6,\dots) \\ -(s-1)/s, & (s=3,5,7,\dots) \end{cases} \tag{44}$$

$$a(K,s) = \begin{cases} \frac{s}{s-1}[1-\frac{2}{sK}], & (s=2,4,6,\dots) \\ 1, & (s=3,5,7\dots) \end{cases} \tag{45}$$

Thus an upper bound for rectangular $h(t)$ on P_b is

$$
\begin{aligned}
P_b &= P_e(1) + \sum_{s=3,5,7,\ldots} sP_e(s) + \sum_{s=2,4,6,\ldots} sP_e(s) \\
&\le e^{-d^2}\{1+(K-1)\sum_{s=3,5,7,\ldots} \frac{s^2-s}{2^{s-1}(K-1)^{s-1}}\} \\
&+(K-1)e^{-2d^2/K}\sum_{s=2,4,6,\ldots} \frac{s^2-s}{2^{s-1}(K-1)^{s-1}} \\
&= \frac{1}{2}e^{-d^2}\left\{2+\left[1-\frac{1}{2(K-1)}\right]^{-3}+\left[1+\frac{1}{2(K-1)}\right]^{-3}\right\} \\
&+\frac{1}{2}e^{-2d^2/K}\left\{1-\left[1-\frac{1}{2(K-1)}\right]^{-3}-\left[1+\frac{1}{2(K-1)}\right]^{-3}\right\}, (d^2 \gg K),
\end{aligned} \tag{46}
$$

$$
P_b < 2e^{-d^2} + \frac{3}{2(K-1)}e^{-2d^2/K}, (d^2 \gg K, K \gg 1) \tag{47}
$$

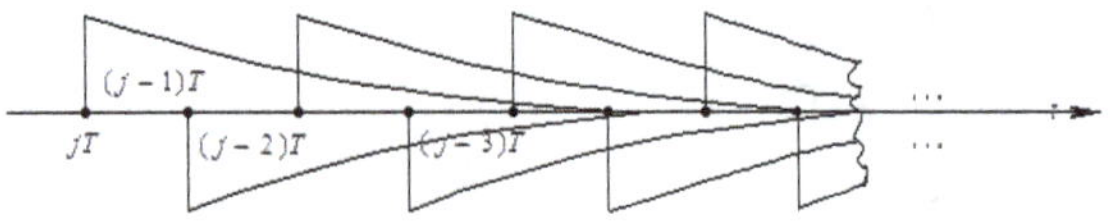

Fig 8. Least favorite $\sum e_n h(t-nT)$ *for truncated exponential* $h(t)$

2. Truncated exponential $h(t)$

When $K \gg 1$,

$$
h_l^0 = 2e^{-2\alpha T} \cong 2(1-2\alpha T), \quad (l=1,2,\ldots,K-1), \tag{48}
$$

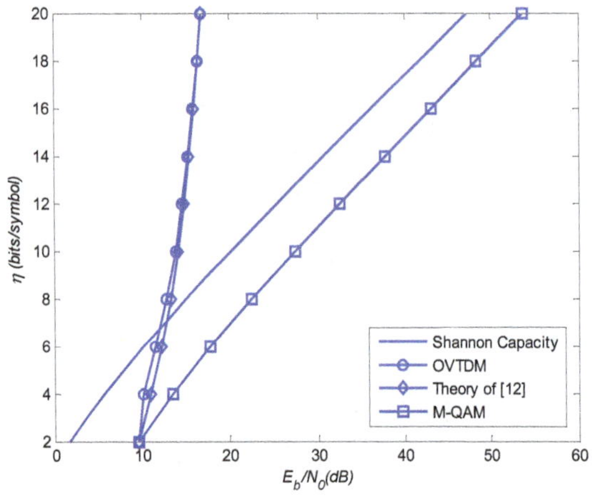

Fig 9. Simulation and theoretical Comparison for rectangular $h(t)$ *{η~ spectral efficiency (bits/symbol),* $d^2 = E_b / N_0$*,* P_b *=10-5 }*

The least favorite error sequence is the continues polar alternative $+-+-+-$.... one (Fig.8)

It can be found:

$$
a(K,s) = \begin{cases} \dfrac{s}{s-1}[1-\alpha T], & (s=1,3,5,\ldots) \\ 1, & (s=2,4,6\ldots) \end{cases}, \tag{49}
$$

Thus when $d^2 \gg K$, an upper bound on P_b for truncated exponential $h(t)$ is

$$
\begin{aligned}
P_b &\le e^{-d^2}\{1+(K-1)\sum_{s=3,5,7,\ldots} \frac{s^2-s}{2^{s-1}(K-1)^{s-1}} \\
&+(K-1)e^{-\alpha T d^2}\sum_{s=2,4,6,\ldots} \frac{s^2-s}{2^{s-1}(K-1)^{s-1}}e^{-(s-1)\alpha T d^2} \\
&= \frac{1}{2}e^{-d^2}\{2+[1-1/2(K-1)]^{-3}+[1+1/2(K-1)]^{-3}\} \\
&+\frac{1}{2}e^{-2\alpha T d^2}\{[1-1/2(K-1)]^{-3}+[1+1/2(K-1)]^{-3}\} \\
&d^2 \gg K
\end{aligned} \tag{50}
$$

Since $\alpha T = b/K$, then when $d^2 \gg K$,

$$
\begin{aligned}
P_b &\le \frac{1}{2}e^{-d^2}\{2+[1-1/2(K-1)]^{-3}+[1+1/2(K-1)]^{-3}\} \\
&+\frac{1}{2}e^{-2bd^2/K}\{[1-1/2(K-1)]^{-3}+[1+1/2(K-1)]^{-3}\}, \\
&d^2 \gg K
\end{aligned} \tag{51}
$$

(5 When $d^2 \gg K, K \gg 1$,

$$
P_b < 2e^{-d^2} + \frac{3}{2(K-1)}e^{-2bd^2/K}, (d^2 \gg K, K \gg 1), \tag{52}
$$

After deeply study (46), (47), (51), (52), It may be find when d^2/K is kept a constant, Pb are roughly the same. That means that spectral efficiency η (or channel capacity) of OVTDM is roughly proportional to the normalized SNR d2. Such conclusion is identical to the conjecture of [12] and also similar to [15], since no mater rectangular or exponential $h(t)$, their bandwidth is defined only by engineering sense rather than Nyquist strictly sense. Regardless the decaying speed, the spectrum tail always extends to infinite. When the strict bandwidth is infinite, for Nyquis $H(f)$ [15] has proved that the channel capacity is linearly to SNR. For any $H(f)$ appendix C get the same conclusion when SNR is high. Simulations shown next in the paper also verify such conclusion. Since among different $h(t)$ with identical η, Rectangular $h(t)$ is the worst due to the Symmetric Principle and shortest constraint length, And among lots of simulated $h(t)$, when η>6, their bit error probability performances all close to such bound (Fig.13). Therefore (46), (47) can be looked upon as an upper bound on bit error probability for any $h(t)$.

4. Two Stage Concatenated OVTDM

A *Two stage Concatenated OVTDM structure and Implementation*

Two stage concatenate OVTDM structure parallel putting in orthogonal I, Q channels is proposed in Fig. 10, where the 1st stage is a K1th order pure OVTDM (no relative shift) changing binary (+1,-1) input into multilevel real input and the 2nd stage is a K2th order shifted OVTDM making its polynomial distribution output approach to Gaussian distribution. The total spectral efficiency of such I, Q parallel concatenate OVTDM structure is $\eta = 2K_1K_2$ bits/symbol, and I, Q real distribution outputs together approach to complex Gaussian distribution.

The multiplexing waveform of the 1st stage OVTDM is a rectangular one with duration K_1T_b. The multiplexing waveform of the 2nd stage OVTDM, denoted within blue line block from transmitter to receiver, is a real waveform $h(t)$ with duration $T_\lambda = \lambda K_1K_2T_b$ $(\lambda \geq 1)$. Where $\lambda \geq 1$ is the waveform duration expanding coefficient corresponding to a rectangular $h(t)$ after filtering. Time duration T_λ of $h(t)$ is the time width of occupying at least 99.9% of the total energy of $h(t)$, depending on the required η. Because the cut off energy of $h(t)$ will become interference, which should be less than the threshold SNR's noise level at least 10-15 dB, larger η should choose larger percentage of the total energy of $h(t)$ (larger λ).

The shift interval of the 2nd stage OVTDM is K_1T_b. In order to keep the spectral efficiency η unchanged in a filtered system. Overlapped folds of the 2nd stage OVTDM would be automatically increased from K_2 to $K_2^\lambda = \lceil \lambda K_2 \rceil$, where $\lceil \bullet \rceil$ is the least integer that is greater or equal to $\bullet$.

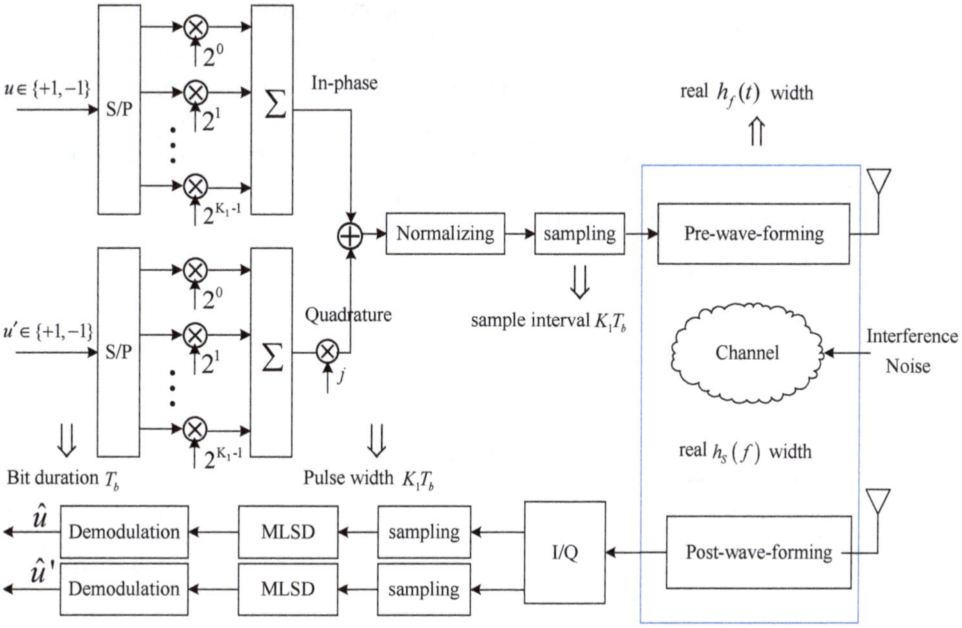

Fig 10. *Concatenate OVTDM structure. The Multiplexing waveform of the 1st stage is a rectangular one of width* K_1T_b*, The Multiplexing waveform of the 2nd stage is* $h(t)$ *of width* $T_\lambda = \lambda K_1K_2T_b$*,* $\eta = 2K_1K_2$.

The 2nd stage of shifted OVTDM with multiplexing waveform $h(t)$ can be also denoted by Fig.10, with shifted interval K_1T_b and overlapping folds K_2^λ. The impulse response of $h_f(t)$ consists of all filters in transmitter, e.g. root transmitter filter, pre-wave-forming, pre-equalizer etc. The impulse response of $h_S(t)$ consists of all filters in receiver, e. g. root receiver filter, post-wave-forming, post-equalizer etc. The total multiplexing waveform of the 2nd stage OVTDM is $h(t) = h_f(t) \otimes h_S(t)$.

MLSD should be employed in such concatenated OVTDM system. From the total $2^{K_1K_2^\lambda}$ possible waveform sequences in each I and Q channel, to select the most possible concatenated OVTDM coded waveform sequence that is nearest to the received signal waveform sequence [2].

Coding steps:

Let the I, Q orthogonal channels' binary (+1,-1) data input sequences be

$$I: \cdots, u_{c,n-1}^T, u_{c,n}^T, u_{c,n+1}^T, \cdots$$

$$Q: \cdots, u_{s,n-1}^T, u_{s,n}^T, u_{s,n+1}^T, \cdots$$

Where $u_{\bullet,n}^T = [\underbrace{u_{\bullet,nK_1}, u_{\bullet,nK_1}, u_{\bullet,nK_1+2}, \cdots, u_{\bullet,(n+1)K_1-1}}_{K_1}]$;

$\bullet \sim$ denote either c or s.

Performing the 1st stage OVTDM operation. That is the K1 folds Pure OVTDM operation for I, Q channel with rectangular multiplexing waveform of width K_1T_b respectively.

Performing the 2nd stage OVTDM operation. That is the $K_2^\lambda(\lambda \geq 1)$ folds shifted OVTDM operation for I, Q channel with $h(t)$ of width $T_\lambda = \lambda K_1K_2T_b(\lambda \geq 1)$ respectively. Where the shifted interval is $K_1T_b, K_2^\lambda = \lceil \lambda K_2 \rceil$.

Finally the complex envelop of the received signal of such concatenate OVTDM is

$$S(t)=\sqrt{2E}\left[\sum_{n}\mathrm{u}_{c,n}^{T}\,\mathrm{H}(t-nK_1T_b)\right.$$
$$\left.+j\sum_{n}\mathrm{u}_{s,n}^{T}\mathrm{H}(t-nK_1T_b)\right]+\tilde{n}(t), \quad (53)$$

Where: E is bit energy of the received signal; $n(t)$ is the complex envelop of the noise.

$$\mathrm{H}(t)=\left(2^{-K_1+1}\sum_{i=1}^{2^{K_1-1}}(2i-1)^2\right)^{-\frac{1}{2}}\begin{bmatrix}h(t)\\2h(t)\\\vdots\\2^{K_1-1}h(t)\end{bmatrix}, \quad (54)$$

$\left(2^{-K_1+1}\sum_{i=1}^{2^{K_1-1}}(2i-1)^2\right)^{-\frac{1}{2}}$ is the normalized coefficient;

The multiplexing waveform vector of the 2nd stage OVTDM is $\mathrm{H}(t)$;

The multiplexing waveform of the 2nd stage OVTDM is $h(t)$.

Summary: The spectral efficiency of the proposed concatenate OVTDM is $\eta=2K_1K_2$ bits/symbol. The system bandwidth is determined by $T_\lambda=\lambda K_1K_2T_b$ $(\lambda\geq 1)$. Power spectrum is determined by $h(t)$. The larger the K_1K_2 or λ, the higher the decoding complexity as well as the narrower the system bandwidth. The number of its output levels is $2^{2K_1}(K_2^{\lambda}+1)^2$, However QAM signal's level with the same spectral efficiency is $2^{2K_1K_2}$, which is much more than the concatenate OVTDM.

B *Decoding Algorithms and Complexity of Concatenate OVTDM*

Based on the complex convolutional coding model of Fig.2 and the orthogonal property between I, Q channels, Decoding procedure can be done independently in I and Q channel. The optimum decoding algorithm is well known the MLSD, i.e. the Viterbi algorithm with state (node in trellis) number $2^{K_1(K_2^{\lambda}-1)}$ and input level number 2^{K_1} [2][3][12]. The decoding complexity will be increased exponentially with $K_1K_2^{\lambda}$. Fast algorithms like Fano, Stack or Sphere decoding algorithm can also be employed when $K_1K_2^{\lambda}$ is large.

C *Overlapping Parameter Selection of Concatenate OVTDM*

The spectral efficiency η depends on the production of $K_1K_2^{\lambda}$. How to put allocation on K_1, K_2^{λ}? In fact, K_1 only offer a transform from binary data to multilevel data, it offers no coding gain. All the coding gain is offered by K_2. For rectangular $h(t)$, According to the symmetric principle, indeed $K_1=1$ is the best allocation. However for uneven $h(t)$, the situation is different, it is still an open problem and can only be determined by simulation.

It is well known that under the uncertainty principle time duration limited signal has infinite bandwidth. Frequency bandwidth limited signal has infinite duration. Both time duration and bandwidth limited signal is physically unrealizable. If a smaller λ (or K_2^{λ}) is chosen, the decoding complexity is reduced. However the signal power outside its processing "bandwidth" (time duration or frequency bandwidth) is larger, that will become interference. On the contrary, when a larger λ (or K_2^{λ}) is chosen, the signal power outside its processing "bandwidth" is smaller, but the decoding complexity would be increased. There should be some "compromise" chosen. Surely "compromise" will reduce the system performance and an "error floor" will be appeared, duo to that the cut off signal's power (interference) is fixed. The smaller power that outside its processing "bandwidth", the lower the "error floor". "Compromise" should be considered among system "bandwidth", error floor, λ (or K_2^{λ}). Error floor is determined by the cut off power and the spectral efficiency η. In engineering the following considerations should be noticed:

1. Employing FIR filter with finite time duration $h(t)$;
2. Choosing suitable λ (or K_2^{λ});
3. Employing a multiplexing waveform $h(t)$ that is robust from filtering;
4. Employing "Minimum error probability equalizer" [2] to equalize the real multiplexing waveform $h'(t)$ approach to the required $h(t)$.

5. Simulation Results

Some summary simulations of OVTDM employing different multiplexing waveforms are given from Fig.11 to Fig.12. Where the system parameters respectively are K1=1, K2=1, 2, 3, 4, 5, 6, 7, the corresponding spectral efficiency η respectively are 2, 4, 6, 8, 10, 12, 14bits/symbol etc. The spectral efficiency η measured by equivalent noise bandwidth [12] for different multiplexing waveform $h(t)$ is given by Fig.13.

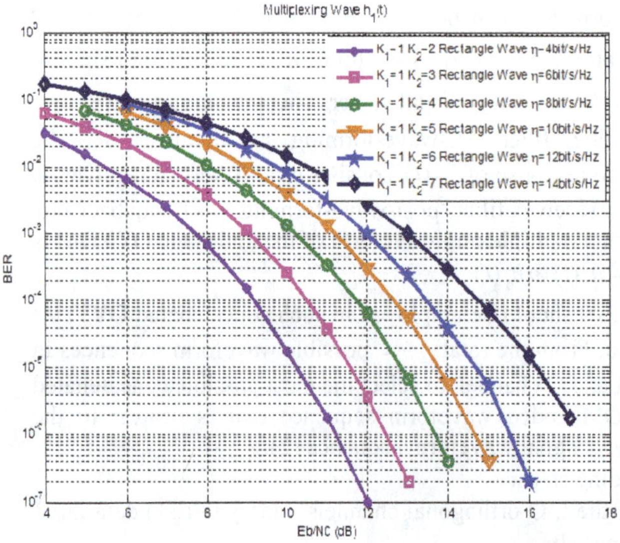

Fig 11. Bit error probability performances of OVTDM with different η, ($h_1(t)$)

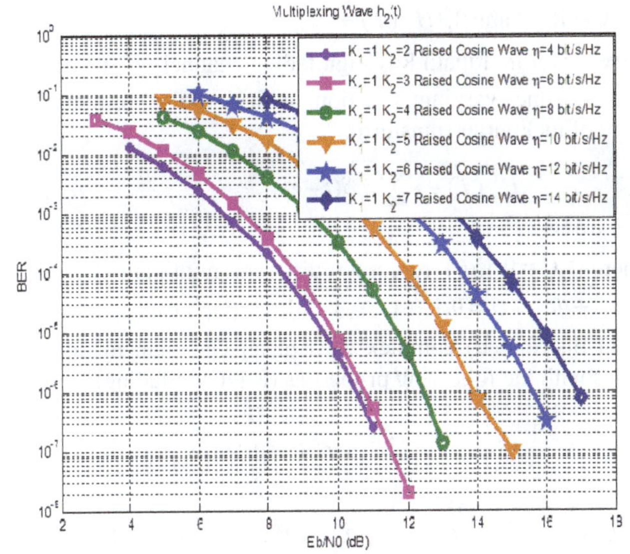

Fig 12. Bit error probability performances of OVTDM with different η, ($h_2(t)$)

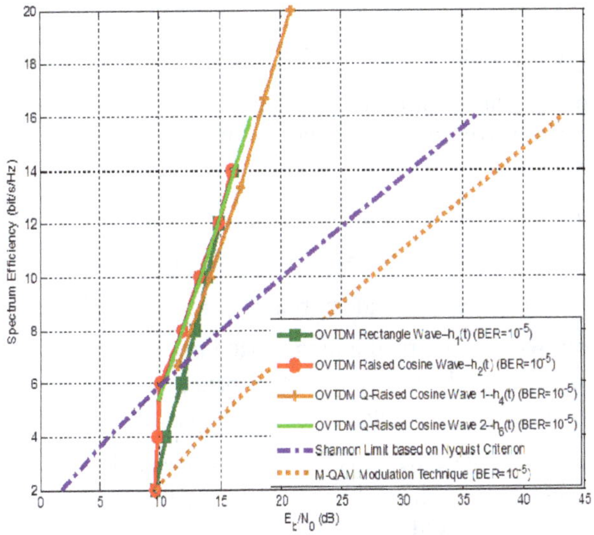

Fig 13. Simulation comparisons of (η, d2) relations for OVTDM employing different $h(t)$ [12] ($P_b=1\times10^{-5}$, equivalent noise bandwidth)

It can be seen that when $\eta \ge 6$, Relation between η and the threshold $d^2 \triangleq E_b/N_0$ (at BER $P_b=1\times10^{-5}$) is roughly a linear relation, which is identical to the conjecture of [12].

Conclusion: The performance of OVTDM is much better than the M-QAM, Especially when spectral efficiency $\eta \ge 6$, Performance of OVTDM begin to go beyond Shannon limit. The lager the η, the higher the gain over Shannon limit.

6. OVTDM with Pre-Coding

In a pre-coded OVTDM system, the binary (+1,-1) input of Fig.14 is not directly from the source but from a pre-coded output. Fig.14 are some simulations employing TPC (64, 51) and TPC (64, 57) respectively. TPC is the abbreviation of Turbo Product Code or Turbo Array Code. The row and column codes of them are BCH (64, 57) and BCH (64, 51) respectively. The code rate respectively are (57/64)2=0.7932 , and (51/64)2=0.6350.

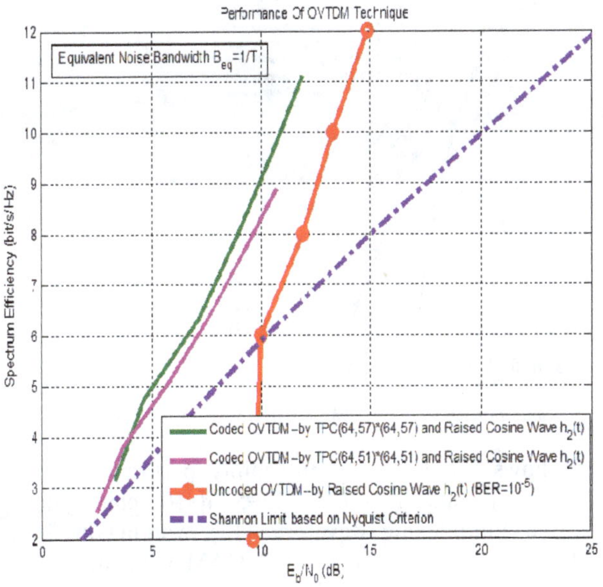

Fig 14. The spectral efficiency of TPC pre-coded OVTDM (equivalent noise bandwidth, h2(t) [12], $P_b=1\times10^{-5}$)

It can be seen that the pre-coding gain is larger when the spectral efficiency of OVTDM is low. However when the spectral efficiency of OVTDM is getting higher pre-coding gain will become lower. For other kind pre-coding we believe that the same conclusion will be obtained.

7. OVTDM in Random Time Varying Channel

A *OVTDM performance in Flat Rayleigh Fading Channel*

The performance of OVTDM in flat Rayleigh fading (i.e. no time dispersion) channel is studied in this section. At the receiver, Optimum detection is still the Matched filtering + MLSD. As comparison a high order QAM of the same spectral efficiency η with different explicit multiple-diversity is employed. Because of that OVTDM itself already have some implicit diversity gain, that is similar to the so called "multipath diversity", the larger the K2, the higher the implicit diversity gain of the OVTDM. There is never need adding extra diversity for the system employing OVTDM. However the high order QAM can't be worked well in flat Rayleigh fading channel.

Fig.15 shows the bit error probability performance of OVTDM with parameter (K1=1, K2=3, η=6) employing raised cosine multiplexing waveform [12], working in flat fading channel. As comparison a 64QAM with different explicit multiple diversity are also given. The slope of the error probability curve of such OVTDM is about the same with the 7th order explicit diversity of 64QAM, and has about 4dB gain (at BER $P_b=1\times10^{-5}$).

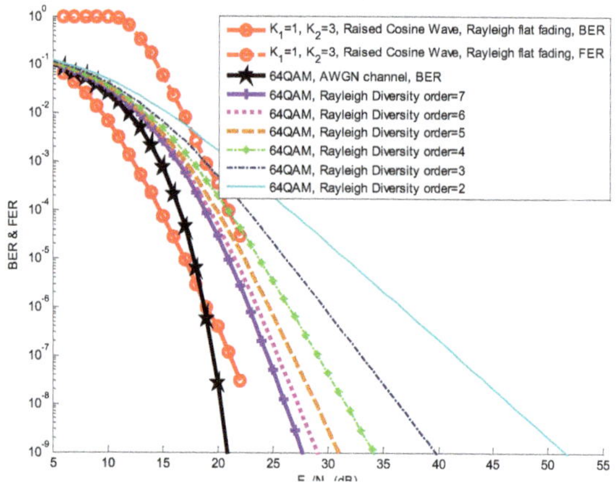

(K1=1, K2=3, η=6, $h_2(t)$, Frame length 1000)

Fig 15. Simulation result of OVTDM in flat Rayleigh fading channel

Fig.16 shows the bit error probability performance of OVTDM with parameter (K1=1, K2=5, η=10) employing raised cosine multiplexing waveform [12], working in flat fading channel. As comparison a 1024QAM with different explicit multiple diversity are also given. The slope of the error probability curve of such OVTDM is about the same with the11th order explicit diversity of 1024QAM, and has about 8dB gain (at BER $P_b = 1\times10^{-5}$).

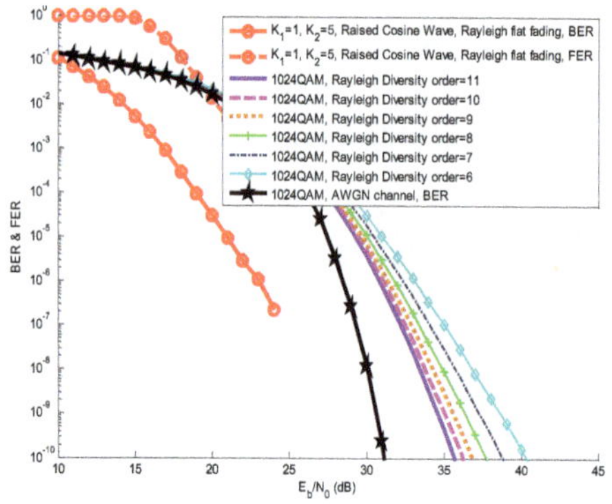

(K1=1, K2=5, η=10, $h_2(t)$, Frame length 1000)

Fig 16. Simulation result of OVTDM in flat Rayleigh fading channel

Conclusion: The performance of OVTDM working in flat Rayleigh fading channels is far beyond the high order QAM with the same η. When η>6, Performance of OVTDM working in flat Rayleigh fading channel is even better than the high order QAM with the same η working in AWGN channel.

B *OVTDM Multiplexing Waveforms in Multipath Fading Channel*

The designed multiplexing waveform and the corresponding spectrum respectively are $h(t)$ and $H(f)$. That is a pair of Fourier transform.

$$h(t) \leftrightarrow H(f) = H^F(f)H^C(f)H^S(f) \quad (55)$$

Where $H^F(f)$, $H^S(f)$, $H^C(f)$ respectively are the transfer functions of transmitter, receiver as well as the channel

For AWGN channel, $H^C(f) \equiv 1$.

However for multipath Rayleigh fading channel, $H^C(f)$ is a random time varying function with impulse response function of its Fourier transform.

$$H^C(f) \leftrightarrow \sum_i \tilde{a}_i \delta(t-\tau_i) \quad (56)$$

Where $\tau_i (i = 0,1,\cdots)$ represent different multipath delay; $\tilde{a}_i = (a_{c,i} + ja_{s,i}),(i = 0,1,\cdots)$ are all i.i.d. (independent identical distributed) complex zero mean Gaussian random variables with the following properties (Rayleigh fading)

$$\frac{1}{2}E\left(\tilde{a}_i\tilde{a}_{i'}^*\right) = \bar{a}_i^2\delta_{i,i'}(\forall i,i' = 0,1,\cdots);$$

$$E\left(a_{c,i}a_{c,i'}\right) = E\left(a_{s,i}a_{s,i'}\right) = \bar{a}_i^2\delta_{i,i'}\forall i,i'$$, $\bar{a}_i^2$ denotes the ith path's normalized average power (variance);

$$E\left(a_{c,i}a_{s,i'}\right) = E\left(a_{c,i}\right) = E\left(a_{s,i'}\right) = 0, \forall i,i';$$

Multipath spread $\Delta \triangleq \max_{i\neq j}\left(\tau_i - \tau_j\right)$.

Obviously in multipath Rayleigh fading channel, multiplexing waveforms of I and Q orthogonal channels respectively are

$$h^I(t) = \sum_i a_{c,i}h(t-\tau_i), h^Q(t) = \sum_i a_{s,i}h(t-\tau_i) \quad (57)$$

$h^I(t)$, $h^Q(t)$ are all random time varying multiplexing waveforms with waveform duration expanding coefficient λ_C related to Δ. Spectral efficiency η only depends on the original designed $h(t)$ and is unrelated to Δ. Larger Δ can cause larger $K_2^{\lambda_C}$ making decoding complexity increased. However larger additional implicit diversity and coding gain are simultaneously achieved.

Where

$$K_2^{\lambda_C} = \lceil (T_\lambda + \Delta)/K_1T_b \rceil, \lambda_C = (T_\lambda + \Delta)/K_1K_2T_b. \quad (58)$$

C *OVTDM performance in Multipath Fading Channel*

The bit error probability performance of OVTDM employing Raised Cosine multiplexing waveform in multipath Rayleigh fading channel is studied in this section. At the receiver, Optimum detection is still the Matched filtering + MLSD. However M-QAM can't be worked well in such channel.

In multipath Rayleigh fading channel, according to (57) multiplexing waveform of I and Q channel are no longer the same $h(t)$ but respectively become $h^I(t)$, $h^Q(t)$ which are all random time-varying waveforms with larger waveform duration expanding coefficient λ_C related to the multipath spread Δ. Although spectral efficiency η of OVTDM only depends on the original designed $h(t)$ and is unrelated to Δ. Larger Δ can cause larger λ_C and larger $K_2^{\lambda_C}$ making decoding complexity increased. However larger additional implicit diversity and coding gain are simultaneously achieved. The larger the λ_C / λ, the larger the additional coding and implicit diversity gain.

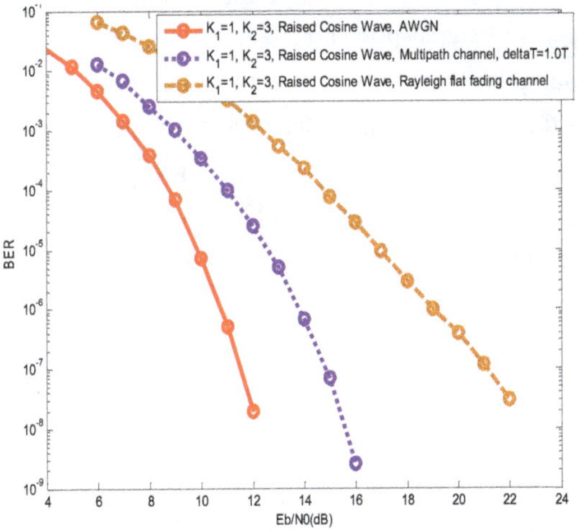

Fig 17. *Simulation result of OVTDM in Multipath Rayleigh fading channel (K1=1, K2=3, η=6,* $h_2(t)$*,* $\Delta = T$ *)*

Fig.17 is one of such examples, which shows the bit error probability performance of OVTDM in multipath Rayleigh fading channel with parameter (K1=1, K2=3, η=6) employing raised cosine multiplexing waveform [12]. Under conditions of the following:

1） Ideal channel estimation achieved;
2） The time dispersion power spectrum of the channel is uniformly distributed with multipath spread Δ equal to T;

It can be seen that the performance of OVTDM working in Multipath Rayleigh fading channel is much better than that of in flat Rayleigh fading channel. According to [2], comparing with the uniform distributed time dispersion power spectrum of the Multipath Rayleigh fading channel, for uneven time dispersion power spectrum channel, the bit error probability curve will shift to right with the same slope at the high SNR region. The shifted value is depended on its uneven degree. However no matter which shape of the time dispersion power spectrum of the channel, when $\Delta \gg T$ the performance of OVTDM working in such Multipath Rayleigh fading channel will converge to that of in AWGN channel [2].

8. Conclusion

Nyquist criterion and level division are the two obstacles of communications. After leaving them a novel waveform coding OVTDM by waveform division is revealed in the paper. The required $d^2 \triangleq E_b / N_0$ is roughly proportional to its spectral efficiency (system capacity). That is an amazing result! So far we never find any coded modulation scheme with better performance than OVTDM. However OVTDM is only at its initial stage. OVTDM still has some unsolved open problems. We do hope there appears better waveform coding latter by other people.

Acknowledgment

The author would like to thank The NNSF (National Nature Science Foundation) of China for their continuously long term support by key projects. Thanks also to my colleagues in Daoben Lab. of Research Institute of Tsinghua Univ. China for their testifying simulations by hardware platform. Special thanks also to Dr. J.K. Omura, and Prof. Chen Junbi, Prof. Tong Zhipeng for their helpful suggestions. Thanks also to my hundreds graduate students they had been working hard successively for over 10 years under the guidance of Overlapped Multiplexing Principle, No their simulations, feedback discussions, a simple principle could not become a systematic OVTDM theory. Specially thank to my Ph.D. Students Wang Ying and Liu Bingchao who offered the final simulations of the paper, and Dong Xuan, Gong Peizhu, Zhao Yujie etc. for their final independent testifying simulations.

Appendix

A *Some simple waves'* $a(K,s)$

To evaluate $a(K,s)$ or equivalently the minimum energy of the least favorite waveform $\sum_n e_n h(t-nT)$ is of importance.

If $h(t) \geq 0,\ \forall t \in [0,\Delta)$, Then in general, the least favorite error sequence is continues polar alternative $+-+-+-....$ one, except rectangular $h(t)$. Any no error 0's between the adjacent errors, or no polar alternative will make $\sum_n e_n h(t-nT)$ with larger energy. Thus

$$Min\ \varepsilon_S = -\frac{1}{s}\sum_{l=1}^{Min(K,s)-1}(-1)^{l-1}(s-l)h_l^0,\ (s \geq 2, K \geq 2), \quad \text{(A-1)}$$

$$Max\ \varepsilon_S = \frac{1}{s}\sum_{l=1}^{Min(K,s)-1}(s-l)h_l^0,\ (s \geq 2, K \geq 2), \quad \text{(A-2)}$$

$$a(K,s) = \frac{s}{s-1}Min\ \varepsilon_S = -\frac{1}{s-1}\sum_{l=1}^{Min(K,s)-1}(-1)^{l-1}(s-l)h_l^0, \quad \text{(A-3)}$$

B *B) The distribution of* σ_l

The definition of σ_l is

$$\sigma_l \triangleq \frac{1}{s}\sum_n e_n e_{n-l},\ (l = 1,2,...,K-1),\ \sum_n e_n^2 = s, \quad \text{(B-1)}$$

σ_l can also be represented as

$$\sigma_l \triangleq \frac{1}{s}\sum_{i=0}^{n_l} x_i,\ (l = 1,2,...,K-1); n_l = 1,2,...s-1; s \geq 2; K \geq 2), \quad \text{(B-2)}$$

Where: binary $x_i \in \{-1,+1\}$, $P_r(x_i = 1) = P_r(x_i = -1) = 0.5$;

n_l is the number of gap length equal to $l-1$ between adjacent errors ;

n_l , x_i are independent random variables;

σ_l is compound binomial distributed with conditional characteristic function as

$$\theta_{\sigma_l}(\omega/n_l) \triangleq E\{e^{j\omega\sigma_l}/n_l] = Cos^n(\omega/s), \quad \text{(B-3)}$$

Since in a node error event, following the first error the second error can occur equal likely at the past K-1 positions. That means

$$P_r(n_l) = (K-1)^{-n_l}\binom{s-1}{n_l}\left(1-\frac{1}{K-1}\right)^{s-n_l-1}, \quad \text{(B-4)}$$

Then the characteristic function of σ_l is

$$\theta_{\sigma_l}(\omega) = \sum_{n_l=0}^{s-1}\theta_{\sigma_l}(\omega/n_l)P_r(n_l)$$
$$= \frac{1}{(K-1)^{s-1}}[Cos(\omega/s)+K-2]^{s-1}, \ (s \geq 2, K \geq 2), \quad \text{(B-5)}$$

(B-5) is independent on l, therefore σ_l are identical distributed, and so the subscript σ_l of l can be ignored.

Since $\theta_\sigma(\omega)$ is an even function of ω, σ only exists even number of moments as

$$E\{\sigma^{2k}\} = (-1)^k\frac{d^{2k}}{d\omega^{2k}}\theta_\sigma(\omega)\Big|_{\omega=0}$$

$$= \frac{1}{(K-1)^{s-1}}\sum_{l=0}^{s-1}\sum_{i=0}^{l}\binom{s-1}{l}\binom{l}{i}\frac{(K-2)^{s-i-1}(l-2i)^{2k}}{2^l s^{2k}}, \quad \text{(B-6)}$$
$$(k=1,2,...;s\geq 2;K\geq 2).$$

Especially

$$E\{\sigma\}=0, \quad E\{\sigma^2\} = \frac{s-1}{s^2(K-1)}, \quad \text{(B-7)}$$

$$E\{\sigma^4\} = \frac{s-1}{s^2(K-1)^2}[(K-1)+3(s-2)], \quad \text{(B-8)}$$

When $s \gg 1$,

$$\lim_{s\to\infty} Log\ \theta_\sigma(\omega) = \frac{s-1}{2s^2(K-1)}\omega^2, \quad \text{(B-9)}$$

It is just the logarithm characteristic function of a Gaussian random variable with mean 0 and variance $(s-1)/s^2(K-1)$.

Although $n_l\ (l=1,2,...)$ are not independent, $x_i\ (i=0,1,2,...)$ are independent, $\sigma_l\ (l=1,2,...)$ are uncorrelated.

$$E\{\sigma_l\sigma_m\} = \frac{s-1}{s^2(K-1)}\delta_{m,l}, \ (m,l=1,2,...,K-1), \quad \text{(B-10)}$$

C *The capacity of channel of non-strictly limited bandwidth with high SNR*

OVTDM employs physical realizable rather than strictly band-limited $h(t) \leftrightarrow H(f)$. In engineer people are interested in $H(f)$ with a long tail. What is the channel capacity C for such $H(f)$ in AWGN channel with high SNR?

When received signal power $P_S \gg 1$, we cut off $H(f)$ by a fictive rectangular filter $G_{B_f}(f)$ with bandwidth $B_f > B_e$, where B_e is the equivalent noise bandwidth or other engineer sense bandwidth, and divide $G_{B_f}(f)$ into many rectangular sub-channels $G_{B_i}(f)$ with band-width $B_i \to 0$, they are all Nyquist ones such that

$$\sum_i G_{B_i}(f) = G_{B_f}(f) = \begin{cases} 0, |f| > B_f \\ 1, \ |f| \leq B_f \end{cases}$$

Basic assumptions:

$$P_S^i \triangleq P_S\int_{-\infty}^{\infty}\left|H(f)G_{B_i}(f)\right|^2 df \gg N_0B_i, \ \forall i; \ \sum_i B_i = B_f > B_e.$$

Where: P_S^i is the signal power of the ith sub-channel; $\sum_i P_S^i \cong P_S$, $\sum_i C_i = C$.

If both B_f, P_S are sufficiently large, $H(f)G_{B_f}(f)$ close to $H(f)$ well under MMSE criterion. The ith sub-channel's capacity Ci is

$$C_i = B_i Log_2(1+P_S^i/N_0B_i), \quad \text{(B-11)}$$

Thus $2^{C_i} = (1+P_S^i/N_0B_i)^{B_i} \xrightarrow{B_i\to 0} e^{P_S^i/N_0}$.

$$2^C = \prod_i 2^{C_i} = 2^{\sum_i C_i} \xrightarrow{B_i\to 0} e^{\sum_i P_S^i/N_0} = e^{P_S/N_0}. \quad \text{(B-12)}$$

Finally we have

$$C = \frac{P_S}{N_0}Log_2 e, \quad \text{(B-13)}$$

When SNR is high, (B-13) is identical to [15], but [15] is only for Nyquist $H(f)$ with infinite bandwidth. Any way when $H(f)$ strictly bandwidth extends to very wide, no matter $H(f)$ is engineer sense band-limited or not, Channel capacity will be always linearly to SNR.

References

[1] Viterbi A J, Omura J K. Principle of Digital Communication & Coding. McGraw-Hill, 1979.

[2] Li Daoben. The statistical Theory of Signal Detection & Estimation. China Academic Press, 2-nd edition, Sept. 2005.

[3] Li Daoben. An Overlapped Time Division Multiplexing Transmission Scheme. PCT application number: PCT/CN2006/001585.

[4] Li Daoben. An Overlapped Frequency Division Multiplexing Transmission Scheme. PCT application number: PCT/CN2006/002012.

[5] Li Daoben. An Overlapped Blocked Time, Frequency and Space Division Multiple Accesses Transmission Scheme. PCT application number: PCT/CN2006/000947.

[6] Li Daoben. A Coded Division Multiple Accesses Transmission Scheme. PCT application number: PCT/CN2007/000536.

[7] Forny G.D. and Jr. Ungerboeck G. Modulation and coding for linear Gaussian channels. IEEE Trans. on Information Theory. Vol. 44(6). 1998: 2348.

[8] Simon, M. K., and Alouini, M. S. Digital Communication over Fading Channels – A Unified Approach to Performance Analysis, 1st ed. Wiley, 2000.

[9] X. Ma and Ping Li, Coded modulation using superimposed binary codes. IEEE Trans. Information Theory, Vol.50, No.10, Oct. 2004.

[10] U Erez, R Zamir. Achieving 1/2 log (1+ SNR) on the AWGN channel with lattice. IEEE Trans. on Information Theory, Vol.50, No.12, Dec. 2004.

[11] John G. Proakis. Digital Communications 4th edition. New York: McGraw Hill, 2001.

[12] Li Daoben. A High Spectral Efficient Waveform Encoding OVTDM Theory and Applications. China Academic Press, Nov. 2013.

[13] Cho, K., and Yoon, D., "On the general BER expression of one- and two-dimensional amplitude modulations", IEEE Trans. Commun.Vol. 50, Number 7, pp. 1074-1080, 2002.

[14] Robert G. Gallager, Principles of Digital Communication. 2010.

[15] Thomas M. Cover, Joy A. Thomas "Elements of Information Theory" John Wiley & Sons, Inc. 2006.

Permissions

All chapters in this book were first published by Science Publishing Group; hereby published with permission under the Creative Commons Attribution License or equivalent. Every chapter published in this book has been scrutinized by our experts. Their significance has been extensively debated. The topics covered herein carry significant findings which will fuel the growth of the discipline. They may even be implemented as practical applications or may be referred to as a beginning point for another development.

The contributors of this book come from diverse backgrounds, making this book a truly international effort. This book will bring forth new frontiers with its revolutionizing research information and detailed analysis of the nascent developments around the world.

We would like to thank all the contributing authors for lending their expertise to make the book truly unique. They have played a crucial role in the development of this book. Without their invaluable contributions this book wouldn't have been possible. They have made vital efforts to compile up to date information on the varied aspects of this subject to make this book a valuable addition to the collection of many professionals and students.

This book was conceptualized with the vision of imparting up-to-date information and advanced data in this field. To ensure the same, a matchless editorial board was set up. Every individual on the board went through rigorous rounds of assessment to prove their worth. After which they invested a large part of their time researching and compiling the most relevant data for our readers.

The editorial board has been involved in producing this book since its inception. They have spent rigorous hours researching and exploring the diverse topics which have resulted in the successful publishing of this book. They have passed on their knowledge of decades through this book. To expedite this challenging task, the publisher supported the team at every step. A small team of assistant editors was also appointed to further simplify the editing procedure and attain best results for the readers.

Apart from the editorial board, the designing team has also invested a significant amount of their time in understanding the subject and creating the most relevant covers. They scrutinized every image to scout for the most suitable representation of the subject and create an appropriate cover for the book.

The publishing team has been an ardent support to the editorial, designing and production team. Their endless efforts to recruit the best for this project, has resulted in the accomplishment of this book. They are a veteran in the field of academics and their pool of knowledge is as vast as their experience in printing. Their expertise and guidance has proved useful at every step. Their uncompromising quality standards have made this book an exceptional effort. Their encouragement from time to time has been an inspiration for everyone.

The publisher and the editorial board hope that this book will prove to be a valuable piece of knowledge for researchers, students, practitioners and scholars across the globe.

List of Contributors

Arnold Adimabua Ojugo
Dept. of Math/Computer, Federal University of Petroleum Resources Effurun, Delta State, Nigeria.

Fidelis Obukowho Aghware
Dept. of Computer Science Education, College of Education, Agbor, Delta State, Nigeria.

Rume Elizabeth Yoro
Dept. of Computer Sci., Delta State Polytechnic, Ogwashi-Uku, Delta State, Nigeria.

Mary Oluwatoyin Yerokun4, Andrew Okonji Eboka4, Christiana Nneamaka Anujeonye4,
Dept. of Computer Sci. Education, Federal College of Education (Technical), Asaba, Delta State, Nigeria.

Fidelia Ngozi Efozia
Prototype Engineering Development Institute, Fed. Ministry of Science Technology, Osun State, Nigeria.

Jai Prakash Prasad
Visvesvaraya Technological University, Research Resource Centre, Belgaum, Karnataka, India.

Suresh Chandra Mohan
Department of ECE, Bapuji Institute of Engineering & Technology, Davangere, Karnataka, India.

Deepak K. Chy
Department of Electrical & Electronic Engineering, University of Information Technology & Sciences (UITS), Dhaka, Bangladesh.

Md. Khaliluzzaman
Department of Computer Science & Engineering, University of Information Technology & Sciences (UITS), Dhaka, Bangladesh.

Harish Kumar and Jai Prakash Gupta
Department of Computer Science Sharda University, Greater Noida, Uttar Pradesh, India.

Prashant Singh
Department of Information Technology, Northern India Engineering College, Delhi.

Shammi Farhana Islam
Department of Material Science and Engineering, Rajshahi University, Rajshahi, Bangladesh.

Mahmudul Haque Kafi and Sk. Sifatul Islam
Department of Applied Physics and Electronic Engineering, Rajshahi, Bangladesh.

Md. Sarwar Hosain
Department of Information and Communication Engineering (ICE), Pabna University of Science and Technology, Pabna, Bangladesh.

Mousumi Haque
Department of Information and Communication Engineering (ICE), University of Rajshahi, Rajshahi, Bangladesh.

Shaikh Enayet Ullah
Department of Applied Physics and Electronic Engineering (APEE), University of Rajshahi, Rajshahi, Bangladesh.

Manato Fujimoto and Yukio Iida
Department of Electrical and Electronic Engineering, Faculty of Engineering Science, Kansai University, Osaka, Japan.

Bourdillon Odianonsen Omijeh
Department of Electronic & Computer Engineering, University of Port Harcourt, Port Harcourt, Nigeria.

Ejioeto Evans Ibara
Centre for Information and Telecommunications Engineering, University of Port Harcourt, Port Harcourt, Nigeria.

Ojo Festus Kehinde
Department of Electronic and Electrical Engineering, Ladoke Akintola University of Technology, Ogbomoso, Nigeria.

Fagbola Felix Adetunji
Department of Works & Physical Planning, University of Lagos, Yaba, Lagos, Nigeria.

Yanzhong Yu
College of Physics & Information Engineering, Quanzhou Normal University, Fujian, China

2Key Laboratory of Information Functional Materials for Fujian Higher Education, Fujian, China.

Jizhen Ni1, Zhixiang Xu
Key Laboratory of Information Functional Materials for Fujian Higher Education, Fujian, China.

Zhu Hong-xiu, Sun Zhi-yuan and Hu Yuan-zhou
School of Mechanical Electronic & Information Engineering, China University of Mining & Technology (Beijing), Beijing, China.

Li Zhi, Hu Jun and Shen Yu
Department of Communication and Command, Chongqing Communication College, Chongqing, China.

Zhang Jianwen
Department of Clinical Medicine, Chongqing Medical and Pharmaceutical College, Chongqing, China.

Liela Khobanizad1,
1Telecommunication of Non-profit Institution of Higher Education, ABA, Abyek, Qazvin, Iran.

Mahmood Khobanizad
Electrical Engineering, Abhar Branch, Islamic Azad University, Abhar, Iran.

Ahmad Houssien Bieg
Non-profit Institution of Higher Education, ABA, Abyek, Qazvin, Iran.

Behrouz Vaseghi
Electrical Engineering, Abhar Branch, Islamic Azad University, Abhar, Iran.

Hamid Chegini
Non-profit Institution of Higher Education, ABA, Abyek, Qazvin, Iran.

Sk. Shifatul Islam, Mahmudul Haque Kafi and Shaikh Enayet Ullah
Department of Applied Physics and Electronic Engineering, University of Rajshahi, Rajshahi, Bangladesh.

Shammi Farhana Islam
Department of Material Science and Engineering, University of Rajshahi, Rajshahi, Bangladesh.

Sylvester Hatsu
Computer Science Department, Accra Polytechnic, Accra, Ghana.

Ujakpa Martin Mabeifam
Faculty of Information Technology and Systems Development, International University of Management (IUM), Dorado Park Campus, Windhoek, Namibia.

Philip Carlis Paitoo
Graduate School, Ghana Technology University College, Takoradi Campus, Accra, Ghana.

Gopal M. Dandime and Veeresh G. Kasabegoudar
Post Graduate Department., Mahatma Basveshwar Education Society's, College of engineering, Ambajogai, India.

Akhtar Hussain, Aimel Khan, Abdul Rehman Qaiser, Muhammad Mohsin Akhtar,
Obaidullah Khalid and Muhammad Faisal Khan

Dept. of Electrical Engineering, National University of Sciences and Technology, Islamabad, Pakistan.

Obiyemi Obiseye O.
Department of Electrical and Electronic Engineering, Osun State University, Osogbo, Nigeria.

Adetan Oluwumi
Department of Electrical and Electronics Engineering, Ekiti State University, Ado Ekiti, Nigeria.

Ibiyemi Tunji S.
Department of Electrical and Electronics Engineering, University of Ilorin, Ilorin, Nigeria.

Dah-Chung Chang and Yen-Heng Lai
Department of Communication Engineering, National Central University, Jhongli City, Taoyuan 320, Taiwan.

Sajad Gharaguozloo
Telecommunication, of Non-profit Institution of Higher Education, ABA, Abyek, Qazvin, Iran.

Abdolhamid Zahedi, Mohammad Norouzi and Hamid Chegini
Non-profit Institution of Higher Education, ABA, Abyek, Qazvin, Iran.

Yuan-Fa Lee
Biomedical Technology and Device Research Laboratories, Industrial Technology Research Institute, Taiwan, R.O.C..

Doru Gabriel Balan, Alin Dan Potorac, Radu Cezar Tărăbuță
Ștefan cel Mare University of Suceava / Computers, Electronics and Automation Department, Suceava, Romania.

Mohamed M. El-Nabawy
Modern Academy for Eng. & Tech in Maadi (M.A.M)/ Electronic and Communication Dept., Cairo, Egypt.

Mohamed A. Aboul-Dahab
Arab Academy for Science and Technology and Maritime Transport (AAST)/ Electronic and Communication Dept., Cairo, Egypt.

Khairy El-Barbary
Canal University, Electronic and Communication Dept., Cairo, Egypt.

R. Sathesh Raaj and J. Kumarnath
Department of Electronics and Communication Engineering, PSNA College of Engineering and Technology, Dindigul, India.

A. Ali
TRENDS Lab, ITU University, Lahore, Pakistan.

Nour Mansour and Dirk Dahlhaus
Communications Laboratory, University of Kassel, Kassel, Germany.

Kochkarev Bagram Sibgatullovich
Department of Mathematics and Mathematical Modeling, Institute of Mathematics and Mechanics Named After Nikolai Ivanovich Lobachevsky, Kazan (Volga Region) Federal University, Kazan, Russia.

Li Daoben
School of Information and Communication, Beijing Univ. of Posts & Telecomm, Beijing, China.

Index

www.ingramcontent.com/pod-product-compliance
Lightning Source LLC
Chambersburg PA
CBHW070153240326
41458CB00126B/4505